Lecture Notes in Physics

Springer-Verlag Berlin Heidelberg GmbH

Hans Elsässer

Steven Beckwith Jakob Staude
Axel Quetz Antonella Natta (Eds.)

Disks and Outflows Around Young Stars

Proceedings of a Conference
Honouring Hans Elsässer
Held at Heidelberg, Germany,
6–9 September 1994

Springer

Editors

Steven Beckwith
Jakob Staude
Axel Quetz
Max-Planck-Institut für Astronomie
Königstuhl 17
D-69117 Heidelberg, Germany

Antonella Natta
Osservatorio di Arcetri
Largo Enrico Fermi, 5
I-50125 Firenze, Italy

Cataloging-in-Publication Data applied for.

Die Deutsche Bibliothek - CIP-Einheitsaufnahme

Disks and outflows around young stars : proceedings of a conference held at Heidelberg, Germany, 6 - 9 september 1994 / Steven Beckwith ... (ed.). - Berlin ; Heidelberg ; New York ; Barcelona ; Budapest ; Hong Kong ; London ; Milan ; Paris ; Santa Clara ; Singapore ; Tokyo : Springer.
ISBN 978-3-662-15720-6
NE: Beckwith, Steven

ISBN 978-3-662-15720-6 ISBN 978-3-540-68511-1 (eBook)
DOI 10.1007/978-3-540-68511-1

Originally published by Springer-Verlag Berlin Heidelberg New York in 1996
Softcover reprint of the hardcover 1st edition 1996

Typesetting: Camera-ready by the authors/editors
Cover design: *design & production*, Heidelberg
SPIN: 10520010 55/3142-543210 - Printed on acid-free paper

Scientific Organizing Committee:

Steven Beckwith, *Heidelberg, Germany (Chair)*
Claude Bertout, *Grenoble, France*
Charles Lada, *Cambridge, USA*
Antonella Natta, *Florence, Italy*
Tom Ray, *Dublin, Ireland*
Steve Strom, *Amherst, USA*

Local Organizing Committee:

Rachel Blythe, Josef Fried, Andreas Glindemann, Christoph Leinert, Karl-Heinz Marien, Reinhard Mundt (Chair), Axel M. Quetz, Jakob Staude

Supporting Institutions and Companies:

Deutsche Forschungsgemeinschaft (DFG)
Max-Planck-Gesellschaft zur Förderung der Wissenschaften (MPG)
Ministerium für Wissenschaft und Kunst Württemberg
Max-Planck-Institut für Astronomie (MPIA)

Carl Zeiss, *Oberkochen*
Colordruck Kurt Weber GmbH, *Leimen*
Schott Glaswerke, *Mainz*

Preface

After the upheavals in the first half of the 20th century, astronomical research in Europe, particularly in Germany, was almost destroyed and had to be rebuilt competely. Hans Elsässer was one of a handful of people with the vision, energy, and tenacity to recreate the scientific excellence that existed previously. He founded the Max-Planck-Institut für Astronomie and built the Calar Alto Observatory in southern Spain, an observatory that is the cornerstone for all optical/infrared research in Germany today. He created *Sterne und Weltraum*, a German-language magazine for popular astronomy, wrote several books, and greatly expanded the influence of astronomy within the Max-Planck-Society. He is without doubt one of the most important astronomers for Germany in the 20th century. It is to him and his achievements that this book is dedicated on the occasion of his 65th birthday.

The subjects of the book, the ubiquitous circumstellar disks around very young stars and the corresponding jets of outflowing matter, were the first addressed by Professor Elsässer when the new Max-Planck-Institut was started, and became the basis for much of the research he undertook in the following 25 years. This subject has recently become one of the hottest areas in astrophysics. The disks are thought to be precursors to planetary systems, and the outflows are thought to be a necessary phase in the formation of a young star, helping the star to get rid of angular momentum and energy as it makes its way onto the main sequence. The possible connections to planetary systems and stellar astrophysics make these topics especially broad, appealing to generalists and specialists alike.

The book is divided into five parts, each emphasizing one aspect of the conference. Parts I and II concern observations and the theory of disks, respectively. Part III concentrates on the young stars and their immediate environments apart from the disks. Parts IV and V discuss the observations and theory of outflows. With the exception of Part IV, each part starts with a general review of the subject followed by a number of topical papers. There were a large number of excellent contributions that could not be printed in the main book. These are included on the accompanying CD-ROM. The CD-ROM contains more than just the papers; it allows the authors to include a fair amount of original data, often displayed as color images. The reader is encouraged to use these data to stimulate new research in a way that was difficult before the advent of compact storage media.

As indicated by the reviews by Sargent and Adams, there is no viable alternative to disk theory to explain the wide range of phenomena seen around young stars. The papers in Part I even give examples of disks that can now be imaged directly with instruments such as HST (McCaughrean) and millimeter-wave interferometers (Dutrey et al., Ohashi and Hayashi). The association of young stars with disks is sufficiently secure in many cases for research to have moved into the next phases: the effect of binary stars (Herbst et al.; Artymowics and Lubow; Whitworth et al.) and the non-stationary aspects of disk accretion (Bell).

Few stars are born in isolation. The presence of other stars – as in clusters – and the residue of gas and dust not in the form of a disk are important both for the understanding of the star formation process and as diagnostics to how it proceeds. Part III emphasizes the importance of stellar evolution theory and the presence of stellar companions to the understanding of observational phenomena related also to disks and outflows. Palla's article shows that our theory for early stellar evolution is converging, just as the theory for main sequence stars did half a century ago. We are reminded by Simon that most young stars are parts of multiple star systems and how this fact affects our thinking about disks (McDonald and Clarke). Several other contributions discuss special aspects of the stars and their environs, including a prediction of the stellar mass function (Nakano et al.).

Outflows were first observed as spectacular, highly collimated jets of hot gas extending large fractions of a pc from their origins. Part IV presents the many new tools used to analyse these jets: the transverse motion of small knots (Eislöffel); the bright molecular hydrogen emission (Zinnecker et al.); the exquisite spectra of the lines and continua providing cooling for the outflowing gas (Staude and Neckel; Hirth et al.; Corcoran and Ray); and high-resolution maps of lines in the millimeter-wave region using the new generation of interferometers (Hirano et al.).

One of the prevailing beliefs is that the outflows and disks are intimately connected, that, without the disks, the outflows would not occur, and it is possible that the disks need the outflows for their own evolution. Königl's review and the subsequent articles by Fendt and Camenzind, Sauty et al., and Bacciotti et al. all rely on the presence of a disk to provide the energy and material for the observed jets and outflows. Almost everyone believes magnetic fields are an integral part of the outflow physics, although the details are in some dispute.

This conference was possible only because of the hard work of many people connected with the MPIA working on behalf of the organizing committees. Reinhard Mundt supervised all the local arrangements aided by Rachel Blythe, Josef Fried, Andreas Glindemann, Christoph Leinert, Karl-Heinz Marien, AMQ, and JS. The members of the scientific organizing committee (Claude Bertout, Charles Lada, AN, Tom Ray, Steve Strom, and SB) put in much time and thought to attract the best speakers to Heidelberg. AMQ was responsible for the technical aspects of the production of these proceedings. We are indebted to the Deutsche Forschungs Gemeinschaft and the Max-Planck-Society for the monetary support needed to make the conference possible.

We are grateful most of all to Professor Hans Elsässer for the visionary work that made this celebration so attractive. It is but a small gesture of gratitude from a community to which he gave his career.

Steven Beckwith, Jakob Staude,
Axel M. Quetz, Antonella Natta

Heidelberg,
February 1996

Contents

Disk Observations

Anneila I. Sargent

California Institute of Technology
Pasadena, CA 91125, U.S.A

Summary. A wide variety of observations suggest that solar mass stars are surrounded by disks of gas and dust for much of their pre-main sequence lifetimes and even beyond. Here, techniques by which the small-scale circumstellar material can be detected are described and evidence for disks is briefly reviewed. Because of the general similarity between the disks around solar-mass stars and the nebula that surrounded the sun as the planets formed, disk properties may have strong bearing on our understanding of solar system evolution. Questions which must be addressed include: How frequently are disks found, and does the fraction vary with environment? What effect do companion stars have? How do disks evolve? Are low infrared fluxes necessarily indicative of disk clearing? Is there evidence of chemical evolution? How long do the disks persist? Are dissipation time-scales different for the gas and dust components? Observations addressing such problems will be summarized and discussed.

1. Introduction

This large gathering of people for a conference on "Disks and Outflows around Young Stars" emphasises how very rapidly the field has advanced since Elsässer & Staude (1978) pointed out that the polarization of scattered light around very young stars is readily explained if the stars are surrounded by disks. Heidelberg is a particularly appropriate venue for this meeting; the polarization studies were carried out at the Max Planck Institut für Astronomie, and supporting evidence for the presence of circumstellar disks, based on analyses of T Tauri star forbidden line profiles, was provided by Appenzeller and collaborators at the Landessternwarte (Appenzeller 1983; Appenzeller et al. 1984). And, largely as a result of Professor Elsässer's foresight in the construction and instrumentation of the Calar Alto observatory, Heidelberg is a world-reknowned center for the study of outflows from young stars.

Numerous talks and posters at this meeting reflect the abundance of observations pertaining to disks. As a reviewer, I find the challenge is not to prove that circumstellar disks are present but rather to combine the diverse disk measurements into a coherent picture. Some focus can be achieved by keeping in mind the fact that disks around solar-mass stars are similar in many respects to the primitive solar nebula. It is important to know the fraction of young stars that have associated disks, because this bears strongly on how many other stars are likely to form planetary systems. Since different star-forming environments, or the presence of companion stars, could affect the incidence of disks, an understanding of these effects is also necessary. I will review what we have learned to date about these topics. I will discuss disk properties, and how these can be deduced from a wide range of observational techniques, in some detail. The persistence time-scales for dust and gas in disks are especially interesting because of the implications for theories of planet formation. Finally, I will comment briefly on the expectations and challenges for this field.

2. The Prototype Disks

The kind of circumstellar disks being discussed here are illustrated in Figure 1. Figure 1a shows the molecular gas surrounding the pre-main sequence T Tauri star, HL Tauri, which is marked by a cross (Sargent & Beckwith 1991). Figure 1b is a coronagraphic image of a main-sequence star, β Pictoris; direct light from the star itself is obscured, allowing radiation scattered by the surrounding dust particles to be seen (Smith & Terrile 1984). For both stars, disk-like structures of radius $\sim$ 1000 AU are evident.

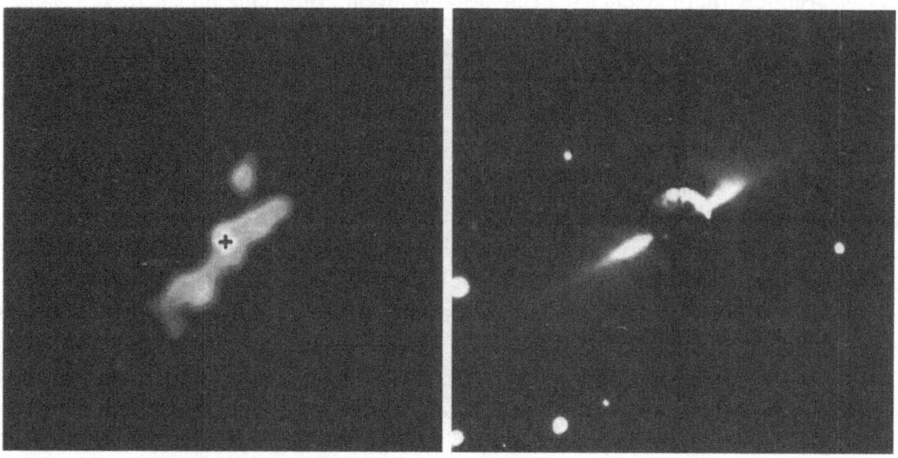

Fig. 1. (a) An image of the ^{13}CO (J=1$\rightarrow$0) emission from HL Tauri at 2″ resolution, reproduced from Sargent & Beckwith (1991). A disk of gas appears centered on the star and extends to radii of order 1000 AU. Emission is unresolved perpendicular to the elongation axis, implying disk thickness $\lesssim$ 100 AU. **(b)** The coronagraph image of scattered light from β Pictoris, reproduced from Smith & Terrile (1984), is rotated by 90° to emphasize its morphological similarity with HL Tau. The scattered light distribution is elongated, indicating the presence of a dust disk of radius about 1000 AU.

Although both disks extend to radii of order 1000 AU, the bulk of the matter around HL Tau is confined to a central region, approximately 100 AU in radius (Beckwith et al. 1990, hereafter BSCG; Beckwith & Sargent 1991; Sargent & Beckwith 1991). By contrast, there appears to be a dearth of material within 40 AU of β Pic (Lagage & Pantin 1994; c.f. Backman & Paresce 1993). HL Tau is a very young star, $\lesssim 10^5$ yrs old (Stapelfeldt et al. 1995), and its disk is substantial, $\sim 0.1\ M_\odot$ (BSCG; Sargent & Beckwith 1987; 1991). Observations of the velocity field in the circumstellar gas are consistent with continuing infall in an outer disk and rotation in an inner disk (Hayashi et al. 1993; Sargent et al. 1995), as might be anticipated for the early stages of star formation (c.f. Galli & Shu 1993a, 1993b). Overall, the properties of the HL Tau system are similar to those of the primitive solar nebula (c.f. Safronov 1991; Beckwith & Sargent 1993a) suggesting that this star/disk system may be the precursor of a planetary system like our own. On the other hand, the mass of circumstellar material around β Pic, a 1.5 $M_\odot$ star about 10^8 years old (Smith & Terrile 1984), is much less than expected for a disk from which Jupiter-like planets may be created (Zuckerman & Becklin 1993). A number of obervations, including the existence of a cleared inner disk, support the view that larger bodies may already have formed (c.f. Backman & Paresce 1993, and references therein) and suggest that the β Pic dust disk may be the detritus of planet formation (c.f. Lagage & Pantin 1994).

The presence of accretion disks around very young stars is expected on theoretical grounds (Shu et al. 1987; Shu et al. 1993 and references therein). Like HL Tau, many solar-mass, pre-main sequence stars of the T Tauri class display features such as strong Balmer lines, asymmetric forbidden emission lines, unusual infrared excesses, and sometimes veiling of the stellar spectrum, that are readily accounted for if they are surrounded by disks (c.f. Appenzeller & Mundt 1989; Bertout 1989). 'Debris' disks may also be prevalent; the IRAS survey detected circumstellar dust around more than 30% of main sequence A-type stars like β Pic and Vega, with ages $\sim 10^8$ years (Aumann et al. 1984; Backman & Gillett 1987; Walker & Wolstencroft 1988). It seems plausible that the history of the solar system could be reflected in the way disks like that associated with HL Tau evolve to become β Pic disks.

The presence of an energetic outflow is intimately linked to accretion of material through a circumstellar disk during the initial stages of star formation (c.f. Shu et al. 1993; Basri & Bertout 1993; Edwards et al. 1993a,b). As a result, the focus of the conference and this review is disks around *young* stars. However, especially in the context of disk evolution, possible end-products such as the β Pic systems must be kept in mind. The nature of Vega-type disks are discussed at length in recent reviews by Backman & Paresce (1993), Lagrange (1995) and Zuckerman (1995).

3. The Incidence of Disks

3.1 Observational Evidence for Disks Around Young Stars

Observational support for the theoretical expectation that young stars are surrounded by disks has accrued relatively recently. Apart from the polarization measurements of Elsässer & Staude (1978; see also Bastien 1982; Hodapp 1984; Bastien & Ménard 1990; Piirola et al. 1992), the strength of the 10 μm silicate absorption feature (Cohen 1983) and the detection of well-collimated optical jets (Mundt & Fried 1983) and bipolar molecular outflows (Snell et al. 1980) emanating from many pre-main sequence stars suggest flattened distributions of circumstellar particles. The preferentially blue-shifted emission line profiles seen in the spectra of very young stars also indicate the presence of circumstellar disks that occlude red-shifted line emission (Appenzeller 1983; Appenzeller et al. 1984; Edwards et al. 1987). Excess infrared emission is one obvious signpost of circumstellar material (c.f. Rydgren & Cohen 1985; Rucinski 1985), and for many young stars the spectral energy distributions between wavelengths of 1 μm and 1 mm are best fit by disk models (Adams et al. 1987; 1988; Hartmann & Kenyon 1987; Hartigan et al. 1989; Strom et al. 1989; Cohen et al. 1989; BSCG; Adams et al. 1990; Weintraub et al. 1991; Mannings & Emerson 1994).

For the most part, the observations cited above are consistent with the presence of disks with radii of order 0.1 to 100 AU. Strong ultraviolet continuum emission that could arise in the boundary layer between the fast-moving inner regions of a Keplerian accretion disk and a more slowly spinning star argue for disks as close to the star as 0.01 AU (Lynden-Bell & Pringle 1974; Bertout 1989; Basri & Bertout 1993). Indeed, a study of optical and infrared spectral line emission in two young objects demonstrates that material very close to the stars is moving in Keplerian orbits (Kenyon et al. 1988; Kenyon & Hartmann 1989). Sensitive, high resolution observations of the first overtone bands of CO at 2.3 μm also provide insights into the behavior of circumstellar material within a few stellar radii of young stars (Carr & Tokunaga 1992; Carr et al. 1993; Chandler et al. 1993). Although the number of objects studied remains small, the observed spectra are often best fit by accretion disk models (Chandler et al. 1995).

The high spatial resolution that can be attained through millimeter-wavelength interferometry (c.f. Sargent & Welch 1993) makes possible *direct* observations of disks of gas and dust that extend to 1000 AU from young stars. At millimeter wavelengths, the low-level, rotational transitions of numerous molecular species 'trace' the dominant molecular hydrogen gas and enable estimates of disk masses, as well as studies of the velocity fields. Images of the molecular line emission show elongated, disk-like structures associated with a significant number of pre-main sequence stars (e.g. Mundy et al. 1986; Sargent & Beckwith 1987; 1989; 1991; Sargent et al. 1988; Weintraub et al. 1989a; Ohashi et al. 1991; Mundy et al. 1992; Kawabe et al. 1993; Koerner et al. 1993a, 1993b; Dutrey et al. 1994).

Alternative interpretations of the observations are, of course, possible for individual cases. Nevertheless, the fact that there is evidence for disks on linear scales ranging from 0.01 to 1000 AU, is very persuasive (cf. Beckwith 1994).

3.2 Disk Frequency

Because of their possible link to forming planetary systems, it is important to determine how often disks exist around other stars. In principle, any of the observational techniques mentioned can be used to ascertain if a young stellar object has an associated disk. Methods which allow many objects to be sampled in a short time are obviously preferable for studies of disk frequency. Measuring infrared, sub-millimeter, and millimeter continuum fluxes from young stars to determine if there is an excess of radiation over that expected from the stellar photospheres has proved to be particularly effective (e.g. Strom et al. 1989;

BSCG; Adams et al. 1990; Cohen et al. 1989; Weintraub et al. 1991; Mannings & Emerson 1994). Model fits to the spectral energy distributions (SED's) that result from these dust continuum observations show that at least 50% of pre-main sequence stars in the Taurus-Auriga star-forming cloud are surrounded by disks (Strom et al. 1989; BSCG). Similar statistics are inferred from millimeter-wave continuum surveys of pre-main sequence objects in Ophiuchus and Chamaeleon (André et al. 1990; André & Montmerle 1994; Henning & Thamm 1993).

A recent $\lambda = 1.3$ mm survey of the continuum emission from young stars by Osterloh & Beckwith (1994) demonstrates that the disk frequency is different for classical Tauri stars (cTT's) and weak-line T Tauri stars (wTT's). First detections of wTT's have in general been made at X-ray wavelengths (e.g. Feigelson et al. 1987; Stahler & Walter 1993; Montmerle et al. 1993). Subsequent spectroscopic and photometric observations identified the optical counterparts as pre-main sequence stars with ages similar to those of cTT's (Walter et al. 1988; Strom et al. 1990). The wTT's have relatively weak Hα emission, little infrared excess, and rarely display signs of accretion activity; it has been inferred that they are not surrounded by active disks (c.f. Edwards et al. 1993a; Stahler & Walter 1993). Although the BSCG 1.3 mm survey of 20 wTT's indicated that 30% supported disks, Osterloh & Beckwith (1995) detect only 6 of an extended sample of 55 wTT's. The dramatic difference in detection rates, 11% for wTT's and 52% for cTT's, strongly supports the contention that relatively few wTT's have associated disks (Skinner et al. 1991; Montmerle et al. 1993).

On the basis of the Einstein X-ray observations, it has been suggested that the number of wTT's in the Taurus-Auriga cloud exceeds that of cTT's by a factor of more than 10 (Walter et al. 1988). This could significantly reduce the overall disk frequency rate. Proper motion studies imply that few of these wTT's are in fact cluster members (Hartmann et al. 1991). However, similarly high detection rates of wTT's have resulted from more complete X-ray surveys of the region carried out with ROSAT (Neuhäuser et al. 1995a; 1995b). The new surveys extend to greater distances from the cluster core so that the newly-identified wTT's may indeed be field stars, but further proper motion studies are needed to clarify the situation.

Herbig Ae/Be stars are somewhat more massive (3 - 10 $M_{\odot}$) analogs of T Tauri stars (Strom et al. 1972). They differ from classical Be stars in that they are confined to star formation regions and display associated nebulosity (Herbig 1960). Infrared and millimeter-wave continuum observations show that the fraction of Herbig Ae/Be stars surrounded by circumstellar material is much greater than 50% (Hillenbrand et al. 1992; Henning et al. 1994; Mannings 1994). Some SED's can be fit by models of geometrically flat, optically thick accretion disks but in many cases an additional circumstellar envelope around the star/disk system is required to explain the observations (Hillenbrand et al. 1992; Natta et al. 1993; Prusti et al. 1994; Mannings 1994; but see also Natta 1994).

Large samples of objects can also be examined for evidence of circumstellar material using plots of J-H versus H-K colors that distinguish stars with intrinsic infrared excesses (Lada & Adams 1992). Relatively few wTT's, but a significant fraction of cTT's and Herbig Ae/Be stars, display infrared excesses consistent with the accretion disk statistics derived from SED's. Using this diagnostic and the new generation of large-format infrared cameras, hundreds of stars in nearby clusters are now being surveyed. To date, about 50% of the stars in each group sampled appear to support disks (Ladd et al. 1993; Lada et al. 1993; Lada & Lada 1995).

Recently, disk-like structures of ionized gas have been detected in Hubble Space Telescope (HST) images of the Orion nebula (O'Dell et al. 1993; O'Dell & Hu 1994). Several compact ionized regions had previously been identified in 2 cm VLA maps of this region and proposed as proto-stellar disks (Felli et al. 1993). The HST images allow a complete census and demonstrate that disks are associated with betweeen 25 and 75% of the pre-main sequence stars in the nebula (Prosser et al. 1994; Stauffer et al. 1994; McCaughrean & Stauffer 1994). In one striking case, presumably far from the ionizing star, a dark disk-like silhouette is seen against the bright nebular background. Its size and mass are consistent with pre-solar nebula values, confirming that here too the fraction of potential proto-planetary disks is high.

Disks appear to be associated with over 50% of young stellar objects in a number of low-mass star forming clouds and the incidence does not seem to be diminished in dense cluster environments like Orion. Nevertheless, there are clear differences between cTT's and wTT's that are not obviously correlated with stellar age and need to be investigated further.

3.3 The Effect of Binary Stars

Within the last decade, it has become clear that many pre-main sequence stars are components of binary or multiple systems (e.g. Simon et al. 1992; Ghez et al. 1993; Leinert et al. 1993; Richichi et al. 1994). In star-forming regions like Taurus-Auriga, the fraction of the pre-main sequence population in multiple systems may be as high as 80% (e.g. Ghez et al. 1993; Leinert et al. 1993; Reipurth & Zinnecker 1993), much higher than the 60% of main-sequence solar-type stars which have companions (Duquennoy & Mayor 1991). In dense clusters like the Trapezium, binary frequency appears similar to the main-sequence, field star value (Prosser et al. 1994), but the binary fraction may well vary among different groups of stars (Durisen & Sterzik 1994). In any case, a large number of T Tauri stars are likely to have companions that could affect the incidence of disks (c.f. Mathieu 1994, and references therein).

Among the BSCG sample, lower 1.3 mm continuum fluxes were detected from binaries with component separations of less than 100 AU (c.f. Beckwith & Sargent 1993a). Osterloh & Beckwith (1995) have demonstrated that, on average, binary systems emit much less 1.3 mm continuum flux. This is illustrated in Figure 2, which shows plots of flux density as a function of binary separation as well as the distribution of fluxes among single stars in their sample. Based on observations of the 800 μm continuum emission from recently identified binaries in the BSCG sample, Jensen et al. (1994) suggest that millimeter emission is critically suppressed at component separations below 50 AU. Figure 2a affords some support for this view; 50 AU corresponds to a separation of −0.35 on the log scale shown. However, the 800 μm measures are largely upper limits. Based on formal tests, Osterloh & Beckwith again find the critical separation of binary components is 100 AU.

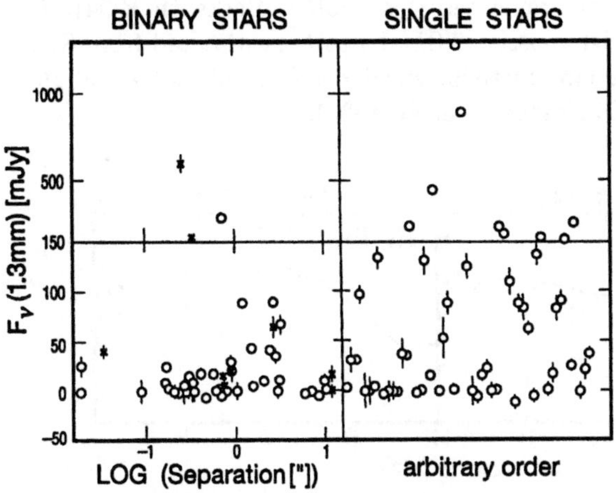

Fig. 2. **(a)** A plot of the projected separations of members of multiple systems as a function of their flux densities at 1.3 mm. Open circles denote binary members and crosses represent objects in multiple systems. From Osterloh & Beckwith 1995). **(b)** Flux densities of single stars in the same sample.

For multiple systems, the disks detected to date appear to be circumbinary (Weintraub et al. 1989; Simon & Guilloteau 1992; Kawabe et al. 1993; Koerner et al. 1993b, Ménard et al. 1993; Dutrey et al. 1994; Ghez et al. 1994). Theory predicts that a circumbinary disk will be truncated at an inner radius defined by the binary orbit (c.f. Pringle 1991; Lin & Papaloizou 1993; Artymowicz & Lubow 1994). Such truncation will substantially reduce the disk surface area, and by implication the infrared/millimeter flux. It is worth noting that 10 μm images of the binary UZ Tau (Ghez et al. 1994) are consistent with clearing of material in the circumbinary disk (c.f. Simon & Guilloteau 1992) out to 50 AU, the orbital radius of the system. A similar result, based on high-resolution mm-wave observations of the molecular gas, has been reported for the GG Tau binary system (Dutrey et al. 1994). Sensitive observations of larger samples of pre-main sequence binaries at a variety of separations are obviously necessary to clarify our understanding of how circumbinary and circumstellar disks in binary systems interact and evolve.

4. General Characteristics of Disks

4.1 Disk Properties from Dust Continuum Surveys

All the observing techniques used to identify circumstellar disks provide information about their gross properties. However, flux measurements from infrared to millimeter wavelengths are most frequently used to constrain disk characteristics. The large numbers of objects that can be sampled no doubt contributes to this bias. As described above, disk masses, sizes, and temperature distributions can be inferred by fitting models to the SED's. For cTT's, disk sizes are typically of order 100 AU and masses range from 0.001 to 0.1 $M_{\odot}$ with an average value of about 0.01 $M_{\odot}$ (c.f. Beckwith & Sargent 1993a). Any disks associated with wTT's are significantly less massive, not merely cooler, and there is a tendency for disk and stellar masses to be correlated (Osterloh & Beckwith 1994). This accords well with the higher disk masses, 0.01 –1.14 $M_{\odot}$, estimated from SED's for Ae/Be stars with extended circumstellar envelopes (Hillenbrand et al. 1992; Mannings 1994; but see also Di Francesco et al. 1994).

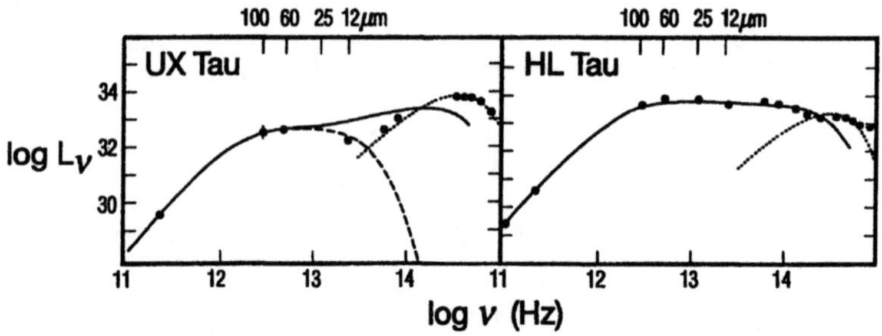

Fig. 3. (a) The SED for the wTT UX Tau from BSCG. Observed fluxes are represented by filled circles. The model of a 100 AU radius disk that extends to the stellar surface is shown as a solid line. The dashed line is a fit that includes a central hole of radius ~ 0.3 AU. The dotted line is a Planck function at the stellar effective temperature fitted to the near-infrared data. **(b)** The SED for the cTT HL Tau. All representations are as in Figure 3a.

Mid-infrared fluxes for some stars are much weaker than expected from the radiation at other wavelengths. Model fits to the SED's require a dearth of radiating particles within a few AU of the star (Beckwith et al. 1990; Strom et al. 1993; Bouvier et al. 1993). To illustrate this point, plots of the SED's from BSCG for HL Tau, and for a wTT star, UX Tau, are displayed in Figure 3. The measured fluxes for HL Tau are well-fitted by a model disk of radius ~ 100 AU that continues to the stellar surface, while the mid-infrared fluxes for UX Tau are significantly lower than predicted by such a model and require the presence of a central hole.

Objects with SED's that suggest central "holes" tend to have relatively low levels of accretion activity (Edwards et al. 1993a; Koerner et al. 1993a; Osterloh & Beckwith 1995). It is tempting to speculate that inner clearing takes place as disks age and is associated with the formation of larger bodies. The relative diminution of dust continuum radiation at ~ 10 μm observed for some objects has, indeed, been interpreted as an indication of gaps in the disks at a few AU from the central star (Skrutskie et al. 1990). The presence of multiple gaps that may reflect the formation of planets has also been proposed (Marsh & Mahoney 1992), but Boss & Yorke (1993) have pointed out that a combination of dust grain opacity and temperature gradients could equally well account for the near-infrared fluctuations in the SED's. Nevertheless, the clearing of inner holes seems reasonably well-subtantiated.

For many disks, the SED's imply a power law temperature distribution, T(r) $\sim r^{-q}$, where r is the disk radius and $q \sim 0.5$ (BSCG; Osterloh & Beckwith 1995); q is rarely as high as 0.75, the value anticipated for an accretion disk or one that is heated by the central star (c.f. Lynden-Bell & Pringle 1974; Adams et al. 1988). A variety of hypotheses have been put forward to explain the unexpectedly low value of q. Most recently, Natta (1993) has proposed that a dusty envelope surrounds the star/disk system and scatters stellar radiation back onto the disk. Since the envelope would cause increased extinction at visual wavelengths, the correlation between q and A_v noted by Osterloh & Beckwith (1994) supports this hypothesis. However, Calvet et al. (1994) are also able to account for heating of the outer disk by invoking the presence of a relatively massive envelope with an infall rate onto the disk of $\sim$ few $\times\ 10^{-6}\ M_\odot\ \mathrm{yr}^{-1}$.

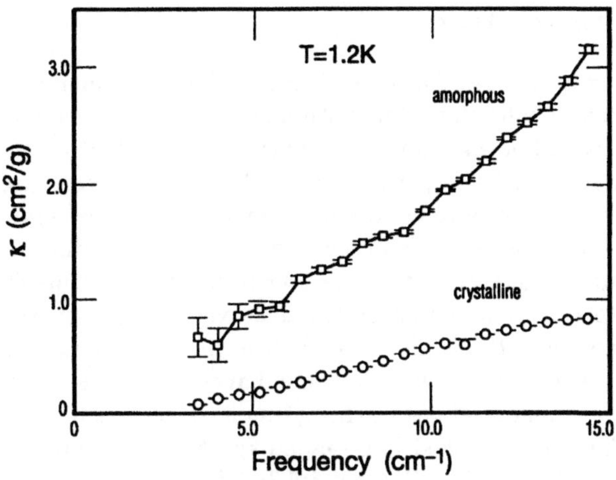

Fig. 4. The absorption coefficient per unit mass for amorphous and crystalline forsterite as a function of frequency at 1.2 K, reproduced from Agladze et al. (1994)

The major uncertainty in disk masses determined from SED's is due to uncertainties in the dust mass opacity coefficient, κ_ν (c.f. BSCG; Beckwith & Sargent 1991; 1993a). Most importantly, the power law index for the dependence of opacity on frequency, β, is not well-known, $\kappa_\nu \propto \nu^\beta$. Draine & Lee (1984) calculated a value of 0.002 cm^2 g^{-1} for $\kappa_{1.1mm}$ in the general interstellar medium; they and others have advocated a value of 2 for β (c.f. Mathis 1990). In principle, β can be established from the ratio of fluxes at different millimeter wavelengths. However, an accurate estimate requires that the emission is optically thin and, even at $\lambda =$ 1 mm, this is not always the case for disks (BSCG). At cm wavelengths, thermal emission from dust in disks is almost certainly optically thin, but the intensity is much lower and there may be contamination by a free-free component. Searches with the VLA have so far been unsuccessful (e.g. Mundy et al. 1993).

Values of β for disks can also be derived from the model fits to SED's and are generally lower than 2 (BSCG; Beckwith & Sargent 1991; Mannings & Emerson 1994). Support for a lower value has recently been provided by laboratory measurements of the absorption spectra of forsterite, Mg_2SiO_4, a form of olivine and one of the main constituents of interstellar dust (Agladze et al. 1994). The measurements were carried out at frequencies between 105 and 350 GHz, and at temperatures below 30 K, very appropriate for sub-millimeter observations of disks. Figure 4 shows the plot of κ_ν as a function of frequency from Agladze et al. (1994). It can be seen that at 230 GHz (1.3 mm), the absorption/unit mass is several times greater than the Draine & Lee (1984) value, with β close to 1.3. Other studies also indicate an increase in the value for κ in disks (Pollack et al. 1994; Krügel & Siebenmorgen 1994; Ossenkopf & Henning 1994). At present, the most appropriate value seems to be a factor of between 3 and 8 times higher than that generally assumed for the interstellar medium. This is quite compatible with the value for $\kappa_{1.3mm}$ adopted by BSCG, Beckwith & Sargent (1991), and Mannings & Emerson (1994) in calculating disk masses.

4.2 Molecular Line Surveys for Disks

Dust continuum radiation from proto-stellar cores is a function of temperature and optical depth and, therefore, falls off rapidly with distance from the core. In contrast, optically thick spectral line emission, particularly the CO (1→0) transition which can be excited at densities as low as 10^3 cm^{-3}, can be pervasive, so that large-scale cloud emission is likely to contaminate molecular line observations of young, solar-mass pre-main sequence objects. As a result, it is difficult to carry out spectral line surveys of young stars in order to identify circumstellar disks. Several other factors can hamper spectral line surveys. At low spatial resolution, beam dilution can make it difficult to distinguish the enhancement in molecular radiation due to a very small source. When enhanced line emission is detectable, self-absorption by foreground gas at the systemic velocity can cause a central dip, mimicking the double-peaked line-shape expected for a rotating disk (c.f. Omodaka et al. 1992; Beckwith & Sargent 1993b). Finally, it is impossible to distinguish the true velocity pattern even when a double peaked profile is seen, since this line shape could result from either rotation or radial infall in a flattened structure; the free-fall velocity scales as $[2GM_*/R]^{1/2}$, while the Keplerian velocity depends on $[GM_*/R]^{1/2}$ (c.f. Koerner 1994).

Successful single-dish molecular line obervations have been carried out for only two disks. The stars GG Tau and DM Tau appear to lie in relatively gas-free regions and the molecular line emission observed at 12 – 15″ spatial resolution can be assumed to be circumstellar (Skrutskie et al. 1993; Guilloteau & Dutrey 1994; Handa et al. 1995). For each star, double-peaked CO and ^{13}CO(2→1) or ^{13}CO and C^{18}O (1→0) lines are seen, consistent with the presence of rotating circumstellar disks with radii ~ 800 AU. However, high resolution imaging is vital to distiguish details of the circumstellar gas emission. C^{18}O aperture synthesis mapping of GG Tau at 2″ resolution confirms that the gas is rotating but is confined to a ring with outer radius ~ 800 AU and inner radius 180 AU (Dutrey et al. 1994). On the other hand, preliminary inspection of CO (2→1) images of

DM Tau at 2″ resolution suggest that a considerable fraction of the circumstellar material is still infalling (Spangler & Sargent 1995).

5. Aperture Synthesis Observations of Disks

The $\sim 1''$ spatial resolution currently attainable through millimeter-wave aperture synthesis observations makes it possible to undertake detailed studies of the disks of gas and dust around young, solar-mass stars in the nearest star forming clouds. Interferometers have the additional advantage that smooth, large-scale cloud emission is resolved out, permitting direct imaging of uncontaminated circumstellar gas. A combination of high spatial and spectral resolution enables investigations of the circumstellar velocity structures of the gas. Independent estimates of disk masses can be obtained from the continuum and molecular line observations and, in principle, different molecular species can be used to probe the physical and chemical properties of the circumstellar environments. For example, imaging the density structure of the inner disks requires observations in an optically thin molecular transition such as $C^{18}O$ (1→0), while outflow or infall in the outer regions can be studied by way of maps in more optically thick transitions (cf. Beckwith & Sargent 1993b).

5.1 Molecular Line Imaging

Aperture synthesis mapping of the $\lambda = 3$ mm dust continuum and CO and $^{13}CO(J = 1\rightarrow0)$ molecular line emission from the HL Tau, T Tau, DG Tau and GG Tau disks has been carried out successfully (Beckwith et al. 1986; Sargent & Beckwith 1987; 1991; Weintraub et al. 1989; Sargent & Beckwith 1994; Kawabe et al. 1993; Dutrey et al. 1994). Disk-like structures with radii $\sim$ 1000 AU were detected in all cases. However, these were also the strongest sources detected in the BSCG survey of 1.3 mm dust continuum emission. To reach the levels of sensitivity required to map molecular line emission from weaker disks requires observations in the $\lambda = 1.3$ mm band. At this wavelength, molecular line emission is stronger than at 3 mm as a result of the Planck function and the larger statistical weights of higher rotational levels of a molecular transition. Several other young stars have been imaged in the J=(2→1) transitions of CO or ^{13}CO (Koerner et al. 1993a; Koerner & Sargent 1995; Sargent et al. 1995; Spangler & Sargent 1995)

Even at high spatial resolution, the interpretation of images of the gas and dust around young stellar objects can be considerably complicated by azimuthal structure in the disks. Thus, early kinematic analyses of $^{13}CO(1\rightarrow0)$ aperture synthesis maps of HL Tau and T Tau suggested that the circumstellar gas was centrifugally supported out to radii of 1000 AU (Sargent & Beckwith 1987; 1991; Weintraub et al. 1989). More recent analyses of position-velocity (p-v) diagrams of ^{13}CO emission from HL Tau indicate that although an inner disk of radius $\sim$ 100 AU may be rotating, a substantial fraction of the circumstellar gas is still infalling (Hayashi et al. 1993). The estimated infall rate, 5×10^{-6} $M_\odot$ yr^{-1}, for HL Tau is much higher than the expected viscous accretion rate, 7×10^{-7} $M_\odot$ yr^{-1}, and it has been suggested that this may be an FU Orionis object in the quiescent state (Lin et al. 1994). Infall onto the circumbinary disk around T Tau has also been inferred from high-resolution HCO^+ aperture synthesis obervations; the mass accretion rate could again lead to FU Orionis activity (van Langevelde et al. 1994). It should be noted, however, that p-v diagrams are not unambiguous and Cabrit et al. (1995) interpret their independent ^{13}CO ($J=1 \rightarrow 0$) observations of HL Tau in terms of an entrained outflow.

The most straightforward way to understand aperture synthesis maps of line emission from disks is by creating synthetic maps at different velocities for comparison with the observations. This method is particularly advantageous when the orientation of the disk is far from edge-on. It has been used to good effect to ascertain the properties of the disks surrounding the cTT GM Aur (Koerner et al. 1993), and the binary cTT system GG Tau (Dutrey et al. 1994). Although GM Aur is only 30° from face-on, synthetic maps of a 0.1 $M_\odot$ disk with radius a few × 100 AU, gravitationally bound to the central star, agree well with the observations at 0.1 km s^{-1} spectral resolution; there is little evidence for infalling material (Koerner 1994). Synthetic representations of the 2″ resolution images of the molecular line emission from GG Tau confirm that the material around this system is also rotating. Dutrey et al. (1994) showed that much of the material within 180 AU of the binary had been cleared, leaving a circumbinary torus of radius about 900 AU.

As mentioned earlier, the spectral line shapes for infalling and rotating circumstellar gas are very similar and can be distinguished only by inspecting the morphology in molecular line maps at high spectral and spatial resolution. The expected emission patterns and velocity gradients associated with gas that is rotating are orthogonal to those expected for radially infalling gas; for rotation, the velocity gradient will be aligned with the disk axis of elongation and, in the case of infall, with the minor axis. Although disk inclination can hamper discrimination between disk major and minor axes, especially if disk size is not much larger than the beam, optical jets or polarization vectors provide an additional diagnostic of disk orientation (c.f. Whitney & Hartmann 1992; 1993). In this way, Koerner and Sargent (1995) have investigated the kinematics of disks around the T Tauri stars DL Tau, RY Tau, AS 209, and DO Tau. The 2″ resolution maps of the CO ($J=2 \rightarrow 1$) emission are displayed in Figure 5, along with gray-scale images of the velocity gradients. In Figure 5a, the orientations of the CO elongation axis (determined from Gaussian fits) and the velocity gradient for DL Tau are in good agreement with the polarization direction, ~ PA 84° (Tamura & Sato 1989), implying the presence of a centrifugally-supported disk of radius 200 AU. Like DL Tau, RY Tau and AS 209 display double-peaked spectra, but alignments in Figures 5c,d and 5g,h are less impressive, although within 20° of one another and the polarization directions, 20° and 134° respectively (Bastien 1982). The analyis suggests in each case a central Keplerian disk of radius ~ 100 AU, with infalling material at larger distances. No clear velocity gradient can be discerned for DO Tau in Figures 5e,f. It seems likely that both infall and rotation are present and detailed kinematic modelling, like that undertaken for GM Aur and GG Tau is required.

5.2 Continuum Emission

Aperture synthesis maps of the continuum emission from disks around about a dozen T Tauri stars have been made at $\lambda = 2.7$ or 1.3 mm (c.f. Sargent & Beckwith 1993, and references therein; Keene & Masson 1990; Simon & Guilloteau 1992; Dutrey et al. 1994; Ohashi et al. 1994; Koerner & Sargent 1995). In all cases, the particle emission is unresolved, even on 2″ scales (equivalent to disk radii of $\sim$ 140 AU), implying lower disk sizes than inferred from molecular line measurements. The difference probably arises from the fact that at wavelengths of 1 mm and longer, emission from dust particles in disks is largely optically thin; the particle emissivity depends on both the temperature and surface density, and thus falls off rather rapidly with increasing radius. For optically thick CO and ^{13}CO measurements, emissivity scales only with temperature (Beckwith & Sargent 1993b). In images at 1″ resolution, the dust disk surrounding HL Tau appears to be marginally resolved; continuum emission at $\lambda = 1.4$ mm is extended along PA 154° with radius 100 AU (Sargent et al. 1995). The dust emission is clearly resolved along this axis in 850 μm, single-baseline interferometer measurements that used the Caltech Submillimeter Telescope (CSO) and the James Clerk Maxwell Telescope (JCMT) on Mauna Kea, Hawaii (Lay et al. 1994). Fits to the visibilities establish a radius of 60 AU. At both 850 μm and 1.4 mm, emission is unresolved along the disk minor axis.

To date, the only other objects where the dust continuum emission has been resolved are T Tau, GG Tau and UZ Tau, all binary systems (Weintraub et al. 1989; Simon & Guilloteau 1992). For T Tau, ^{13}CO emission appears to be circumbinary, but spatial resolution is quite low $\sim$ 6″. As already noted, model fits to the $\lambda = 2.7$ mm line and continuum radiation from GG Tau are consistent with the presence of a circumbinary ring of outer radius 800 AU and inner radius 180 AU (Dutrey et al. 1994). The situation for UZ Tau is less clear-cut. In 1.4 mm maps at 1″ resolution, unresolved continumm radiation appears confined to one component of the wide binary system; the extended emission observed at 2.7 mm is not seen (D. W. Koerner, private communication).

6. Disk Evolution

A knowledge of how the disks around protostars evolve has implications for theories of solar system development (c.f. Podosek & Cassen 1994). For example, an understanding of the formation of giant planets like Jupiter and Saturn requires information on the quantity of nebular gas available when the gaseous envelopes collapse onto the rock-ice cores (c.f. Stevenson 1982; Bodenheimer 1985). Moreover, since aggregation of particulate matter is probably the nucleating event

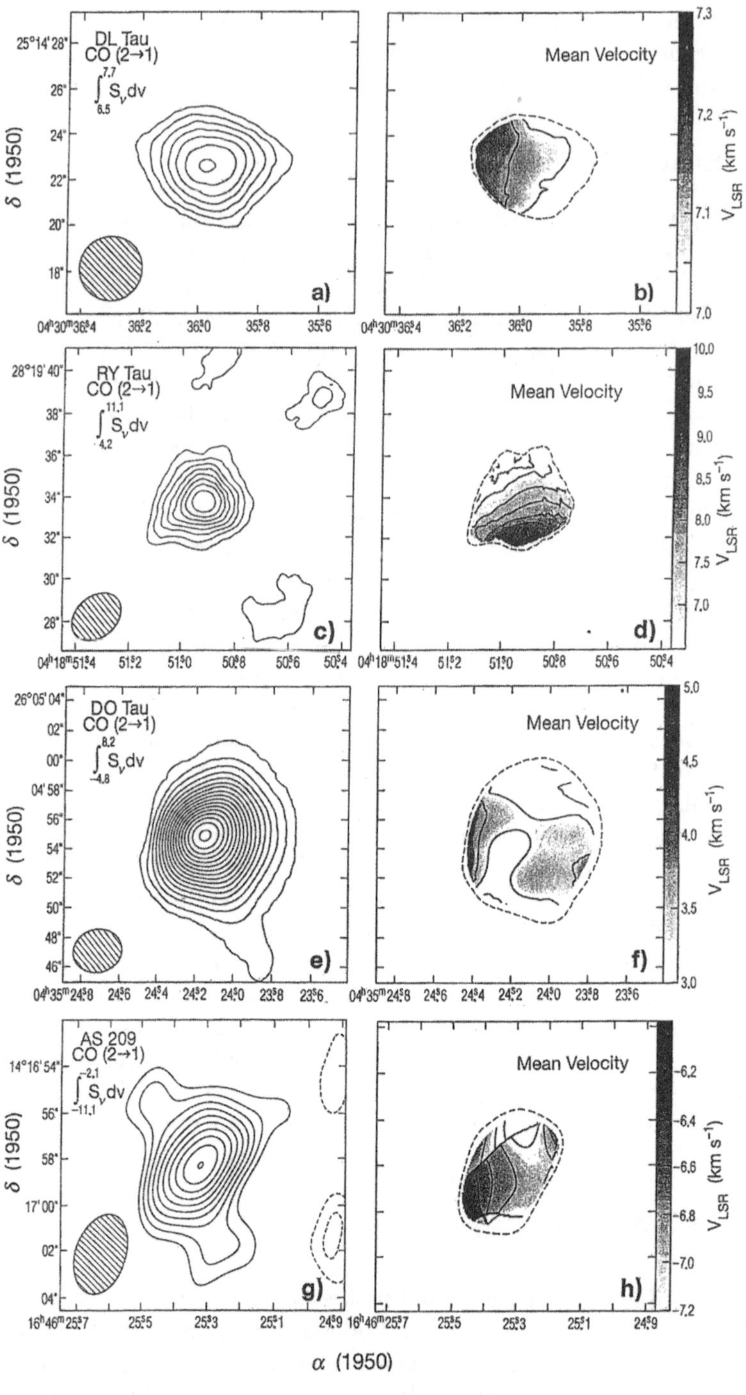

Fig. 5. Contour maps of the integrated CO(2→1) emission towards DL Tau, RY Tau, DO Tau, and AS 209, with corresponding grey-scale maps of the mean velocity variations. All are reproduced from Koerner & Sargent (1995).

for all planets (c.f. Weidenschilling 1989), observational indications of when this takes place are critical. Thus, it is important to determine when the signature of emission from small dust particles disappears and when disk gas disperses. It is of particular interest to establish if these timescales are different.

6.1 Chemical Evolution

The behavior of molecular gas in disks is usually probed through observations of readily-detectable species such as CO and its isotopes. The mass of gas is determined by assuming that the abundance ratios between the observed species and the main constituent of the disks, H_2, are the same as in typical interstellar clouds (e.g. Frerking et al. 1982). In most cases, the observed molecular line emission is optically thick and only lower limits to the disk masses can be derived (c.f. Beckwith & Sargent 1993b). Nevertheless, the estimated masses are often much lower than inferred from dust continuum observations – sometimes by two orders of magnitude (c.f. Koerner et al. 1993a; Dutrey et al. 1994; Koerner 1994; Guilloteau & Dutrey 1994; Handa et al. 1995; Koerner & Sargent 1995). The discrepancy still exists even when conservatively high values of the dust mass opacity coefficient, κ_ν, are used to determine disk masses. In addition, masses derived from kinematic models of high resolution spectral line maps are compatible with those derived from dust continuum measurements (Koerner et al. 1993a; Dutrey et al. 1994). The most plausible explanation of the discrepancy is that the CO isotopes are markedly depleted in disks relative to their canonical values, possibly due to condensation onto cold (≤ 20 K) grains.

Notable exceptions to the pattern are HL Tau and L1551 where spectral line and dust continuum masses are consistent (Sargent & Beckwith 1991; Hayashi et al. 1993; Sargent et al. 1988; Keene & Masson 1990). Both objects are still highly obscured and appear to be significantly younger than the other objects (c.f. Stapelfeldt et al. 1995) so that depletion may be an evolutionary effect. However, there are already strong indications that other species are depleted in the HL Tau disk. From aperture synthesis maps in the CS ($J = 2 \rightarrow 1$) transition, and follow-up excitation studies based on multi-transitional CS observations, Blake et al. (1992) find that CS must be depleted by factors of 25 – 50. They note that this could be a manifestation of the conversion of gas-phase CS into other sulfur-bearing molecules as predicted for the inner part of the primitive solar nebula (Barshay & Lewis 1976). In high resolution maps of several other cTT's the position of peak CS ($2 \rightarrow 1$) intensity is, as for HL Tau, significantly different from that of the continuum emission (Ohashi et al. 1990). The implication is that CS emission again originates in ambient core material and is depleted in the disks. Recent observations and excitation calculations for two, very young, proto-stellar cores in NGC133/IRAS4 also indicate depletion, by factors of 10-20, of all molecules including CO (Blake et al. 1995). The onset of depletion can evidently begin early in the formation process.

Other explanations of the discrepancy between gas- and dust-derived masses are possible. The bulk of the line emission could arise in an optically thick core of radius only a few $\times$ 10 AU, for example. Whatever the explanation, it is clear that disk masses that rely on molecular line measurements may well underestimate the total mass of material present. From the point of view of disk evolution, diminishing disk masses that are based on molecular line measurements cannot be interpreted as unambiguous evidence for gas dispersal. Supporting mass estimates, perhaps from rotation curves, are vital to determine disk-dispersal time-scales.

6.2 Persistence Timescales

Diminution of the dust emission around T Tauri stars can be accounted for by the aggregation of grains to form planetesimals (c.f. Weidenschilling 1989; Boss 1993). Observational studies indicate that some form of dust dispersal, including the clearing of the inner disk and perhaps the creation of gaps, takes place by stellar ages of about 10^7 years (Walter et al. 1988; Skrutskie et al. 1990; BSCG, Skinner et al. 1991; Beckwith & Sargent 1993a; Strom et al. 1993; Walter et al. 1994). Unsuccessful searches for dust around possible 'post-T Tauri' stars with ages up to 1.5×10^7 yrs support this dissipation time (Gahm et al. 1994; Jewett 1994; Ray et al. 1995).

Slightly greater timescales are suggested by the observed range of rotational velocities for solar-mass stars on the main-sequence, $10 - 200$ km s^{-1}. Based on the premise that the presence of disks causes lower rotational velocities (Bouvier et al. 1993; Edwards et al. 1993b), Bouvier (1995) argues that very fast rotation, ~ 200 km s^{-1}, will result if the associated disk has a lifetime of only about 10^6 yrs, while slow rotators, $\lesssim 10$ km s^{-1}, require disks that persist for $\gtrsim 2 \times 10^7$ yrs. In this context, it is interesting to recall that although cTT's and wTT's occupy more or less the same region of the Hertzsprung-Russell diagram, wTT's on average rotate faster than cTT's (c.f. Stahler & Walter 1993). Initial angular momentum considerations may be the key to disk evolution timescales.

From solar system studies, it seems likely that the time required to dissipate nebular gas is $\sim$ a few $\times 10^6$ years at Saturn and more than 10^7 years at Neptune (Lissauer et al. 1994). This accords well with the time-scales indicated by photo-evaporation models of disk evolution (Shu et al. 1993; Hollenbach et al. 1993). In these models, the high UV flux of a newly-formed central star irradiates the upper levels of the disk causing an outflow that begins at larger radii as the star ages. Gas dispersal in this way can take up to a few $\times 10^7$ yrs.

Observations that address the gas persistence times are limited. Skrutskie et al. (1991) suggest an upper limit of 3×10^8 yrs, based on 45″ observations of eight stars in the Ursa Major stream. This is considerably higher than the limit recently derived by Zuckerman et al. (1995), a few $\times 10^6$ yrs. The latter authors searched for CO ($2 \rightarrow 1$) and ($1 \rightarrow 0$) emission from field stars at distances ranging from 40 to 140 pc that showed evidence of circumstellar dust emission. At stellar ages of a few $\times 10^6$ yrs, they find gas masses that are less than the mass of Jupiter. However, as discussed earlier, the effects of beam dilution in their 11″ beam or CO depletion in proto-planetary disks could cause these low estimates. Rotation curves imply disk masses well in excess of Jupiter's mass around the stars GM Aur and DM Tau, both of which are at least 2×10^6

yrs old (Koerner et al. 1993; Guilloteau & Dutrey 1994). Statistical samples involving very sensitive molecular line observations with high spectral and spatial resolution are clearly required.

6.3 Kinematic Evolution

Galli & Shu (1993a; 1993b) have postulated a transitional stage in the very early evolution of a T Tauri star, when a flattened, accreting "pseudo-disk" of radius ~ 1000 AU still surrounds the forming star and its much smaller accretion disk. HST images and ground based images (Stapelfeldt et al. 1995; Beckwith & Birk 1995; Weintraub et al. 1995) lend support to the suggestion that HL Tau may be such a transitional object (Hayashi et al. 1993). They demonstrate that, far from being a visible star, HL Tau is highly obscured and therefore much younger than hitherto thought (c.f. BSCG), with an age less than 10^5 yrs. Kinematic modelling of a statistical sample of disks with a range of ages could clarify if there is an evolutionary sequence.

Initial model fits to spectral line maps of RY Tau, DL Tau, and DO Tau indicate that their surrounding disks are much smaller than those surrounding the much younger HL Tau (Koerner 1994). These are essentially two-component models that allow for an infalling outer disk with radius up to ~ 1000 AU and a much smaller rotating inner disk. For each star, the radius R_c within which gas is centrifugally supported is given in Table 1, as is the 'outer' radius of the infalling disk, R_d. Stellar ages and disk masses are also shown and parameters for HL Tau are included for comparison. There is a tendency for the infall region to decrease in size with stellar age but the sample is obviously extremely small. Moreover, GM Aur, with an age of 2×10^6 yrs, seems to support a rather extended disk (Koerner et al. 1993a).

Table 1. Radius as a function of stellar age.

star	age [yrs]	M_D [$M_\odot$]	M_* [$M_\odot$]	R_C [AU]	R_{out} [AU]
HL Tau	$< 10^5$	0.1	~ 1.1	150	1000
RY Tau	2×10^5	0.04	~ 1.25	150?	200
DO Tau	6×10^5	0.02	1.6	350	500
DL Tau[1]	1.2×10^5	0.08	0.25	200	200
GM Tau[2]	2×10^5	0.08	0.72	1000?	1000

[1,2] No outflow/accretion; clearing inside $R = 0.3$ AU – transition objects?

In fact, more realistic models of star/disk systems must be developed. The two-component model fails to reproduce some critical details of the high resolution spectral line maps. This is not surprising since there will undoubtedly be some component of rotation combined with the free-fall motion in the outer disk due to the fact that cloud cores rotate at ~ 2 km s^{-1} pc^{-1} (Kleiner & Dickman 1984). Three-component models (e.g. Stahler et al 1994; D'Alessio et al. 1995) that include a contribution from rotation to the free-fall motion in an outer disk, an intermediate region where the free-fall component falls off, and an interior region where material moves in Keplerian orbits appear to achieve better fits to ^{13}CO (J = 2→1) images of HL Tau at $\sim 1''$ resolution (Sargent et al. 1995).

7. Future Directions

The observations summarized here demonstrate that many questions concerning the nature of disks remain open. Disks appear to be numerous, but if wTT's are as prevalent as indicated by X-ray surveys, they may constitute a smaller fraction of the pre-main sequence population of any one group than is currently accepted. Proper motion studies that define cluster membership could resolve this question. Where companion stars are present, detected disks are circumbinary. Theories of binary formation would benefit from information about the prevalence of disks around individual elements of multiple systems. It is also important to understand the reasons why disk emission is inhibited in some binary systems. Near infrared spectroscopy of CO bandhead emission could be helpful here, as would high resolution molecular line and continuum imaging. The problem of depletion of some molecular species compared to their usual abundances in the interstellar medium deserves considerable attention, particularly because of the difficulties inherent in establishing the dissipation time-scale for the disk gas. The most tantalizing questions involve disk evolution and the time-scales for gas and dust dispersal. High resolution imaging of the millimeter and sub-millimeter dust continuum emission will establish inner disk sizes. For nearby, older stars it can perhaps be determined if there is clearing as indicated by the SED's. Likewise, high resolution molecular line imaging will allow kinematic modeling of disks around stars of different ages to examine the possibility that an evolutionary sequence from infall to pure rotation can be discerned. Disk rotation curves would provide an independent measure of disk mass and enable a better understanding of dispersal times. In addition, an independent mass determination is one certain way to constrain κ_ν and β in disks (Beckwith & Sargent 1991).

We can expect to see some of these questions resolved in the near future. As a result of the recent upgrades to all the operating millimeter-wave arrays, higher resolution imaging at millimeter wavelengths is becoming more routine. The CSO-JCMT results hint at the power of sub-millimeter interferometry that will be realized with the construction of the Smithsonian array. The launch of ISO in 1995 will enable sensitive searches for dust around many stars with ages intermediate between those of the pre-main sequence objects described here and the β Pic-like stars. Follow-up aperture synthesis mapping should begin to provide clues to disk dissipation times. However, these older circumstellar disks are likely to be concomitantly fainter and their investigation may have to await the construction of the national millimeter-wave arrays planned for early in the next century. Near infrared spectroscopy with the 8-meter and 10-meter telescopes that are just coming on-line promises to be very fruitful for studies of the inner disks. Finally, near infrared interferometry with these large telescopes, perhaps supplemented by smaller outriggers, may provide us with real intimations of the presence of other planets around some of the debris disks.

References

Adams, F.C., Emerson, J.P., & Fuller, G.A. 1990, ApJ, 357, 606

Adams, F.C., Lada, C.J., Shu, F.H. 1987, ApJ, 312, 788

Adams, F.C., Lada, C.J., Shu, F.H. 1988, ApJ, 326, 865

Agladze, N.I., Sievers, A.J., Jones, S.A., Burlitch, J.M., & Beckwith, S.V.W. 1995, Nature, 372, 243

André, P., Montmerle, T., & Steppe, H. 1990, A&A, 240, 321

André, P., & Montmerle, T. 1994, ApJ, 420, 837

Appenzeller, I. 1983, RevMexAA, 7, 151

Appenzeller, I., Jankovics, I., & Ostreicher, R. 1984, A&A, 141, 108

Appenzeller, I., & Mundt, R. 1989, A&A Rev., 1, 291

Artymowicz, P., Clarke, C,J., Lubow, S.H., & Pringle, J.E. 1991, ApJ, 370, L35

Artymowicz, P., & Lubow, S.H. 1994, ApJ, 421, 651

Aumann, H.H., Gillett, F.C., Beichman, C.A., de Jong, T., Houck, J.R., et al. 1984, ApJ, 278, L23

Backman, D.E. & Gillett, F.C. 1987, in Cool Stars, Stellar Systems and the Sun, ed. J.C. Linsky & R.E. Stencel (Berlin: Springer-Verlag), 340

Backman, D.E. & Paresce, F. 1993, in Protostars & Planets III, ed. E.H. Levy & J.I. Lunine (Tucson: University of Arizona Press), 1253

Barshay, S.S. & Lewis, J.S. 1976, ARA&A, 14, 81

Basri, G. & Bertout, C. 1993, in Protostars & Planets III, ed. E.H. Levy & J.I. Lunine (Tucson: University of Arizona Press), 543

Bastien, P. 1982, ApJS, 48, 153

Bastien, P., & Ménard, F. 1990, ApJ, 364, 232

Beckwith, S.V.W., Birk, C.C. 1995, ApJ, 449, L59

Beckwith, S.V.W. 1994, in Theory of Accretion Disks - 2, eds. W. Duschl, J. Frank, F. Meyer, E. Meyer-Hofmeister, & W. Tscharnuter, NATO ASI Series (Kluwer: Dordrecht), 1

Beckwith, S.V.W., & Sargent, A.I. 1991, ApJ, 381, 250

Beckwith, S.V.W., & Sargent, A.I. 1993a, in Protostars & Planets III, ed. E.H. Levy & J.I. Lunine (Tucson: University of Arizona Press), 521

Beckwith, S.V.W., & Sargent, A.I. 1993b, ApJ, 402, 280

Beckwith, S.V.W., Sargent, A.I., Chini, R.S., & Güsten, R. 1990, AJ, 99, 924 (BSCG)

Beckwith, S.V.W., Sargent, A.I., Scoville, N.Z., Masson, C.R., Zuckerman, B. Phillips, T.G. 1986, ApJ 309, 755

Bertout, C. 1989, ARA&A, 27, 351

Blake, G.A., Sandell, G., van Dishoeck, E.F., Groesbeck, T.D., Mundy, L.G., & Aspin, C. 1995, ApJ, 441, 689

Blake, G.A., van Dishoeck, E.F., & Sargent, A.I. 1992, ApJ, 391, L99

Bodenheimer, P. 1985, in Protostars and Planets II, ed. D.C. Black, & M.S. Matthews (Tucson:University of Arizona Press), 873

Boss, A.P., & Yorke, H.W. 1993, ApJ, 411, L99

Bouvier, J. 1995, in The Eighth Cambridge Workshop on Cool Stars, Stellar Systems, and the Sun, ed. J.-P. Caillault (San Francisco: Bookcrafters), in press
Bouvier, J., Cabrit, S., Fernandez, M., Martin, E.L., & Matthews, J.M. 1993, A&A, 272, 176
Cabrit, S., Guilloteau, S., André, P., Bertout, C., & Montmerle, T. 1995, in Circumstellar Disks, Outflows & Star Formation, ed. S. Lizano & J. Torrelles, in press
Calvet, N., Hartmann, L., Kenyon, S.J., & Whitney, B.A. 1994, ApJ, 434, 330
Carr, J.S., & Tokunaga, A.T. 1992, ApJ, 393, L67
Carr, J.S., Tokunaga, A.T., Najita, J., Shu, F.H., & Glassgold, A.E. 1993, ApJ, 411, L37
Chandler, C.J., Carlstrom, J.E., Scoville, N.Z., Dent, W.R.F., & Geballe, T.R. 1993, ApJ, 412, L71
Chandler, C.J., Carlstrom, J.E., & Scoville, N.Z. 1995, ApJ, 446, 793
Cohen, M. 1983, ApJ, 270, L69
Cohen, M., Emerson, J., & Beichman, C.A. 1989, ApJ, 339, 455
D'Alessio, P., Cantó, J., Lizano, S., Calvet, N., & Rodriguez, L. 1995, in Circumstellar Disks, Outflows & Star Formation, ed. S. Lizano & J. Torrelles, in press
Difrancesco, J., Evans, N.J., Harvey, P.M., Mundy, L.G., & Butner, H.M. 1994, ApJ, 432, 710
Draine, B.T., & Lee, H.M. 1984, ApJ, 285, 89
Duquennoy, A., & Mayor, M. 1991, A&A, 248, 485
Durisen, R.H., & Sterzik, M.F. 1994, A&A, 286, 84
Dutrey, A., Guilloteau, S., & Simon, M. 1994, A&A, 286, 149
Edwards, S., Cabrit, S., Strom, S.E., Heyer, I., Strom, K.M., et al. 1987, ApJ, 321, 473
Edwards, S., Ray, T., & Mundt, R. 1993a, in Protostars & Planets III, ed. E.H. Levy & J.I. Lunine (Tucson: University of Arizona Press), 567
Edwards, S., Strom, S.E., Hartigan, P., Strom, K.M., Hillenbrand, L.A., et al. 1993b, AJ, 106, 372
Elsässer, H., & Staude, H.J. 1978, A&A, 127, 290
Emerson, J.P. 1988, in Formation & Evolution of Low Mass Stars, ed. A.K. Dupree & M.T.V. Lago (Kluwer: Dordrecht), 21
Feigelson, E.D., Jackson, J.M., Mathieu, R.D., Myers, P.C., & Walter, F.M. 1987, AJ, 94, 1251
Felli, M., Churchwell, E., Wilson, T.L., & Taylor, G.B. 1993, A&AS, 98, 137
Frerking, M.A., Langer, W.D., & Wilson, R.W. 1982, ApJ, 262, 590
Gahm, G.F. Zinnecker, H., Pallavicini, R., & Pasquini, L. 1994, A&A, 282, 123
Galli, D., & Shu, F.H. 1993a, ApJ, 417, 220
Galli, D., & Shu, F.H. 1993a, ApJ, 417, 243
Ghez, A.M., Emerson, J.P., Graham, J.R., Meixner, M., & Skinner, C.J. 1994, ApJ, 343, 707
Ghez, A.M., Neugebauer, G., & Matthews, K. 1993, AJ, 106, 2005
Guilloteau, S., & Dutrey, A. 1994, A&A, 291, L23
Handa, T., Miyama, S., Yamashita, T., Omodaka, T., Kitamura, Y., et al. 1995, ApJ, 449, 894
Hartigan, P., Hartmann, L., Kenyon, S., Hewett, R., & Stauffer, J. 1989, ApJS, 70, 899
Hartmann, L., & Kenyon, S. 1987, ApJ, 312, 243
Hartmann, L., Jones, B.B., Stauffer, J.R., & Kenyon, S. J. 1991, AJ, 101, 1050
Hayashi, M., Ohashi, N., Miyama, S.M. 1993, ApJ, 418, L71
Henning, T., & Thamm, E. 1994, Ap&SS, 212, 215
Henning, T., Launhardt, R., Steinacker, J., & Thamm, E. 1994, A&A, 291, 546
Herbig, G.H. 1960 ApJS, 4, 337
Hillenbrand, L.A., Strom, S.E., Vrba, F.J., & Keene, J. 1992, ApJ, 397, 613
Hodapp, K.W. 1984, A&A, 141, 255

Hollenbach, D.D., Johnstone, S., & Shu, F. 1993, in Massive Stars: Their Lives in the Interstellar Medium, ed. J. Cassinelli & E.B. Churchwell (San Francisco: BookCrafters), 26
Jensen, E.L.N., Mathieu, R.D., & Fuller, G.F. 1994, ApJ, 429, L29
Jewett, D. C. 1994, AJ, 108, 661
Kawabe, R., Ishiguro, M., Omodaka, T., Kitamura, Y., & Miyama, S.M. 1993, ApJ, 404, L63
Keene, J., & Masson, C.R. 1990, ApJ, 355, 635
Kenyon, S.J., & Hartmann, L. 1989, ApJ, 342, 1134
Kenyon, S.J., Hartmann, L., & Hewett, R. 1988, ApJ, 325, 231
Kleiner, S.C., & Dickman, R.L. 1984, ApJ, 286, 255
Koerner, D.W. 1994, Ph. D. Dissertation, California Institute of Technology
Koerner, D.W., Sargent, A.I., & Beckwith, S.V.W. 1993a, Icarus, 106, 2
Koerner, D.W., Sargent, A.I., & Beckwith, S.V.W. 1993b, ApJ, 408, L93
Koerner, D.W., & Sargent, A.I. 1995, AJ, 109, 2138
Krügel, R., & Siebenmorgen, R. 1994, A&A, 288, 929
Lada, C.J., & Adams, F.C. 1992, ApJ, 393, 278
Lada, C.J., & Lada, E.A. 1995, AJ, 109, 1682
Lada, C.J., Young, E.T., & Greene, T.P. 1993, ApJ, 408, 471
Ladd, E.F., Lada, E.A., & Myers, P.C. 1993, ApJ, 410, 168
Lagage, P.-O. & Pantin, E. 1994, Nature, 369, 628
Lagrange, A.M. 1995, Ap&SS, 223, 19
Langevelde, H. J. van, van Dishoeck, E.F., & Blake, G. A. 1994, ApJ, 425, L45
Lay, O.P., Carlstrom, J., Hills, R.J., & Phillips, T.G. 1994, ApJ, 434, L75
Leinert, C., Zinnecker, H., Weitzel, N., Christou, J., Ridgway, S.T. et al. 1993, A&A, 278, 129
Lin, D.N.C., & Papaloizou, J. 1993, in Protostars & Planets III, ed. E.H. Levy & J.I. Lunine (Tucson: University of Arizona Press), 749
Lin, D.N.C., Hayashi, M., Bell, K.R., & Ohashi, N. 1994, ApJ, 435, 821
Lissauer, J.J., Pollack, J.B., Wetherill, G.W., & Stevenson, D.J. 1994, in Neptune and Triton, ed. D.P. Cruickshank & M.S. Matthews (Tucson: University of Arizona Press), in press
Lynden-Bell, D. & Pringle, J. E. 1974, MNRAS, 168, 603
Mannings, V. 1994, MNRAS, 271, 587
Mannings, V., & Emerson, J.P. 1994, MNRAS, 267, 361
Marsh, K.A., & Mahoney, M.J. 1992, ApJ, 395, L115
Mathieu, R.D. 1994, ARA&A, 32, 405
Mathis, J.S. 1990, ARA&A, 28, 37
McCaughrean, M.J., & Stauffer, J.R. 1994, AJ, 108, 1382
Ménard, F., Monin, J.L., Angelucci, F., Rouan, D. 1993, ApJ, 414, L117
Montmerle, T., Feigelson, E.D., Bouvier, J., & André, P. 1993, in Protostars & Planets III, ed. E.H. Levy & J.I. Lunine (Tucson: University of Arizona Press), 689
Mundt, R., & Fried, J. W. 1983, ApJ, 274, L83
Mundy, L.G., McMullin, J.P., Grossman, A.W., & Sandell, G. 1993, Icarus, 106, 11
Mundy, L.G., Wilking, B.A., & Myers, S.T. 1986, ApJ, 311, L75
Mundy, L.G., Wootten, H.A., Wilking, B.A., Blake, G.A., & Sargent, A.I. 1992, ApJ, 385, 306
Natta, A. 1993, ApJ, 412, 761
Natta, A. 1994, Infrared Physics & Technology, 35, 375
Natta, A., Palla, F., Butner, H.M., & Harvey, P. M. 1993, ApJ, 406, 674
Neuhäuser, R. Sterzik, M.F., Schmitt, J.H.M.M., Wichman, R., & Krautter, J. 1995a A&A, 295, L5
Neuhäuser, R. Sterzik, M.F., Schmitt, J.H.M.M., Wichman, R., & Krautter, J. 1995b A&A, 297, 391

O'Dell, C.R., & Wen, Z. 1994, ApJ, 436, 194
O'Dell, C.R., Wen, Z., & Hu, X. 1993, ApJ, 410, 696
Ohashi, N., Kawabe, R., Hayashi, M., & Ishiguro, M. 1991, AJ, 102, 2054
Ohashi, N., Kawabe, R., Ishiguro, M., & Hayashi, M. 1994, Ap&SS, 212, 239
Omodaka, T., Kitamura, Y., & Kawazoe, E. 1992, ApJ, 396, L87
Ossenkopf, V., & Henning, T. 1994, A&A, 291, 943
Osterloh, M., & Beckwith, S.V.W. 1995, ApJ, 439, 288
Piirola, V., Scaltriti, F., & Coyne, G. V. 1992, Nature, 359, 399
Podosek, F.A., & Cassen, P. 1994, Meteoritics, 29, 6
Pollack, J.B., Hollenbach, D., Beckwith, S.V.W., Sandelli, D.P., Roush, T., et al. 1994, ApJ, 421, 615
Pringle, J.E. 1991, MNRAS, 248, 754
Prosser, C.F., Stauffer, J. R., Hartmann, L., Soderblom, D. R., Jones, B. F., et al. 1994, ApJ, 421, 517
Prusti, T., Natta, A., Palla, F. 1994, A&A, 292, 593
Ray, T.P., Sargent, A.I., Beckwith, S.V.W., Koresko, C., & Kelly, P. 1995, ApJ, 440, L89
Reipurth, B., & Zinnecker, H. 1993, å, 278, 81
Richichi, A., Leinert, C., Jameson, R., Zinnecker, H. 1994, A&A, 287, 145

Rydgren, A.E. & Cohen, M. 1985, in Protostars & Planets II, ed. D.C. Black & M.S. Matthews, (Tucson:University of Arizona Press), 371
Safronov, V.S. 1991, Icarus, 94, 260
Sargent, A.I., & Beckwith, S., 1987, ApJ, 323, 294
Sargent, A.I., Beckwith, S.V.W. 1989, in I.A.U. Colloquium No. 120, Structure and Dynamics of the Interstellar Medium, ed. G. Tenorio-Tagle, J. Melnick, & M. Moles (Berlin: Springer-Verlag), 221
Sargent, A.I. & Beckwith, S.V.W. 1991, ApJ, 382, L31
Sargent, A.I., & Beckwith, S.V.W. 1993, in IAU Colloquium 140, Astronomy with Millimeter & Submillimeter Wave Interferometry, ed. M. Ishiguro & Wm. J. Welch (San Francisco: BookCrafters), 203
Sargent, A.I., & Beckwith, S.V.W. 1994, Ap&SS, 212, 181
Sargent, A.I., Beckwith, S.V.B., Keene, J., & Masson, C.R. 1988, ApJ, 333, 936
Sargent, A.I., Koerner, D.W., & Beckwith, S.V.W. 1995, in preparation
Sargent, A.I., & Welch, W.J. 1993, ARA&A, 31, 297
Shu, F.H., Adams, F.C., & Lizano, S. 1987, ARA&A, 25, 23
Shu, F.H., Johnstone, D., & Hollenbach, D. 1993, Icarus, 106, 92
Shu, F.H., Najita, J., Galli, D., & Ostriker, E. 1993, in Protostars & Planets III, ed. E.H. Levy & J.I. Lunine (Tucson: University of Arizona Press), 3
Simon, M., Chen, W.P., Howell, R.R., Benson, J.A., & Slowik, D. 1992, ApJ, 384, 212
Simon, M., & Guilloteau, S. 1992, ApJ, 397, L47
Skinner, S.L., Brown, A., Walter, F.M. 1991, AJ, 102, 1742
Skrutskie, M.F., Dutkevitch, D., Strom, S.E., Edwards, S., Strom, K.M., et al. 1990, AJ, 99, 1187
Skrutkie, M.F., Snell, R.L., Strom, S.E., Edwards, S., Fukui, Y., et al. 1993, ApJ, 409, 422
Smith, B.A., & Terrile, R.J. 1984, Science, 126, 1421
Snell, R.L., Loren, R.B., & Plambeck, R.L. 1980, ApJ, 239, L17
Spangler, E.C., & Sargent, A.I. 1995, in preparation
Stahler, S.W., Korycansky, D.G., Brothers, M.J., & Touma, J. 1994, ApJ, 431, 341
Stahler, S.W., & Walter, F.M. 1993, in Protostars & Planets III, ed. E.H. Levy & J.I. Lunine (Tucson: University of Arizona Press), 405
Stapelfeldt, K.R., Burrows, C.J., Krist, J.E., Trauger, J.T., Hester, J.J., et al. 1995, ApJ, 449, 888

Stauffer, J.R., Prosser, C.F., Hartmann, L., & McCaughrean, M.J. 1994, AJ, 108, 1375
Stevenson, D.J. 1982, Pl&SS, 30, 755
Strom, K.M., Strom, S.E., Edwards, S., Cabrit, S., & Skrutskie, M.F. 1989, AJ, 97, 1451
Strom, K.M., Strom, S.E., Wilkin, F.P., Carrasco, L., Cruzgonzalez, I. et al. 1990, ApJ, 362, 168
Strom, S.E., Edwards, S., & Skrutskie, M.F. 1993, in Protostars & Planets III, ed. E.H. Levy & J.I. Lunine (Tucson: University of Arizona Press), 837
Strom, S.E., Strom, K.M., Yost, J., Carrasco, L., & Grasdalen, G. 1972, ApJ, 173, 353
Tamura, M., & Sato, S. 1989, AJ, 98, 1368
Walker, H.J. & Wolstencroft, R.D. 1988, PASP, 100, 1509
Walter, F.M., Brown, A., Mathieu, R. D., Myers, P. C., & Vrba, F. J. 1988, AJ, 96, 297
Walter, F.M., Vrba, F. J., & Mathieu, R. D. 1994, AJ, 107, 692
Weidenschilling, S. J. 1989, in The Formation & Evolution of Planetary Systems, ed. H. A. Weaver & L. Danly (Cambridge: Cambridge University Press), 131
Weintraub, D.A., Kastner, J.H., Whitney, B.A. 1995, ApJ 452, L141
Weintraub, D.A., Masson, C.R., & Zuckerman, B. 1989a, ApJ, 344 915
Weintraub, D.A., Sandell, G., & Duncan, W. D. 1989b, ApJ, 340 L69
Weintraub, D.A., Sandell, G., & Duncan, W.D. 1991, ApJ, 382, 270
Whitney, B.A., Hartmann, L. 1992, ApJ 395, 529
Whitney, B.A., Hartmann, L. 1993, ApJ 402, 605
Zuckerman, B. 1995, in Circumstellar Dust Disks and Planet Formation ed. R. Ferlet & A. Vidal-Madjar (Editions Frontières), in press
Zuckerman, B. & Becklin, E.E. 1993, ApJ, 414, 793
Zuckerman, B., Forveille, T., & Kastner, J.H. 1995, Nature, 373, 494

Keplerian Circumstellar Disks in Taurus: The GG Tau Ring and the DM Tau Disk

A. Dutrey[1], S. Guilloteau[1], and M. Simon[2]

[1] Institut de Radio Astronomie Millimétrique, 300 Rue de la Piscine, F-38406 Saint Martin d'Hères, France
[2] Astronomy Program, State Univ. of New-York, Stony Brook, NY 11794 U.S.A.

1. Introduction

We report here the detection and the analysis of two circumstellar keplerian disks around the T Tauri stars GG Tau and DM Tau which are located in the Taurus clouds at about 150 parsecs. The GG Tau system contains a pair of close binaries (Leinert et al. 1991), GG Tau and GG Tau/c, which are separated by 10″. GG Tau itself is a binary with an angular separation on the sky plane of 0.25″ ($\sim$ 38 AU) and, regarded as an unresolved system, GG Tau is an emission line ("classic") T Tauri star. This system is the second strongest 1.3 mm source in the Beckwith et al. (1990) survey. So far, it is the only pre-main sequence object around which dust emission has been clearly resolved (Simon and Guilloteau 1992).

DM Tau is a single T Tauri star with a spectral type M0.5, an age $\sim 5\ 10^6$ years and a mass $\sim 0.65\ M_\odot$ from Mazzitelli (1989) isochrones. Like GG Tau, it is located inside a hole of the surrounding Taurus molecular clouds. In the Beckwith et al. survey, its flux at 1.3 mm is 109 mJy.

2. Observations and Data Reduction

For GG Tau, we used the IRAM Plateau de Bure Interferometer to perform high resolution (~2") and high sensitivity imaging in the ^{13}CO $J = 1 \rightarrow 0$ line and in the continuum at 2.7 mm. The observations were carried out between August 1992 and February 1993. The observing frequency was 110.2 GHz in the lower side-band and 113.2 GHz in the upper side-band. The spectral resolution is 1.6 times the channel spacing (78 kHz), or about 0.35 km·s^{-1}. The continuum images were obtained summing the upper and lower side-band of the continuum correlator. We estimate that the flux density calibration is accurate to 10% while the rms phase noise (between 10 and 25° on all baselines) implies a positional uncertainty better than 0.5″. All maps were produced using natural weighting. Since the source is small and located near ($\sim$ 4″) the beam pointing center, no correction for primary beam attenuation was made.

For DM Tau, we used the IRAM 30m radio telescope to observe in position switching mode several isotopes of CO (^{13}CO and C^{18}O) in the J=1-0 and J=2-1 transitions. The observations were made in June 1994 with an excellent weather for the season. The pointing was good (better than 2″ rms). The line intensity was regularly checked on known sources and their accuracy are estimated to be $\sim 10\%$.

All these data were reduced using the GILDAS software package.

3. Dust and CO Models

In order to analyse the physical parameters of disks, we developed a LTE radiative transfer model for the dust and gas component. We solve the radiative transfer equation by a simple step by step integration along the line of sight and we assume the gas is in hydrostatic equilibrium. For the dust, we assume a mass absorption coefficient $\kappa_d(2.7\,mm) = 0.01$ cm^2/g (Beckwith et al., 1990). The kinetic temperature, the H_2 density and the orbital velocity in the disk have a power law dependence with radius. Finally the computed lines and continuum maps were convolved with the clean beam of the observations and smoothed to the effective velocity resolution to provide a direct comparison with the observations.

4. The GG Tau Disk

We show in Figures 1 and 2 the continuum emission pattern due to cold dust. This feature is consistent with what would be expected from a ring inclined by about 40° along the line of sight and definitely implies that the inner part of the disk is free of dust. We estimate the radius of the dust ring to be about $R_{\rm min} \sim 160 - 180$ AU, for an assumed distance of 150 pc. The spectral image presented in superimposition to the continuum emission in Figure 1, is strikingly similar to those expected from a rotating disk tilted along the line of sight. Further evidence that this rotation is keplerian comes from the agreement with models of keplerian disks (see Figure 2). The velocity-position diagram of Figure 2(2) agrees with the rotation curve around a 1.2 $M_\odot$ central object, in perfect agreement with the total mass of the binary (1.15 $M_\odot$). Finally, comparison between continuum and line emissions (Figure 1) shows that they are centered at the same position (as expected for a rotating disk) and are clearly resolved, with two main peaks coincident with the extreme velocity emission.

4.1 A Circumbinary Disk

The detection of a dusty ring is a strong support for theories of disk-binary interactions. Artymowicz et al. (1991) performed a Smooth Particle Hydrodynamics simulation of the evolution of such a disk around a binary. In the model, the disk surface density peaked near $r = 2.7\ a$ ($a =$ semimajor axis). The main evolution is a change in orbital parameters of the stars, mostly in the eccentricity. The results can be simply understood by considering tidal effects at apastron: the

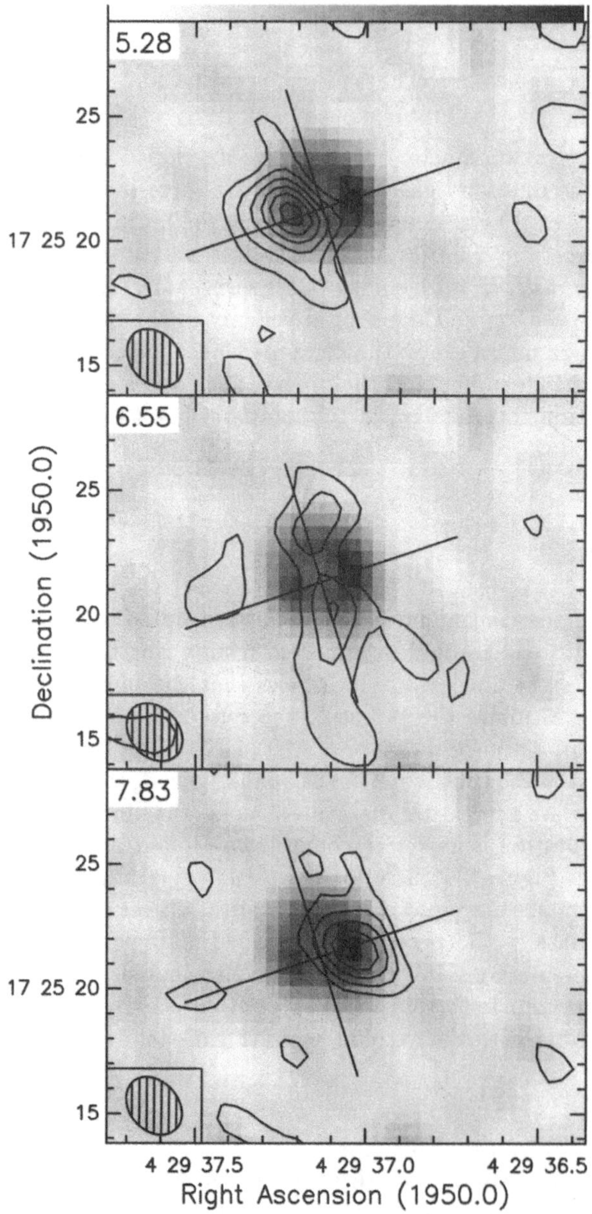

Fig. 1. Superposition of the high resolution continuum map (greyscale) to three velocity channels at LSR velocities: 7.83, 6.55 and 5.28 $\mathrm{km \cdot s^{-1}}$ (contour maps; contour step is 1 K). The cross, tilted by $\sim$ 18° from the R.A. axis, indicates the apparent disk plane and axis. The restoring CLEAN beam for the spectral data is 2.6" $\times$ 2.0" oriented at $P.A. = 45^o$ and 2.1" $\times$ 1.4" with $P.A. = 26^o$ for the continuum data.

binary always rotates faster than the disk which tends to delay the binary. This causes angular momentum and energy transfer from the binary to the disk associated with elongation of the binary orbit and creation of the central hole. Using this model, we predict that the GG Tau binary should have an eccentricity of ~ 0.25 and a semi-major axis a ~ 70 AU.

4.2 Disk Model

We used the model described in the section 3 to reproduce the ^{13}CO $J = 1 \rightarrow 0$ emission and the 2.7mm dust emission maps of GG Tau. Although such a disk model has many parameters, some of them can be estimated relatively directly from the observations or from knowledge of the GG Tau system. The CO abundance and the density law can only be derived from a comparison of model results with the observed images. Because the GG Tau disk is resolved by the interferometer beam, our model does not suffer from beam dilution and these parameters have been determined with a good accuracy.

We present in Table 1 the best model whose results are displayed in Figure 2. This fit can only be obtained if we concentrate about 90% of the disk mass inside a narrow ring of width ~ 200 AU and radius ~ 400 AU.

Table 1. Parameters of the best model for GG Tau used in Figure 2.

Parameter	Law or Value [Unit]
Size	$R_{\text{min}} \sim 180 - R_{\text{max}} \sim 800$ [AU]
Total Mass	$\sim 0.15[\mathrm{M}_{\odot}]$
Inclination	$\sim 43°$, $0°$ means face-on
$T_k(r)$	$30 \cdot (\frac{r}{100\ \mathrm{AU}})^{-0.5}$ [K]
$H(r)$	$14.55 \cdot (\frac{r}{100\ \mathrm{AU}})^{1.25}$ [AU]
$V(r)$	$3.3 \cdot (\frac{r}{100\mathrm{AU}})^{-0.5}[\mathrm{km \cdot s^{-1}}]$
$n(r)$	$2.0 \cdot 10^8 \cdot (11.0\ \exp(-(\frac{r-R_{\text{min}}}{150\ \mathrm{AU}})^2) + (\frac{r}{100\ \mathrm{AU}})^{-2})$
$^{13}CO/H_2$	$5.0 \cdot 10^{-8}$

5. The DM Tau Disk

Figure 3 presents emission line spectra of the CO isotopes towards the nominal position of DM Tau (R.A. $04^h33^m48.70''$ and Dec. $18°10'10.7''$ J2000.0). The velocity component centered on DM Tau, broad and weak, is at $V_{lsr} \simeq 6.1\,\mathrm{kms^{-1}}$

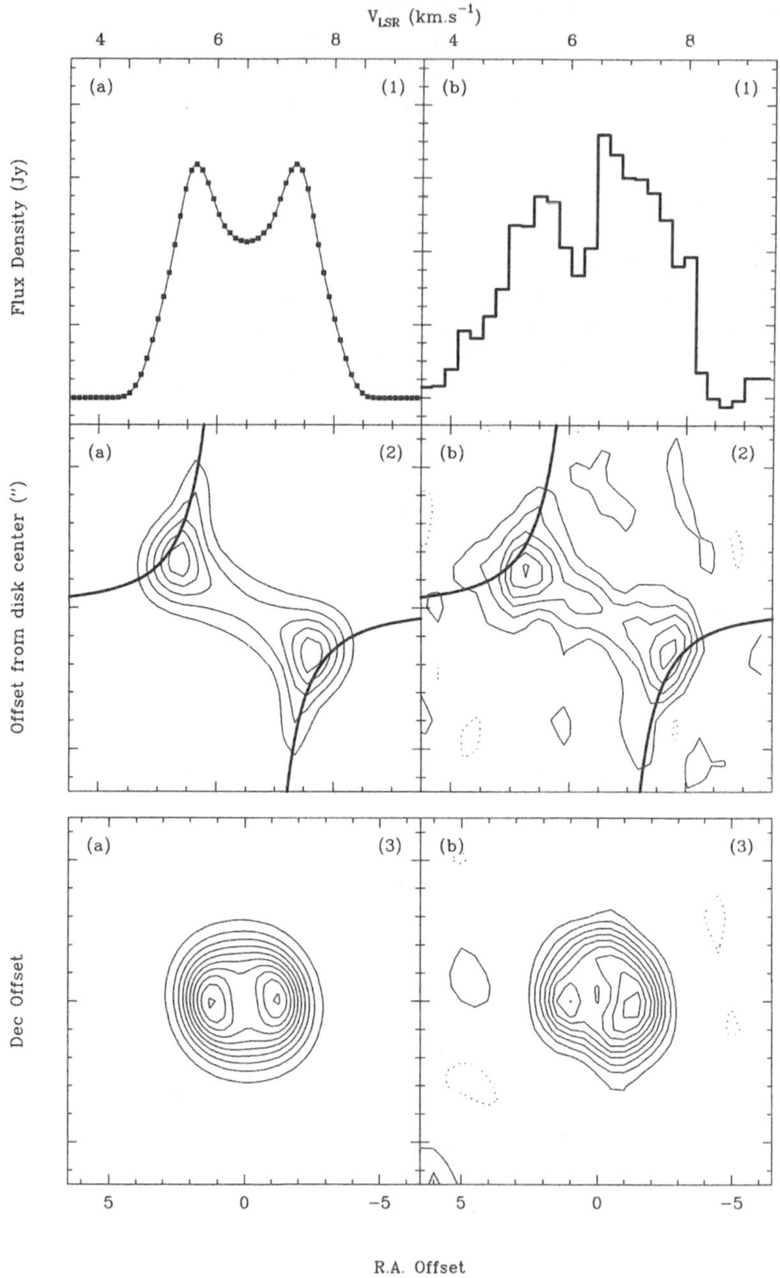

Fig. 2. Results from the best model (a), compared to the observations (b) of GG Tau. (1) ^{13}CO $J = 1 \to 0$ integrated flux density (Jy). (2) Velocity-Position diagram in the plane of the disk. The curve is the theoretical Keplerian rotation curve for 1.2 $M_\odot$ and 43° inclination. Contour step is 1 K (50 mJy/beam). (3) Continuum maps. Contour step is 2 mJy per beam. The model has been convolved with the clean beam used for the observations and smoothed to the effective velocity resolution in order to provide a direct comparison with the observations.

while the relatively narrow line near $V_{lsr} \simeq 9.0\,\mathrm{kms}^{-1}$ corresponds to the residual emission from a rather diffuse cloud along the line of sight also mentioned by Skrutskie et al., (1993). The ^{13}CO J=2-1 emission clearly shows a double peaked profile, as would be expected in the case of a rotating disk, while the high ^{13}CO /^{12}CO ratio indicates that the ^{12}CO emission is optically thick. Moreover, the ^{12}CO J $= 2 \rightarrow 1$ map (Figure 2 in Guilloteau and Dutrey 1994) shows that the disk is marginally resolved by the 30m beam and elongated north/south; since weak ^{12}CO J $= 2 \rightarrow 1$ emission is detected at offsets (0,+12″) and (0,−12″), this implies a disk radius of about $\sim$ 750 AU.

We summarize in Figure 4 the velocity measurements deduced from the ^{12}CO J $= 2 \rightarrow 1$ map and the ^{13}CO J $= 2 \rightarrow 1$ central spectrum: we have superimposed Keplerian curves for several central masses assuming the disk is seen 30° from face-on. Evidence for rotation comes from the fact that all lines north of DM Tau are blueshifted while all lines south of DM Tau are redshifted with respect to the centroid of the double peak profiles towards DM Tau, located at 6.05 ± 0.03 km·s^{-1}.

Table 2. DM Tau: parameters of the model fit

Parameter	Law or Value [Unit]
Size	$R_{max} \sim 750$ [AU]
Total Mass	* ~ 0.03 [$M_\odot$]
Total Mass	** $\sim 1.4 \cdot 10^{-3}$[$M_\odot$]
Inclination	$\sim 30^\circ, 0^\circ$ means face-on
$T_k(r)$	$20 \cdot (\frac{r}{100\ \mathrm{AU}})^{-0.25}$ [K]
$H(r)$	$14.7 \cdot (\frac{r}{100\ \mathrm{AU}})^{1.375}$ [AU]
$V(r)$	$2.42 \cdot (\frac{r}{100\ \mathrm{AU}})^{-0.5}$ [km·s^{-1}]
$n(r)$	$2.5 \cdot 10^7 \cdot (\frac{r}{100\ \mathrm{AU}})^{-2.875}$ [cm^{-3}]
$^{13}CO/H_2$	10^{-6}

Note to Table 2
* Derived from 1.3mm continuum flux.
** Assuming the mentionned ^{13}CO /H_2 ratio (standard)

Then we used our radiative transfer code to compute the expected line profiles from the disk. In the model, we considered that the molecular emission might be produced by i) a Keplerian rotating disk or ii) a non rotating and accreting disk at near free-fall velocities. In this last hypothesis, the observed line intensity ratios are only marginally reproduced by a highly tilted disk (~70°) at low

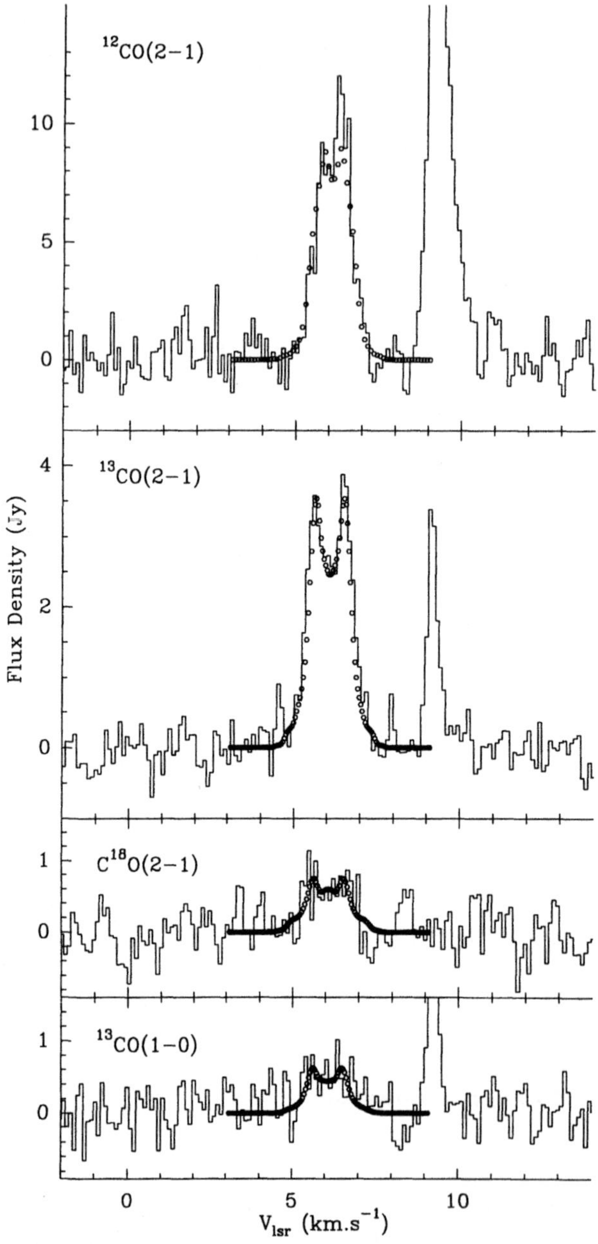

Fig. 3. Spectra of several lines for CO isotopes toward DM Tau. The open circles are synthetic profiles from a Keplerian disk model convolved with the telescope beam.

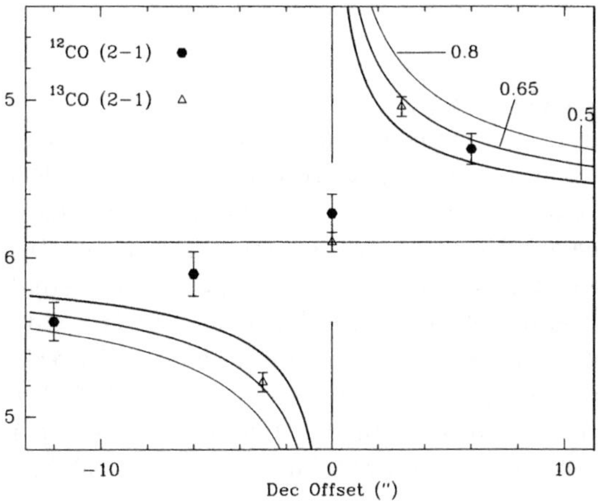

Fig. 4. Velocity diagram for DM Tau, with 3 theoretical Keplerian rotation curves superimposed. The open tirangles indicate velocities derived from the ^{13}CO J=2-1 spectrum towards DM Tau, and were assigned to offsets $\pm3''$ since for Keplerian rotation the two peaks represent emission from the outer edge of the disk.

kinetic temperature (15 K) but this model fails to fit the double-peak profile observed in ^{13}CO $J = 2 \rightarrow 1$ and the measured velocities at offsets $(0,\pm6'')$.

The best fit model is given by a Keplerian disk rotating around a 0.65 $M_\odot$ central mass and the Table 2 summarizes its main characteristics.

Finally, we would like point out the fact that with an age of ~ 5 Myr, DM Tau is the oldest pre-main sequence star for which a Keplerian gaseous disk has been detected, adding some constraint on the timescale for the formation of planets.

6. CO Depletion?

Detailed analysis of the GG Tau interferometer ^{13}CO data shows that ^{13}CO abundance is about 20 times lower than in "normal" molecular clouds. This limit is imposed by the fact that the ^{13}CO $J = 1 \rightarrow 0$ line must be optically thin in the outer part of the disk to reproduce the observed low brightness temperature. Our data strongly support the simple explanation by CO condensation onto grains in the very dense, cold circumbinary disk. This depletion indirectly argues in favour of the high value we have adopted for the absorption coefficient of the dust in the millimeter range, since lower dust emissivity would imply more mass, and even higher depletion. CO observations of the DM Tau keplerian disk show also some evidence of CO depletion inside the disk by a similar amount. However, because the continuum emission is not resolved by the 30m radiotelescope, the DM Tau result can also be explained by a very compact and optically thick core (with a radius of ~ 30 AU) which is superimposed to a large radius disk (~ 750 AU) and is diluted inside the 30m beam. Only high resolution mm interferometric observations will resolve this uncertainty.

7. Conclusion

From these new observations, we conclude that:

- We have imaged and resolved a circumbinary disk in continuum at 2.7mm and in ^{13}CO J=1-0 around GG Tau. Its velocity pattern is consistent with Keplerian rotation in an inclined disk around a 1.2 $M_{\odot}$ star. The ^{13}CO emission is optically thin in most of the imaged disk. The dust emission is concentrated in a very dense ring of radius about 2.4″, or 400 AU. All results are consistent with the expected behaviour of an external disk around a binary system, with masses 0.5 and 0.65 $M_{\odot}$, on slightly excentric orbit of semi-major axis 70 AU.
- We have detected a new circumstellar disk around the single star DM Tau. Modelling of the line profiles and line ratios from the different CO isotopes show that the data strongly support the existence of a large (radius $\sim$ 750 AU) keplerian disk around this star.
- These observations present strong evidence for CO depletion by a factor of about $\sim$ 20. This result is non ambiguous in the GG Tau case because the disk is resolved. However the discrepancies between the disk masses derived from the DM Tau CO measurements and dust emission will only be solved by high resolution mm interferometric observations.

References

Artymowicz, P., Clarke C.J., Lubow, S.H., Pringle J.E., 1991, ApJ 370, L35
Beckwith, S.V.W., Sargent, A.I, Chini, R.S., and Güsten, R., 1990, AJ 99, 924
Guilloteau S., and Dutrey A., 1994, A&A 291, L23
Leinert, Ch., Haas, M., Richichi, A., Zinecker, H., and Mundt R., 1991, A&A 250, 407
Mazzitelli, I. 1989, in: *Low Mass Star Formation and Pre-Main Sequence Objects*, B. Reipurth, ed., ESO Proc. No. 33.
Simon, M., Guilloteau S., 1992, ApJ 397, L47
Skrutskie, M.F., Snell, R.L., Strom, K.M., Strom, S.E., Edwards, S., Fukui, Y., Mizano, A., Hayashi, M., and Ohashi, N., 1993, AJ 409, 422

Circumstellar Disks in the Trapezium Cluster

Mark McCaughrean[1], John Rayner[2], Hans Zinnecker[3], and John Stauffer[4]

[1] Max-Planck-Institute für Astronomie, Königstuhl 17, D-69117 Heidelberg, Germany
[2] Institute for Astronomy, University of Hawaii, 2680 Woodlawn Drive, Honolulu, HI 96822, USA
[3] Institut für Astronomie und Astrophysik, Universität Würzburg, Am Hubland, D-97074 Würzburg, Germany
[4] Harvard-Smithsonian Center for Astrophysics, 60 Garden St., Cambridge, MA 02138, USA

Summary. We discuss the recent discovery and study of circumstellar disks in the Trapezium Cluster via two techniques. First we discuss the population of objects sometimes called 'proplyds', as revealed by ionizing radiation from the central massive OB stars of the Trapezium, and the fraction of these ionized objects that are associated with stars. Second, we look at the wider cluster population and the statistical occurrence of circumstellar disks detected through the more traditional signature of infrared excess. Both techniques reveal that a large fraction ($\sim$ 50–80%) of the young stars in the Trapezium Cluster are likely surrounded by disks.

1. Introduction

1.1 Background

Over the past ten years, the existence of circumstellar disks around young stellar objects (YSOs) has been firmly established, and substantial efforts have been made to elucidate their properties, both observationally and theoretically (see, *e.g.*, the recent reviews of Beckwith & Sargent 1993 and Strom *et al.* 1993). In the Taurus-Auriga and Ophiuchus star-forming regions, disks are found around $\sim$30–50% of all YSOs. Typical host YSOs have masses ranging from 0.3–3 $M_\odot$ and ages in the range 0.5–5 Myr. The disks have masses on the order of 0.01–0.1 $M_\odot$, radii ranging from 10–1000 AU, and they survive for $\sim$10 Myr before being dissipated to the point of ending significant accretion.

However, there is an important caveat to be considered before we use these properties as the basis of a circumstellar disk paradigm. Both the Taurus-Auriga and Ophuichus clouds are low-mass dark clouds and relatively nearby (100–200 pc). Most stars in them have formed in the so-called 'isolated mode', *i.e.*, alone, or in binary and triple systems (see, *e.g.*, Ghez *et al.* 1993; Leinert *et al.* 1993). It is the proximity and relative isolation of these samples of YSOs and disks that has made them amenable to study with a wide range of ground- and space-based instrumentation, resulting in the parameters listed above.

But the dominant mode of star formation in the galaxy is the so-called 'cluster mode' found in the giant molecular clouds (GMCs) (see, *e.g.*, Zinnecker *et al.* 1993; Lada *et al.* 1993). Most molecular gas (the raw material of star formation) in the galaxy is in GMCs, and recent infrared surveys have shown that most young stars in GMCs are to be found in dense clusters ranging from 10–1000 members in size. Thus, most stars in the galaxy probably form in clusters, not in isolation, and disk formation, evolution, and dissipation may be dramatically affected by the dense cluster environment, for example due to star-disk interactions, winds and outflows from massive OB stars in the cluster, and the apparent near-simultaneous formation of a cluster.

1.2 The Observational Challenge

Clearly then, we need to study the properties of circumstellar disks in clusters before drawing conclusions as to the general nature of disks and issues such as the typical time-scale for planet building. In the nearby dark clouds, such studies typically rely on mid-IR to millimetre wavelength bolometry, imaging, and mapping, where the disks are radiating reprocessed stellar radiation and/or accretion luminosity, leading to an excess above the stellar photospheric emission. Unfortunately, discovering and studying disks in dense clusters embedded in GMCs via these techniques is relatively difficult compared to those in the nearby low-mass dark clouds.

First, the nearest star-forming GMCs with dense clusters are at 300–500 pc, *i.e.*, more than twice the distance to Taurus, and thus their stellar members are relatively faint. Second, the high surface density of stars leads to a crowding problem, making long-wavelength observations with low spatial resolution harder to intepret. Third, massive stars in or near the GMC will heat dust in the cloud, leading to a diffuse background emission at long wavelengths. Fourth, many dense clusters are in the galactic plane towards the inner galaxy, leading to confusion with fore- and background stars. Commensurately, disks around YSOs in a GMC should appear fainter and smaller than their Taurus counterparts, and thus be harder to detect and map. It will be harder to separate individual YSOs and disks in clusters with the limited spatial resolution generally available at long wavelengths, and it will be hard to distinguish long wavelength disk emission from the diffuse GMC background. Finally, statistical studies will require special efforts to account for contamination of the population by field stars.

1.3 The Trapezium Cluster

With these observational difficulties in mind, the Trapezium Cluster in the Orion Nebula turns out to be one of the best hunting grounds for disks in a cluster environment. It is one of the nearest clusters ($\sim$450 pc), is located in the galactic anti-centre and well out of the plane, thus minimizing contamination problems, and contains a large number of young ($\sim$1 Myr) stars across a wide mass range, from OB stars to M dwarfs (Herbig & Terndrup 1986; Prosser *et al.* 1994). While traditional optical studies are hampered by the extinction and the bright H II region nebulosity found in massive star-forming regions, as well as the intrinsic redness of cool low-mass stars, these problems are substantially ameliorated at near-IR wavelengths (Zinnecker *et al.* 1993). Near-IR imaging of the Trapezium

Cluster reveals over 700 stars in a 0.65×0.65 pc region (Figure 1), and within the 0.1 pc diameter core, the density rises to an extraordinary $\sim 5 \times 10^4$ stars/pc^3 (McCaughrean & Stauffer 1994). Thus the Trapezium Cluster is an important region in which to search for the possible effects that a dense cluster environment may have on disk formation and dissipation.

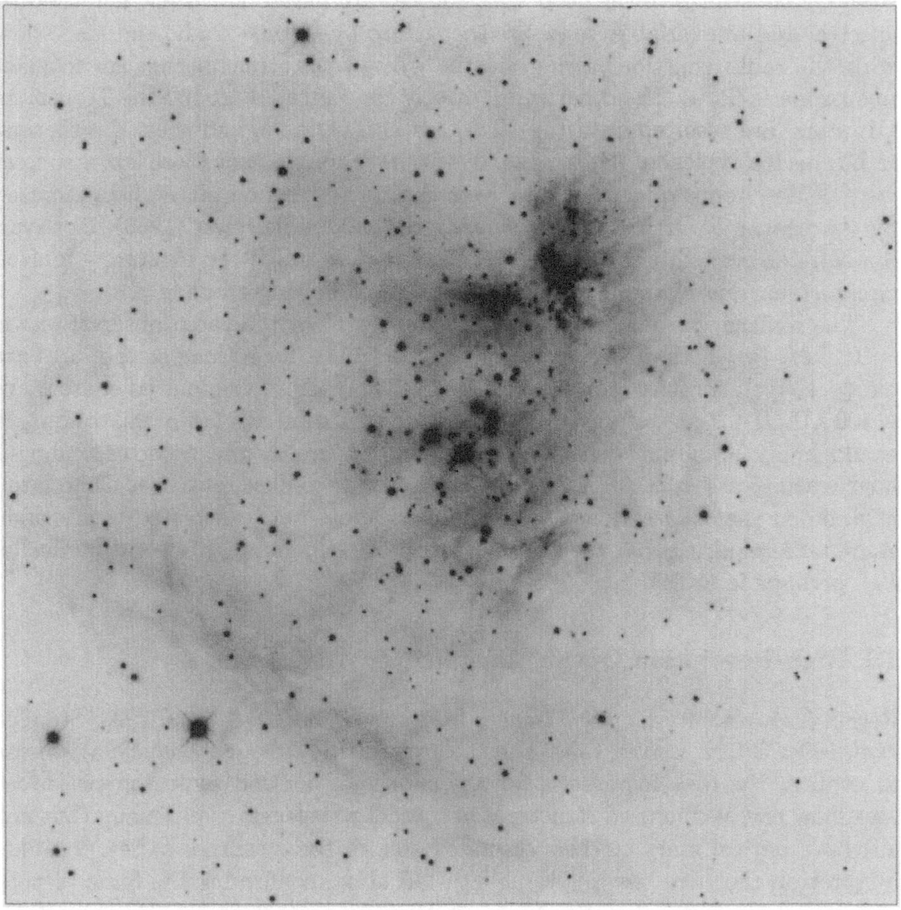

Fig. 1. A near-infrared (2.2 μm, K) image of the Trapezium Cluster taken using MAGIC on the Calar Alto 3.5m telescope in January 1994. The field is $5' \times 5'$ (0.65 pc × 0.65 pc), and the resolution is $0.7''$ FWHM. Over 700 stars are seen, ranging from K=5^{m}–17^{m}. The Trapezium OB stars are at the centre; the BN-KL region is at upper-right, and the Bright Bar is at lower left. North is up and East to the left.

2. Externally Ionized Disks

It has been known for a number of years that the core of the Trapezium Cluster is host to a number of compact ionized sources, first seen in optical emission line images (Laques & Vidal 1979), and later extensively studied in the radio continuum using the VLA (Garay *et al.* 1986; Churchwell *et al.* 1986; Felli *et al.* 1993). Roughly two-thirds of the ~ 50 VLA sources are unresolved, show non-thermal spectra, and are variable: these are thought to be intrinsic radio-emitting stars, with the radio emission arising via the gyro-synchrotron mechanism in giant photospheric flares. The other third, mostly concentrated around the Trapezium OB stars, are resolved with the VLA, are non-variable, and show the thermal radio spectra typical of ionized gas. It was first thought that these latter sources were dense clumps of gas left over from the star-formation phase, being ionized by the nearby θ^1 Ori OB stars following the model of Dyson (1968). However, it was later argued by Churchwell *et al.* that these might be externally ionized circumstellar disks around young stars, on the following grounds.

The ionized gas mass loss rate determined from the radio observations is $\sim 10^{-7}\,M_\odot/\mathrm{yr}$. If the reservoir of material is to last for as long as the stars are old ($\sim 1\,\mathrm{Myr}$), its mass must be $\gtrsim 0.1\,M_\odot$. The radii of the ionized sources are $\sim 100\,\mathrm{AU}$. If $0.1\,M_\odot$ of gas were spherically distributed within this radius, it would imply a column density in excess of $10^{25}\,\mathrm{cm}^{-2}$, and a correspondingly huge extinction ($A_V > 10^3$) assuming a typical gas-to-dust ratio. Yet Churchwell *et al.* noted that at least some of the ionized knots were associated with visible stars, thus requiring that the gas reservoir be distributed highly non-spherically, *i.e.*, perhaps in a disk.

2.1 High-Resolution Optical Imaging

Recent high-resolution optical images from the HST (O'Dell *et al.* 1993; Stauffer *et al.* 1994; O'Dell & Wen 1994) and the ground (McCullough *et al.* 1995) appear to confirm the disk hypothesis for the externally ionized radio sources. Most also show resolved ionized structures at optical wavelengths, and many (but not all) have optical stars at their centre[1]. Some of these sources appear arc-like, brighter on the edge towards the θ^1 Ori OB stars, confirming the basic picture of external ionization by the OB stars. However, rather than seeing the disks directly, models indicate that we are probably seeing ionized material streaming

[1] It was these sources that O'Dell *et al.* (1993) christened 'proplyds', short for proto-planetary disks. The name is somewhat of an overinterpretation in evolutionary terms, as it implies that these structures will necessarily form planets. While this is certainly an attractive idea and drew significant attention from the media, it is by no means clear that planets will be able to form around stars that are in such close proximity to the massive OB stars. At the present time, it seems more appropriate to stay with the strictly physical interpretation implied by the usual name 'circumstellar disk'. Potential confusion between the names proto-planetary disks and proto-planetary nebulae (which have nothing to do with planets) would also be avoided in this way.

off the disk, then interacting with the OB stellar wind at some larger distance (McCullough 1993; McCullough *et al.* 1995; Stauffer *et al.* 1994). In the thermal IR (Figure 2), emission is seen from hot dust in bow-shock structures lying between the ionized radio sources and the OB stars, apparently confirming the basic picture of interaction with the OB star wind (Hayward *et al.* 1994).

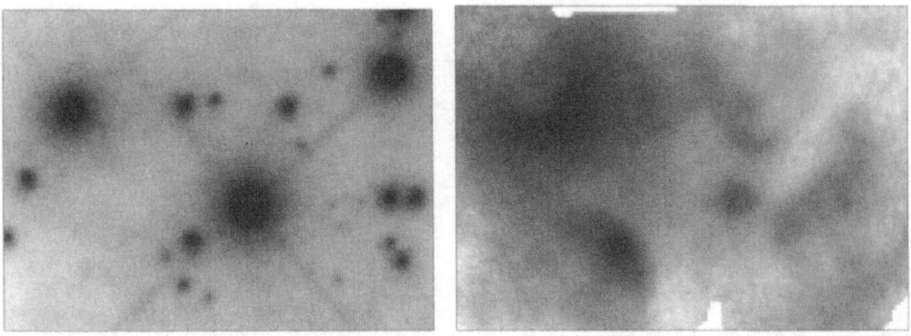

Fig. 2. *Left panel:* The centre of the Trapezium at 2.2 μm, taken using MAGIC and the CHARM tip-tilt compensation system on the Calar Alto 3.5m telescope in September 1993 (McCaughrean & Stauffer 1994). The field is 20×40 arcsec, and the spatial resolution is 0.35″ FWHM. The brightest star near the centre of the image is θ^1Ori C. *Right panel:* The same region at 12 μm, taken using SpectroCam10 on the Hale 5m telescope (Hayward *et al.* 1994). The spatial resolution is 0.5″ FWHM. The stars are no longer seen, only the dust arcs associated with the compact ionized sources. The bright arc at top-left is the Ney-Allen nebula, around θ^1Ori D, a Trapezium OB star.

2.2 A High-Resolution Near-IR Stellar Census

With the discovery of this population of ionized sources in the Trapezium Cluster, some very important questions are raised, namely:

- What fraction of the ionized radio sources can be associated with visible stars? In particular, what fraction of the extended, thermal spectrum radio sources are associated with stars, and are therefore likely to be circumstellar disks, as opposed to isolated dense gas globules?
- Conversely, what fraction of the total stellar population in the core of the cluster is associated with a candidate circumstellar disk?

In order to address these questions, a full census of the stellar population is needed at near-IR wavelengths, where the stars are best detected as described above. As the core of the Trapezium Cluster is crowded and holds a number of very bright OB stars, high spatial resolution imaging techniques are needed to reveal the lower-mass stars. Accordingly, we used shift-and-add and tip-tilt imaging on the Calar Alto 3.5m telescope to obtain 0.6″ and 0.35″ resolution images at broad-band K' (2.1 μm), covering the central 0.2×0.2 pc of the cluster (McCaughrean & Stauffer 1994). Some 123 stars were seen in the region from

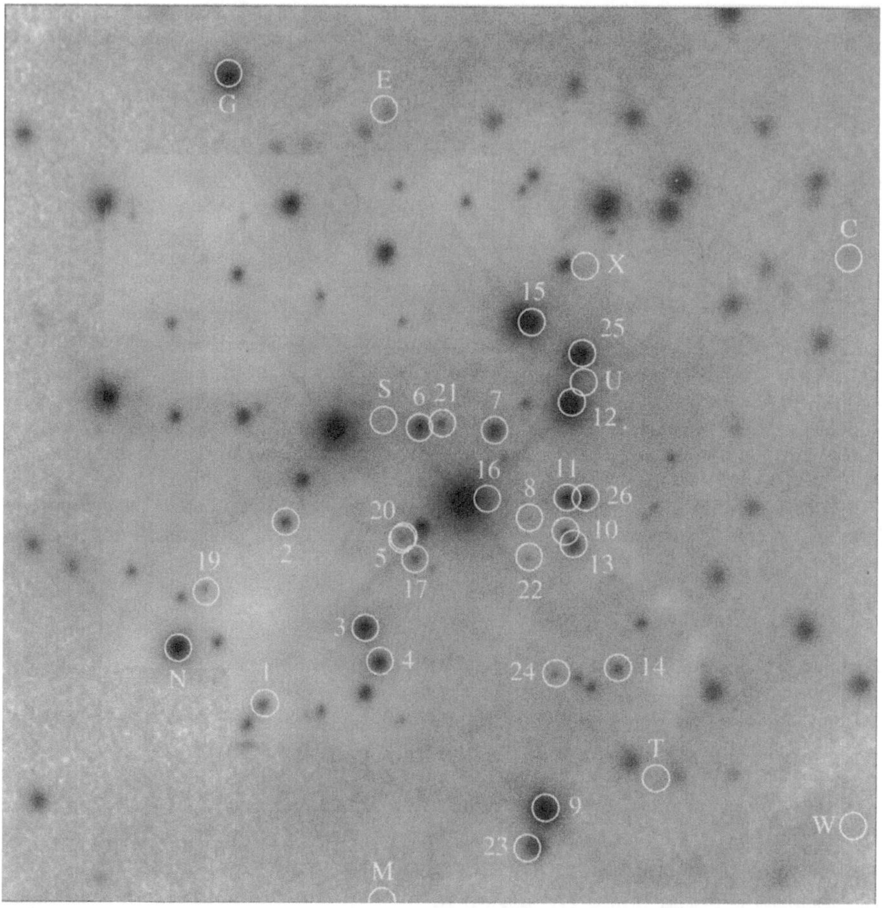

Fig. 3. A high-resolution tip-tilt image of the core of the Trapezium at 2.2 μm with 0.35″ FWHM resolution, taken using MAGIC and CHARM on the Calar Alto 3.5m telescope in September 1993. The outer section is a shift-and-add image with 0.6″ FWHM resolution. Overlaid are the positions and names of the compact radio sources (Felli *et al.* 1993). All but a few are identified with near-infrared stars.

$K \sim 5^m$ down to a limiting magnitude of $K \sim 16^m$, *i.e.*, from OB stars to M dwarfs.

Figure 3 shows a composite of the two images, with the higher-resolution tip-tilt data in the centre. Overlaid are the positions and names of the 35 ionized radio sources which lie within the image, of which 30 are seen to have stellar counterparts. Combined with wide-field near-IR imaging (Figure 1), we find that 40 out of 49 compact radio sources in the Trapezium Cluster are associated with stars, while another 3 have nebular counterparts. Thus, the vast majority of the radio sources are not left-over gas clouds, but are instead star+disk systems.

2.3 The Fraction of Stars with Disks: I

Within a 30″ (0.07 pc) radius of θ^1 Ori C, 20 of the radio sources exhibit the disk-like characteristics of non-variability and thermal radio spectra (Felli *et al.* 1993); all are associated with stars. Within the same radius there are 75 stars, and thus a naive first guess at the fraction of stars with disks might be 20/75, *i.e.*, ~25%. However, some of these stars are actually further than 0.07 pc from the centre, and are simply projected against the core. Using wide-field near-IR images of the Trapezium Cluster to determine the cluster spatial distribution in 3D, we found that a projection correction of ~50% should be made, *i.e.*, roughly half of the 75 stars *seen* within 0.07 pc of the centre really *are* that close to the centre (McCaughrean & Stauffer 1994).

Thus, applying that correction, we find that in the core of the Trapezium Cluster, the fraction of stars with externally ionized disks is 20/38, *i.e.*, ~50% (Stauffer *et al.* 1994).

At smaller radii, the mean density of OB star ionizing photons will increase, and even less favourably oriented disks might be illuminated. Indeed, by repeating the experiment at smaller and smaller radii, we find that the disk fraction increases with decreasing radius, rising to ~100% at 15″ (0.035 pc) radius from θ^1 Ori C.

3. Thermal Excess Emission from Disks

While this core population of ionized disks is very interesting, it might also be 'special' in some way, due to the extreme conditions near the OB stars. The strong ionizing flux and stellar winds that allow us to see the disks may also destroy them more readily, and perhaps sweep ambient cloud material up against them. As such, they might not necessarily be representative of disks around the wider population of stars in the cluster.

To determine the disk fraction over the wider cluster, we must use a method which reveals disks without relying on ionizing photons from a nearby OB star. The classical technique for detecting disks around YSOs in isolated star forming regions is to search for excess IR and millimetre radiation above the stellar photosphere. This emission arises in the disks via local heating, either by the host star or via dissipation as matter is accreted through the disk. Ideally, a search for thermal excess emission should be done at 10 μm and beyond, where a typical young disk strongly dominates the total system luminosity. Unfortunately, the enormous sky background, combined with the relative faintness of systems in clusters at $\gtrsim 500$ pc, and the sheer number of stars to be measured, makes such a survey very difficult from the ground, while existing space-based infrared data (*e.g.*, IRAS) have completely inadequate spatial resolution to resolve the cluster members.

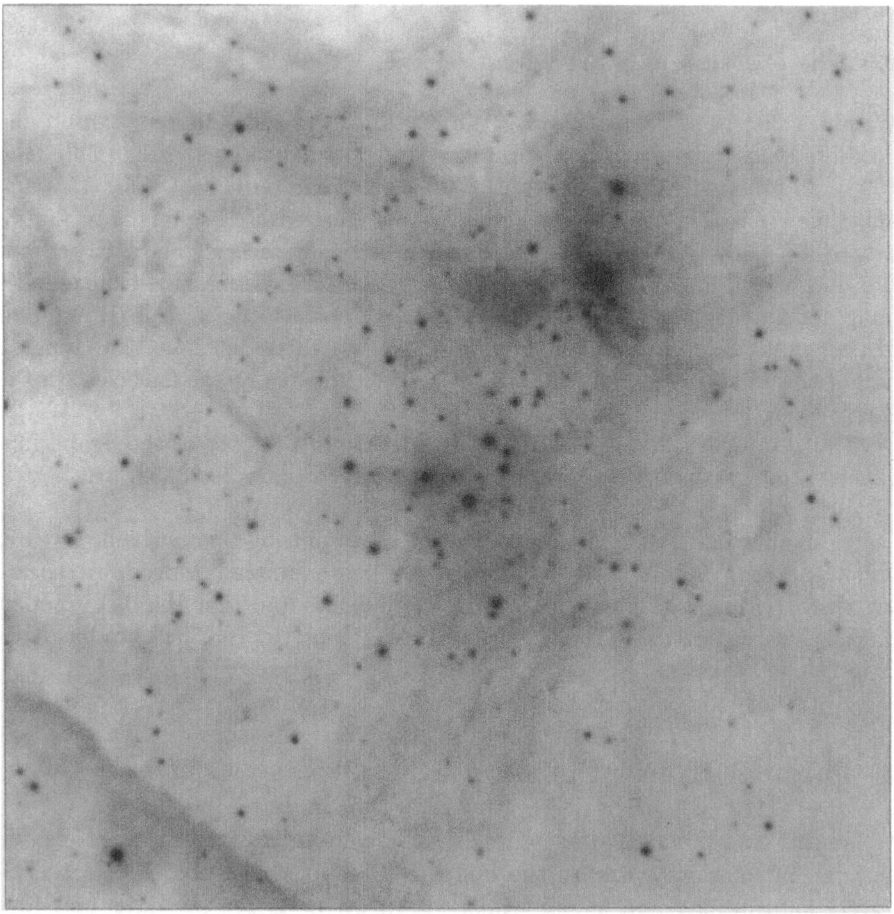

Fig. 4. The central 4×4′ section of the Trapezium Cluster at L (3.4 μm), taken using the NSFCAM on the NASA IRTF in December 1993. Stars as faint as L=13$^{\rm m}$ are detected. Much of the nebular structure is due to the so-called 'dust feature' at 3.3 μm, which lies in the L window. Note the dominance of the BN-KL complex even at these relatively short wavelengths.

However, thermal excess emission from disks can be significant even at 4–5 μm and can be measured by combining imaging photometry taken at 1–5 μm.

We have imaged the central 0.65×0.65 pc of the Trapezium Cluster at 1–2.5 μm (J, H, K) using an HgCdTe array camera on the Calar Alto 3.5m telescope (Figure 1), and a 0.5×0.5 pc sub-section at 3–4 μm (L, L') using an InSb array camera on the NASA IRTF 3.0m telescope on Mauna Kea (Figure 4). In this contribution we present preliminary photometry for some 300 stars which were detected and unsaturated at all wavelengths.

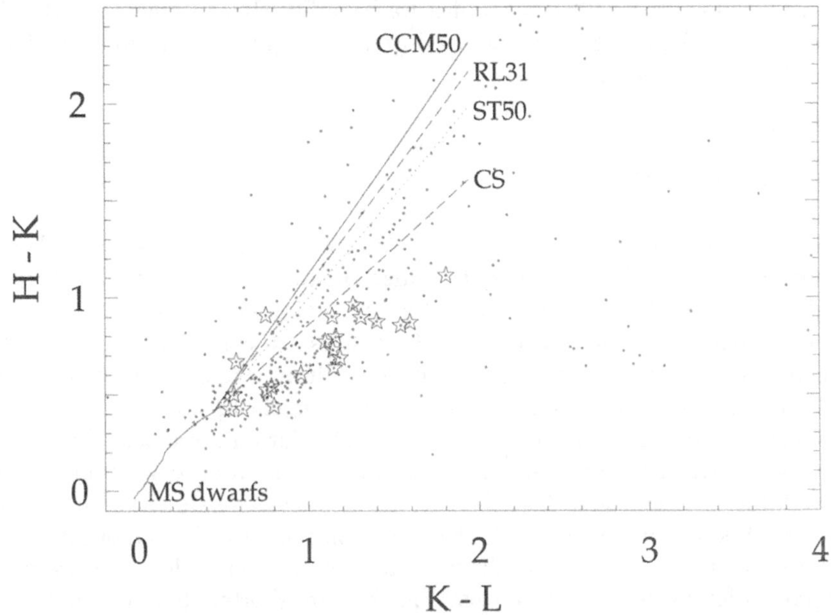

Fig. 5. The (H–K) vs. (K–L) colour-colour diagram for over 300 stars measurable in the wide-field images. Many brighter stars were saturated at J, H, and/or K, while the faintest stars were not detected at L. Several reddening vectors are shown (as described in the text), but regardless of which law is chosen, a great majority of the stars nevertheless exhibit thermal excess emission indicative of disks. The stars are ionized sources from O'Dell & Wen (1994) for which we have full 1–4 μm photometry: almost all appear to show thermal excess emission.

3.1 The Fraction of Stars with Disks: II

The (H–K) vs. (K–L) colour-colour diagram for the present sample is shown in Figure 5. A dwarf main sequence is shown, as are several reddening vectors, including those for normal interstellar extinction (R=3.1; Rieke & Lebofsky 1985; RL31), a model extinction law for larger dust grains (R=5; Steenman & Thé 1989; ST50), and the measured extinction for the Orion Nebula (R=5; Cardelli *et al.* 1989; CCM50). Also shown is a model of extreme scattering in a spherical circumstellar shell (Casali 1994; CS).

Despite the wide area of the diagram covered by these various vectors and the obvious uncertainty that an unknown reddening vector can introduce into the analysis, it is nevertheless clear that the majority of the stars lie *below* all the various vectors, and therefore that they show thermal-IR excess emission.

A rough count of the stars in excess region of the diagram shows that, over the whole Trapezium Cluster, roughly 60–80% show evidence for thermal excess emission, probably best explained by circumstellar disks.

Finally, to tie the two techniques (external ionization and thermal excess) together, we have also plotted in Figure 5 all those externally ionized sources from O'Dell & Wen (1994) for which we have full IR photometry. Virtually all show thermal-IR excess emission, again buttressing the basic circumstellar disk model for these sources.

4. Conclusions

In the core of the Trapezium Cluster, ~50–100% of the stars show evidence for *externally ionized* disks. Across the wider cluster, ~60–80% of the stars show evidence for *locally heated* thermal-IR excess emission from disks.

It is therefore clear that circumstellar disks are common in the Trapezium Cluster. Apparently the very dense cluster environment has had relatively little effect on the disk population, *i.e.*, there are as many (if not more) disks in the Trapezium Cluster as in quieter 'isolated' star-forming regions. However, the studies presented here probe only small parts of the total disk, namely ionized material flowing off the disk or the hotter *inner* disk emission as seen at near-IR wavelengths. It remains plausible that the cooler *outer* disk component has been affected by interactions in the cluster, and that the timescales for accretion and reservoir dissipation in a cluster environment are shorter than 'normal'.

In order to measure (and perhaps map) the outer disks across the entire cluster, mid-IR imaging and millimetre interferometry are needed. While certainly difficult, such projects should soon be possible using the next generation of ground-based thermal-IR cameras, cryogenic IR space observatories (ISO, SIRTF), and mm interferometers with improved *u-v* plane coverage and mosaicing abilities.

Acknowledgement. Thanks to Tom Hayward for permission to use his 12 μm image of the centre of the Trapezium (Figure 2).

References

Beckwith, S. V. W. B. & Sargent, A. I. 1993, in *Protostars & Planets III*, eds. E. H. Levy & J. I. Lunine (Tucson: Univ. Arizona Press), p. 521
Cardelli, J. A., Clayton, G. C. & Mathis, J. S. 1989, *Ap. J.*, **345**, 245
Casali, M. M. 1994, *Astro. Lett. & Comm.*, **29**, 209
Churchwell, E. B., Felli, M., Wood, D. O. S., & Massi, M. 1987, *Ap. J*, **321**, 516
Dyson, J. E. 1968, *Astrophys. Space Sci.*, **1**, 388

Felli, M., Churchwell, E. B., Wilson, T. L., & Taylor, G. B. 1993, *Astron. Astrophys. Suppl.*, **98**, 137
Garay, G., Moran, J. M., & Reid, M. J. 1987, *Ap. J*, **314**, 535
Ghez, A. M, Neugebauer, G., & Matthews, K. 1993, *Astron. J.*, **106**, 2005
Hayward, T. L., Houck, J. R., & Miles, J. W. 1994, *Ap. J*, **433**, 158
Herbig, G. H. & Terndrup, D. M. 1986, *Ap. J*, **307**, 609
Lada, E. A., Strom, K. M., & Myers, P. C. 1993, in *Protostars & Planets III*, eds. E. H. Levy & J. I. Lunine (Tucson: Univ. Arizona Press), p245
Laques, P. & Vidal, J.-L. 1979, *Astron. Astrophys.*, **73**, 97
Leinert, Ch., Zinnecker, H., Weitzel, N., Christou, J., Ridgway, S. T., Jameson, R., Haas, M., & Lenzen, R. 1993, *Astron. Astrophys.*, **278**, 129
McCaughrean, M. J. & Stauffer, J. R. 1994, *Astron. J.*, **108**, 1382
McCullough, P. R. 1993, PhD thesis, University of California-Berkeley
McCullough, P. R., Fugate, R. Q., Ellerbroek, B. L., Higgins, C. H., Spinhirne, J. M., Moroney, J. F., & Cleis, R. A. 1995, *Ap. J*, **438**, 394
O'Dell, C. R., Wen, Z., & Hu, X. 1993, *Ap. J*, **410**, 696
O'Dell, C. R. & Wen, Z. 1994, *Ap. J*, **436**, 194
Prosser, C. F., Stauffer, J. R., Hartmann, L., Soderblom, D. R., Jones, B. F., Werner, M. W., & McCaughrean, M. J. 1994, *Ap. J*, **421**, 517
Rieke, G. H. & Lebofsky, M. J. 1985, *Ap. J.*, **288**, 618
Stauffer, J. R., Prosser, C. F., Hartmann, L. & McCaughrean, M. J. 1994, *Astron. J.*, **108**, 1375
Steenman, H. & Thé, P. S. 1989, *Ap. Space Sci.*, **189**, 189
Strom, S. E., Edwards, S., & Skrutskie, M. F. 1993, in *Protostars and Planets III*, eds. E. H. Levy & J. I. Lunine, (Tucson: Univ. Arizona Press), p837
Zinnecker, H., McCaughrean, M. J., & Wilking, B. A. 1993, in *Protostars & Planets III*, eds. E. H. Levy & J. I. Lunine (Tucson: Univ. Arizona Press), p421

Note: copies of this paper, including figures, can be obtained over the Internet as follows:

- Via anonymous ftp:

Connect to `spitfire.mpia-hd.mpg.de`, login as `ftp`, with your e-mail address as password. `cd` to `pub/preprints`, and `get` the `README` file for more information.

- Via the World-Wide-Web:

Connect to `http://spitfire.mpia-hd.mpg.de/Preprints.html` and follow the links relevant to this paper.

High Resolution Observations of Disks Around Protostellar Sources with the Nobeyama Millimeter Array

Nagayoshi Ohashi[1] and Masahiko Hayashi[2]

[1] Nobeyama Radio Observatory, National Astronomical Observatory, Japan, and Harvard-Smithsonian Center for Astrophysics, 60 Garden st., Cambridge, MA02138, USA

[2] SUBARU project office, National Astronomical Observatory, Japan

Summary. We present high spatial resolution observations of disks around young stellar objects in Taurus with the Nobeyama Millimeter Array. A survey of circumstellar disks at 3 mm shows that the 3 mm continuum emission tends to be weaker for embedded sources than for visible T Tauri stars, which is consistent with our previous interpretation of protoplanetary disk formation. The CS gas around L1489 and IRAS 04365+2535 show disk-like structures of ~1000 AU in radius. A ^{13}CO observation of HL Tau indicates direct evidence of dynamical infalling motion. We discuss the formation of a centrifugally supported viscous accretion disk around HL Tau.

1. Introduction

Disks play a most essential role in star and planet formation. A large number of disks around young stellar objects have been observed, and their properties at the ~10,000 AU scale have been investigated (Kaifu et al. 1984, other references). On the other hand, recent studies have suggested that such large disks commonly have more compact structures, i.e., the protoplanetary disk of ~100 AU scale (e.g., Strom et al. 1989, Beckwith et al. 1990) and the pseudo-disk of ~1,000 AU scale (Galli and Shu 1993a, b). These circumstellar disks should have more direct relationship with star and planet formation than 10,000 AU scale disks. The high spatial resolution of an interferometer is required in order to study such small scale disks. Although some interferometric observations of circumstellar disks were individually made for limited sources (e.g., HL Tau: Sargent and Beckwith 1987, 1991; L1551-IRS5: Sargent et al. 1988, Keene and Masson 1990; IRAS 16293-2426: Mundy et al. 1992; GG Tau: Kawabe et al. 1993, Koerner et al. 1993), a systematic interferometric survey of circumstellar disks has not been previously carried out.

We have completed a survey of 3 mm continuum and CS ($J = 2 - 1$) line emission from protostellar sources associated with the Taurus molecular cloud using the Nobeyama Millimeter Array(NMA) in order to study their associated circumstellar disks. We also started systematic ^{13}CO ($J = 1 - 0$) observations of circumstellar disks with the NMA. The typical spatial resolution of these observations is $\sim 5''$, corresponding to 700 AU at the distance of Taurus ($d =$ 140 pc). In this paper, we present the results of the NMA survey at 3 mm and a ^{13}CO observation of HL Tau. A ^{13}CO observation of T Tau is separately presented in this volume (Momose et al.).

2. NMA Survey in 3 mm Continuum and CS($J = 2 - 1$)

2.1 Summary of the Results

After reporting observations of 11 sources at 3 mm (Ohashi et al. 1991), we completed the survey of 19 young stellar objects.

Table 1. Observational results of the NMA survey at 3 mm.

Observed Source	CS ($J = 2 - 1$)			3mm Continuum		
	I_{CS} [a] [Jy km s^{-1}]	Size [b] [AU]	M_{CS} [c] [$M_\odot$]	F_ν [a] [mJy]	Size [b] [AU]	M_d [d] [$M_\odot$]
Embedded Sources						
L1489	5.6 ± 0.32	2000 · 1700	0.11	< 9.6		< 0.022
04108+2803	1.4 ± 0.32	2700 · 1200	0.026	< 21		< 0.047
04154+2823	< 0.42			< 15		< 0.044
04169+2702	2.6 ± 0.32	2000 · 1200 [e]	0.049	< 14		< 0.032
		1600 · 1500 [e]				
04191+1523	0.80 ± 0.085	1400 · 1400	0.015	< 9.0		< 0.021
04239+2436	2.8 ± 0.20	2700 · 1700	0.054	< 12		< 0.032
04248+2612	1.7 ± 0.16	3800 · 1500 [e]	0.033	< 12		< 0.034
		1800 · 1100 [e]				
04295+2251	< 0.56			< 21		< 0.028
L1551-IRS5	15 ± 0.32	2300 · 1600	0.28	131 ± 4.0	< 430 · 360	0.070
04325+2402	0.86 ± 0.080	1800 · 1400	0.016	< 9.2		< 0.021
04361+2547	1.4 ± 0.32	2100 · 1500 [e]	0.026	< 14		< 0.032
		2100 · 1500 [e]				
04365+2535	2.5 ± 0.14	1500 · 1500	0.048	30 ± 3.8	< 940 · 700	0.068
L1527	11 ± 0.32	2600 · 2000	0.21	< 11		< 0.025
T Tauri Stars						
FS Tau	< 0.34			< 11		< 0.041
T Tau	< 0.60			56 ± 10	< 550 · 420	0.10
DG Tau	1.4 ± 0.32	1600 · 730	0.026	57 ± 4.0	< 800 · 770	0.10
HL Tau	1.5 ± 0.32	1200 · 850	0.028	74 ± 4.0	< 530 · 420	0.15
GG Tau	< 0.44			41 ± 4.0	< 780 · 670	0.16
DL Tau	< 0.37			23 ± 4.0	< 1470 · 880	0.10

[a] Upper limit values are 3s and errors are 1σ [b] FWHM size [c] H_2 mass derived from the CS total integrated intensity [d] H_2 mass derived form the total flux density of the dust emission [e] Two spatially separated components are identified

The 19 sources are consist of 13 optically invisible embedded sources and 6 visible T Tauri stars.

Table 1 summarizes our results. The 3 mm continuum emission was detected toward 5 T Tauri stars out of 6, while only 2 embedded sources out of 13 possess detectable 3 mm continuum. All the continuum detections were not spatially resolved (see Figure 1), implying that the emission arises from compact protoplanetary disks. The implied disk masses of the 5 T Tauri stars are ~0.1 $M_\odot$, and those of the embedded sources are ~0.07 $M_\odot$.

In contrast, CS emission was detected toward 11 embedded sources out of 13, while 2 T Tauri stars out of 6 showed detectable CS emission. The CS emission is extended as compared with the continuum emission: the FWHM extent of the emission is ~2,000 AU around embedded sources and ~1,000 AU around T Tauri stars. The masses of the CS condensations associated with the embedded sources are ~0.01–0.3 $M_\odot$ and those associated with the T Tauri stars are ~0.03 $M_\odot$. The detailed method of the mass estimation is described in Ohashi et al. (1991).

2.2 Formation of Protoplanetary Disks

The most striking result in Table 1 is the strong contrast of the continuum and line detectabilities between the embedded sources and T Tauri stars; embedded sources generally have detectable CS ($J = 2 - 1$) emission without detectable 3 mm continuum emission except for two sources, whereas the opposite trend was observed for T Tauri stars. Both, embedded sources without as well as with detectable continuum emission are considered to be associated with compact disks because (1) their spectral energy distributions are well fitted with a model including a compact disk (e.g., Adams, Lada, and Shu 1987), and (2) a non-spherical mass distribution on the scale of ~100 AU is suggested from their low visual extinctions of ≤100 mag (Myers et al. 1987). The upper limit to the disk mass of the observed embedded sources without detectable continuum is ~0.03 $M_\odot$. This is consistent with the recent result that the disk masses of embedded sources in Taurus are at most ~0.01 $M_\odot$ (Moriarty-Schieven et al. 1994). Moreover, recent observations have shown that ~ 40% of classical T Tauri stars in Taurus have disk masses larger than 0.01 $M_\odot$ (Osterloh and Beckwith 1995), which is also consistent with the high continuum detectability for T Tauri stars in the present study. Such a contrast between the embedded and visible objects means that the disk masses of embedded sources are generally smaller than those of T Tauri stars. This suggests that the disk mass must grow rapidly in evolution from embedded to visible sources, as has been discussed on the basis of a disk formation scenario in a previous paper (Ohashi et al. 1991). In this picture, the main accretion phase continues until the early T Tauri phase, which is consistent with a recent kinematic observation of dynamical accretion toward a young T Tauri star, HL Tau (Hayashi, Ohashi, & Miyama 1993, see also Sect. 3).

Galli & Shu (1993a, b) calculated the inside-out collapse of a magnetized cloud core, and suggested that an infalling extended disk of ~1,000 AU in radius is formed in the embedded phase although a compact circumstellar disk is not yet well developed. Nakamoto & Nakagawa (1994, also presented in this volume) suggested through a numerical calculation of disk formation that protoplanetary disks grow suddenly in the late stage of the infalling (embedded) phase . These results support our scenario of disk formation and evolution

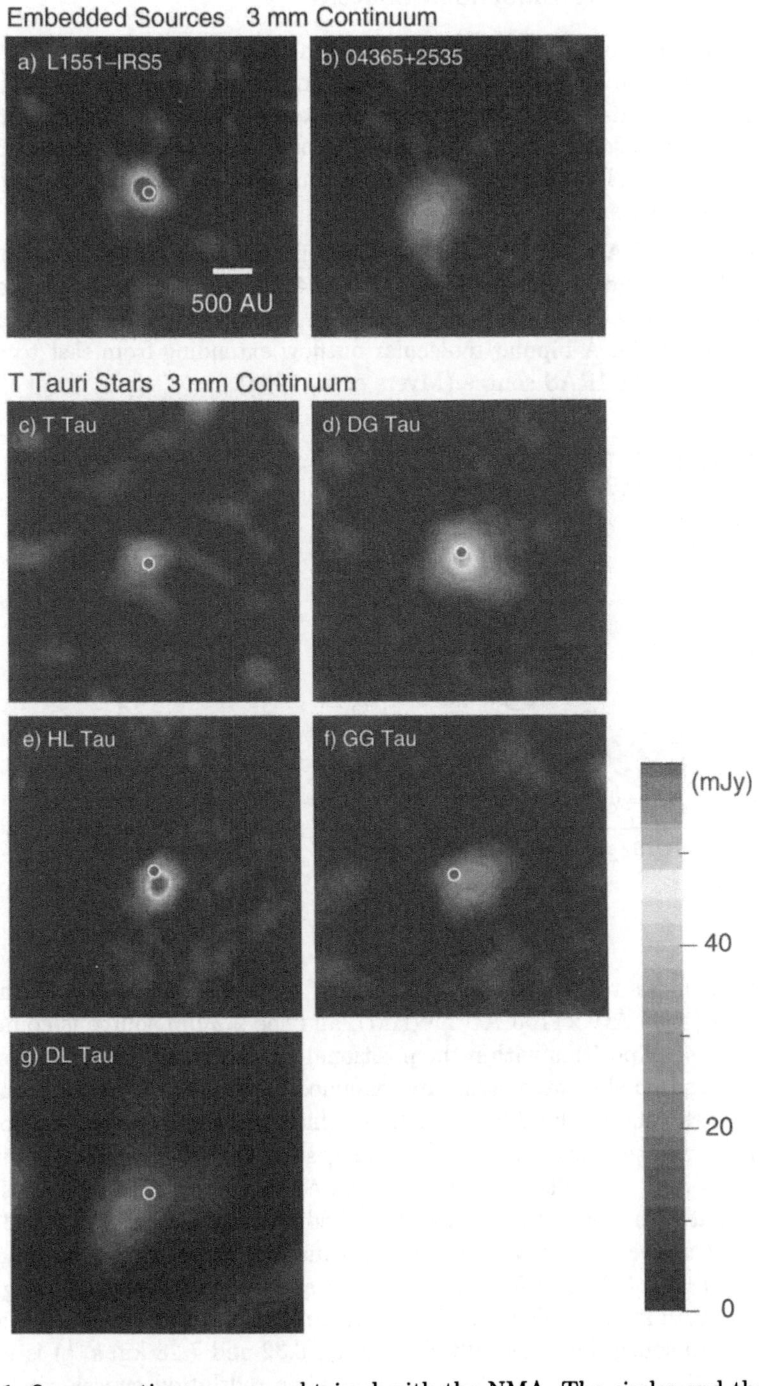

Fig. 1. 3 mm continuum maps obtained with the NMA. The circles and the plus in (b) indicate the position of the 2.2μm sources (Tamura et al. 1991). [This figure is presented in color on CD-ROM.]

2.3 CS Gas Around Embedded Sources

In studying the process of star and disk formation, it is important to investigate molecular line emission as well as dust continuum emission. In particular, information regarding the velocity structure derived from line observations is of great value. We discuss CS gas kinematics around 4 embedded sources (L1489, IRAS 04365+2535, L1527, and L1551-IRS5) that show remarkable velocity structures in the NMA survey.

2.3.1 L1489 (IRAS 04016+2610). L1489 is one of the dense cores observed in NH_3 (Benson and Myers 1989), and IRAS 04016+2610 is located at the western edge of the core. A 2.2 μm point source was detected toward the IRAS source (Myers et al. 1987). A bipolar molecular outflow extending from east to west is associated with the IRAS source (Myers et al. 1988).

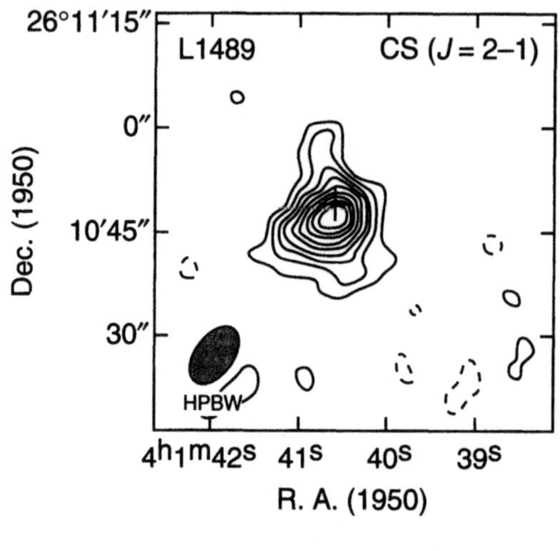

Fig. 2. CS integrated intensity map of L1489. The cross marks the position of the 2.2 μm source (Myers et al. 1987).

Figure 2 shows the CS integrated intensity map of L1489. The CS emission is extended (2000 AU $\times$1700 AU FWHM), and the 2.2 μm source is coincident with the CS peak position within the positional uncertainty of the 2.2 μm source (3″). We found an elongated structure extended by $\sim 3,000$ AU with a position angle of $\sim$ 15° across the 2.2 μm source. This elongated structure is roughly perpendicular to the axis of the outflow, suggesting that the elongated structure is a gaseous disk. The disk radius is estimated to be $\sim 1,500$ AU, consistent with the spatial extent of the pseudo-disk predicted by Galli & Shu (1993a, b).

Figure 3 shows the CS velocity channel maps. CS emission is significantly detected between 6.32 and 8.24 km s^{-1}, and additional emission is marginally detected at 9.20 km s^{-1}. The velocity channel maps indicate a velocity gradient from north to south, i.e., blueshifted emission(6.32 and 7.28 km s^{-1}) is mainly detected to the north of the 2.2 μm source while redshifted emission(8.24 and 9.20 km s^{-1}) is mainly detected to the south. This velocity gradient is along the major axis of the disk structure, implying rotational motion.

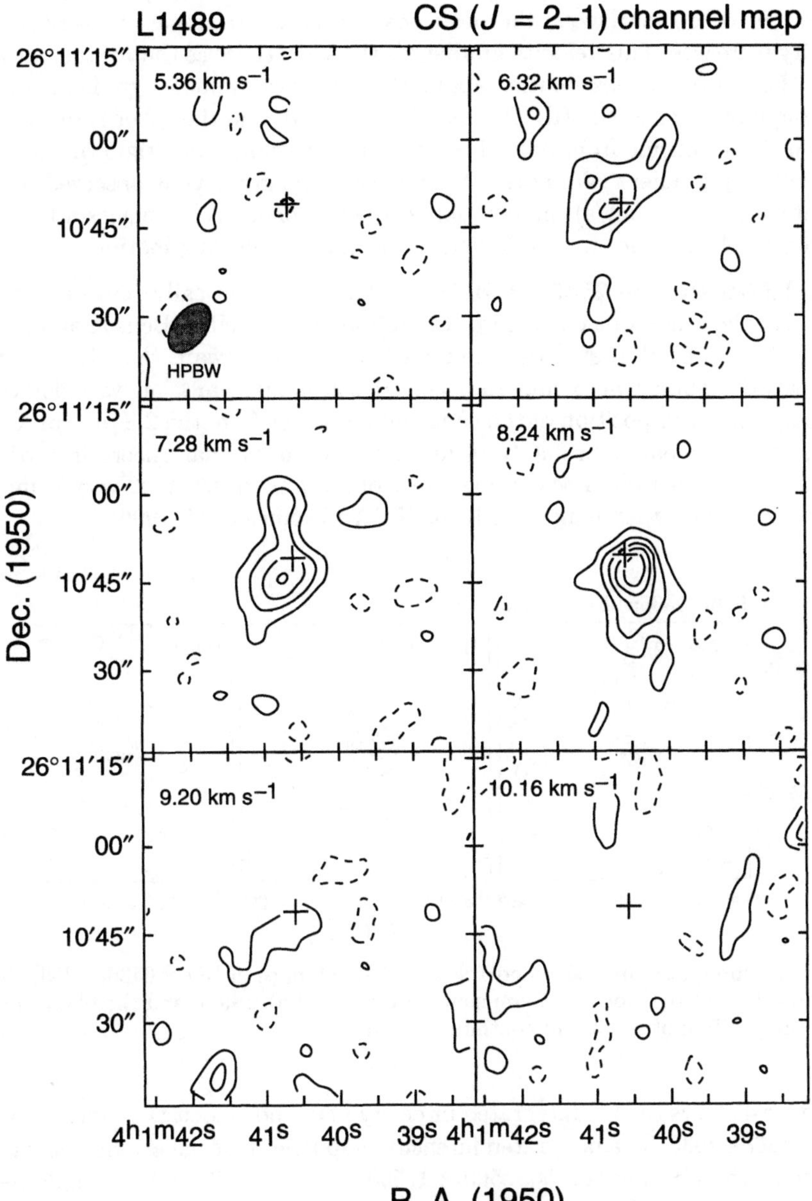

Fig. 3. CS velocity channel maps of L1489. The plus marks the position of the 2.2 μm source (Myers et al. 1987).

Inspection of Figure 3 may indicate that there is also a velocity gradient from east to west. The peaks at the blueshifted velocities are located to the southeast of the 2.2 μm source, and the peak at the redshifted velocity of 8.2 km s^{-1} is located to the southwest; the peak position moves from east to west as the velocity increases. This velocity gradient is the same sense as that of the outflow. Thus there are two possibilities about this east-west velocity gradient; one is infalling motion as in the HL Tau case (see Sect. 3), and the other is outflowing motion. It is difficult from only these data to determine the origin of the east-west velocity gradient. We note, however, that the gradient is observed at low velocities ($\Delta V \sim 1$ km/s), namely line core component. This suggests that the velocity gradient arises from infalling rather than outflowing motion.

2.3.2 IRAS 04365+2535 (TMC1A). This is an optically invisible source located at the center of a dense core (L1543) observed in NH_3 (Benson and Myers 1989). Tamura et al. (1991) detected a 2.2 μm source toward the IRAS source. In the NMA survey at 3 mm, unresolved continuum emission was detected. Although the peak position of the continuum is offset from the 2.2 μm source by $\sim 3''$, this discrepancy probably results from the positional uncertainty of the 2.2 μm source, or from a scattering effect of 2.2 μm emission. A very compact molecular outflow was imaged with the NMA (Tamura et al. 1996).

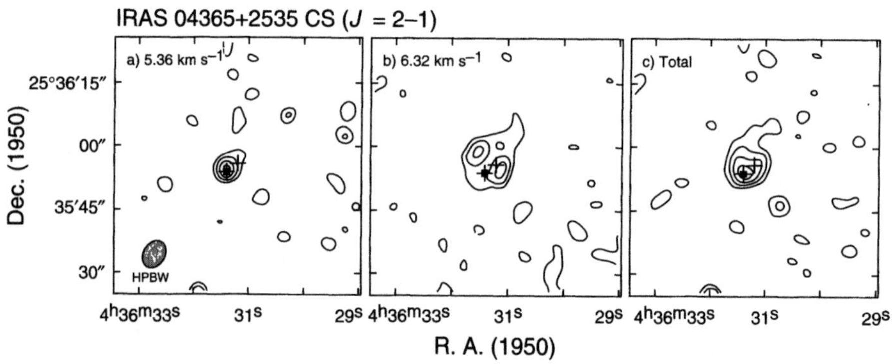

Fig. 4. CS integrated intensity and velocity channel maps of IRAS 04365+2535. The plus indicates the position of 2.2 μm source (Tamura et al. 1991), and the filled square shows the position of the 3 mm continuum source.

Figure 4 shows the CS integrated intensity map and velocity channel maps of IRAS 04365+2535. The integrated intensity map (the right panel) indicates that the size of the CS emission is $1,500 \times 1,500$ AU in FWHM with a single peak that is coincident with the 3 mm continuum position. CS emission is detected in two channels; one is centered at nearly the systemic velocity (6.32 km s^{-1}; the middle panel), and the other is at a blueshifted velocity (5.36 km s^{-1}; the left panel). No significant CS emission was detected at redshifted velocities. The CS emission at 5.36 km s^{-1} has a single peak coincident with the continuum position, while the 6.32 km s^{-1} component shows two peaks; one peak is located to the northeast of the continuum position, and the other is located to the southwest of the continuum source. The two peaks are approximately equidistant from the continuum source.

Moreover the two peaks are symmetrically placed with respect to the compact outflow (Tamura et al. 1994). Figure 5 shows the CS emission at 6.32 km s^{-1} superimposed on the blue- and redshifted lobes. We find from Figure 5 that the intensity peaks of the blue- and redshifted lobes are located between the two peaks, and that the two peaks align perpendicularly to the axis of the outflow. This suggests that the two peaks are the edges of a disk or a ring surrounding the outflow. The radius of the disk/ring is estimated to be ~900 AU, which is again consistent with the predicted size of the pseudo-disk (Galli and Shu 1993a, b). Figure 5 reminds us that the two peaks pinch the outflow so that the outflow is well collimated. The disk/ring may play an important role in the collimation of the outflow at the 1,000 AU scale.

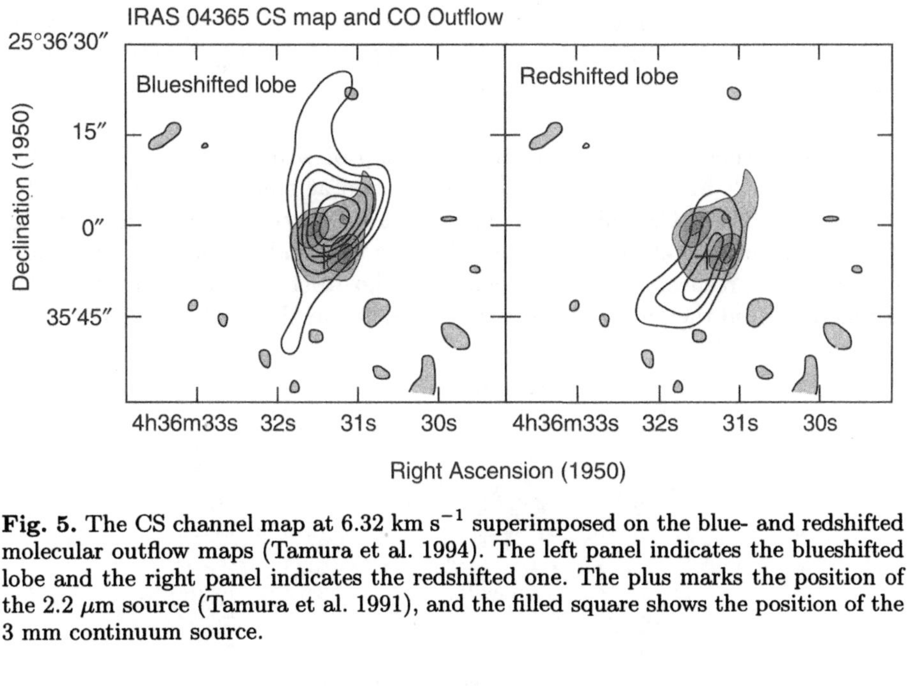

Fig. 5. The CS channel map at 6.32 km s^{-1} superimposed on the blue- and redshifted molecular outflow maps (Tamura et al. 1994). The left panel indicates the blueshifted lobe and the right panel indicates the redshifted one. The plus marks the position of the 2.2 μm source (Tamura et al. 1991), and the filled square shows the position of the 3 mm continuum source.

2.3.3 L1527 (IRAS 04368+2557). L1527 is one of the dense cores observed in NH_3 (Benson and Myers 1989), and IRAS 04368+2557 is located at the center of the dense core. This IRAS source shows the coldest color temperature of the 19 sources in our sample, and its high visual extinction of ~1,000 mag indicates that this source is very deeply embedded in the core (Ladd et al. 1991). Tamura et al. (1991) showed that there is no near-infrared point source toward the IRAS source although extended nebulosity is detected at the east side of it. Spatially unresolved continuum emission was detected at 2.7 mm with OVRO (Terebey et al. 1992) and at 2 mm with the NMA (Ohashi et al. 1995). This continuum position is considered to indicate the position of the central source. A CO outflow, which shows a v-shaped structure centered on the continuum position, was imaged with the NMA (Tamura et al. 1995).

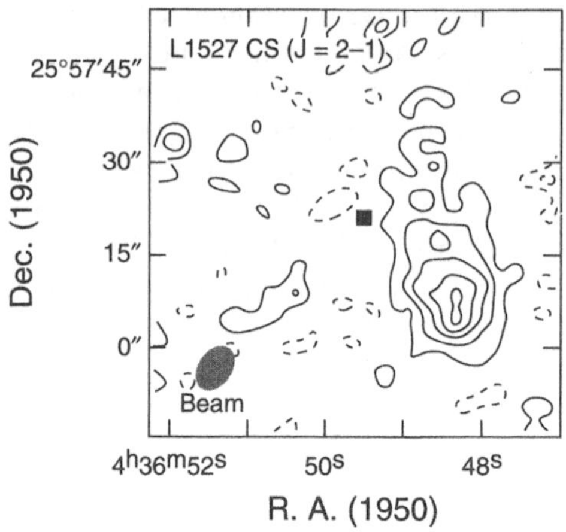

Fig. 6. CS integrated intensity map of L1527. The filled square indicates the position of the 2 mm continuum source (Ohashi et al. 1995).

Figure 6 shows the CS total intensity map of L1527 integrated from 3.44 to 7.28 km s^{-1}. Most of the emission is detected at blueshifted velocities. It is remarkable that no CS emission is detected at the continuum position, although strong and weak CS emission is detected to the west and southeast of the continuum position, respectively. This means that the observed CS gas around L1527 is different from the gas observed toward L1489 and IRAS 04365+2535.

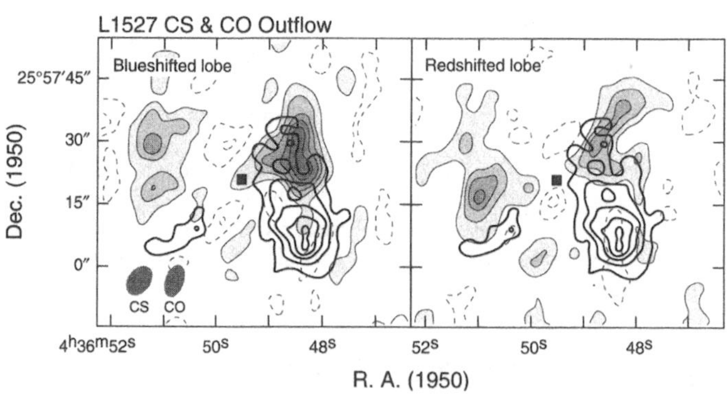

Fig. 7. The 04368+2557 CS map with thick contours superimposed on the CO outflow map in gray scale(Tamura et al. 1995). The filled square indicates the position of the 2 mm continuum source (Ohashi et al. 1995).

Figure 7, which compares the CS map with the v-shaped CO outflow observed with the NMA (Tamura et al. 1994b), shows the anticorrelation between the CS and the CO outflow gas; strong CS emission is located at the southern edge of the western part of the outflow, and weak CS emission delineates the southern edge of the eastern red lobe. This anticorrelation suggests that the CS gas is a dense shell swept up by the outflow.

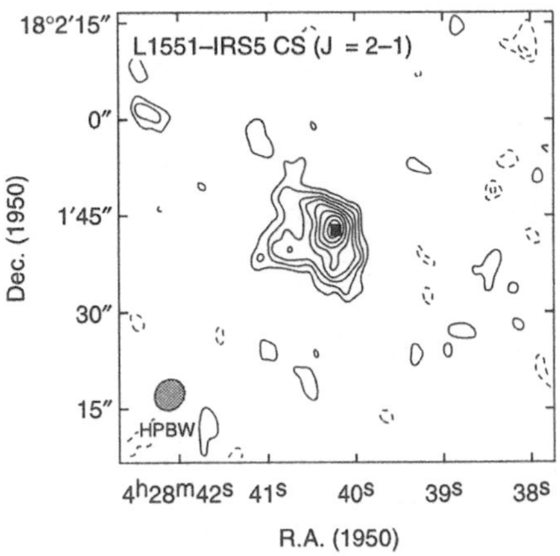

Fig. 8. CS integrated intensity map of L1551-IRS5. The square indicates the position of the 2 cm radio continuum source (Rodríguez et al. 1986).

2.3.4 L1551–IRS5 (IRAS 04287+1801). We present the CS total integrated intensity map in Figure 8. The CS peak position is coincident with the 2 cm radio continuum position (Rodríguez et al. 1986). The size of the CS condensation is $2,300 \times 1600$ AU in FWHM, and the emission is elongated from northeast to southwest. The molecular outflow associated with L1551-IRS5 (Snell, Loren, & Plambeck 1980) is extended in the same direction as the CS emission. The CS emission has an additional component that extends to the southeast. This elongated structure is roughly perpendicular to the axis of the outflow. Sargent et al. (1988) found from $C^{18}O$ observations with OVRO a gaseous disk with a position angle of $\sim 135°$, which is consistent with the CS extension to the southeast.

Figure 9 shows the CS velocity channel map of L1551-IRS5. Strong CS emission is detected at 6.32 km s^{-1} and 7.28 km s^{-1} though the CS emission is detected over 6 channels. The blueshifted CS emission at 6.32 km s^{-1} is mainly detected to the southwest of the central source while the redshifted CS emission at 7.28 km s^{-1} extends to the northeast. This implies a velocity gradient from southwest to northeast. This gradient is in the same sense as that of the outflow. Thus, the extended CS structures to the southwest and northeast may result from outflowing motion. On the other hand, there is no clear velocity gradient along the axis perpendicular to the outflow.

3. Disk Formation Around HL Tau

Recent ^{13}CO observations of HL Tau with the Nobeyama Millimeter Array have revealed direct evidence of dynamical infall in the gas disk around it (Hayashi,

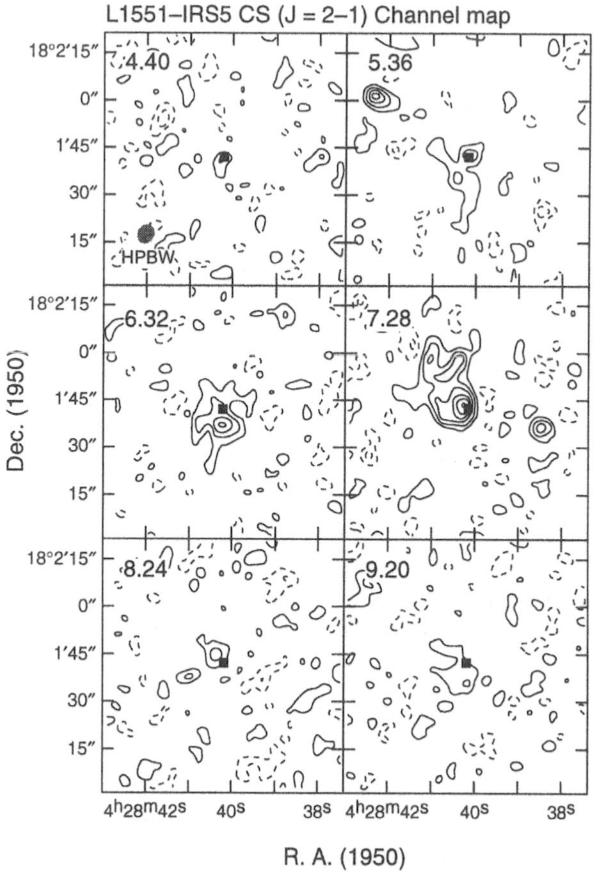

Fig. 9. CS velocity channel maps of L1551-IRS5. The square indicates the position of the 2 cm radio continuum source (Rodríguez et al. 1986).

Ohashi, & Miyama 1993). This result gives evidence for the formation of a centrifugally supported compact disk at the very central part of the infalling disk as discussed below (Lin et al. 1994).

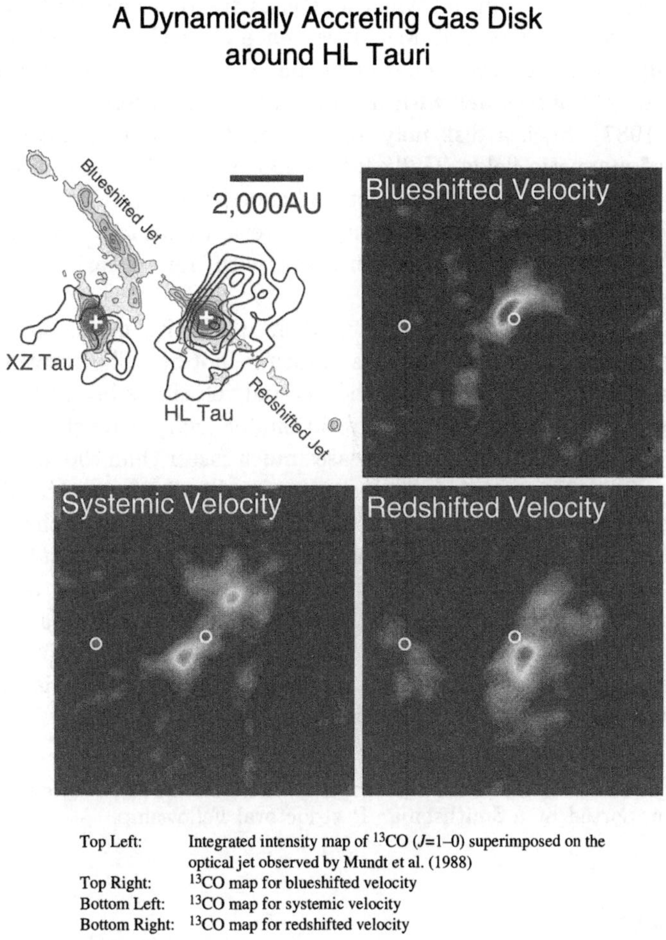

Fig. 10. ^{13}CO integrated intensity map and channel maps of HL Tau. [This figure is presented in color on CD-ROM.]

Figure 10 summarizes the result. The upper left panel shows an integrated intensity map of the $^{13}\mathrm{CO}(J = 1 - 0)$ emission superimposed on emission from the optical jet (Mundt et al. 1988). The other three panels are velocity channel maps at blueshifted (upper right), systemic (lower left), and redshifted (lower right) velocities. The jet is blueshifted at its northeastern part and redshifted at its southwestern part. Because the disk is perpendicular to the jet, the northeastern part of the disk is the far side, and its southwestern side is the near side. The three velocity channel maps show that the disk is approaching us at its far side, and is receding away at its near side, suggesting that the disk is shrinking toward the central star.

The radius and mass of the infalling disk are $\sim 1,400$ AU and 0.03 $M_\odot$, respectively. The measured infall velocity, ~1 km s^{-1} at a radius of 700 AU, is consistent with that acquired when matter is freely accelerated by the central star which has a mass of 0.55 $M_\odot$. It is hence clear that the infalling disk is a dynamically accreting protostellar envelope. The derived mass accretion rate of 5×10^{-6} $M_\odot$ yr^{-1} agrees well with theories of star formation (see e.g. Shu, Adams & Lizano 1987). Such a disk may form through dynamical collapse under the influence of magnetic fields (Galli & Shu 1993a, b). The derived accretion rate is comparable to those estimated for embedded sources from their bolometric luminosities. This means that the main accretion phase still continues even at the early T Tauri phase. This is consistent with our scenario of compact disk formation discussed in Sect. 2.2.

The disk also shows a velocity gradient along its major axis, which implies that the infalling disk also has rotating motion. The rotational velocity is ~0.2 km s^{-1} at 700 AU. Because the free-fall velocity is inversely proportional to the square root of radius and the rotational velocity is inversely proportional to radius, the rotational velocity increases much faster than the infalling velocity as matter infalls. The rotational velocity eventually dominates the kinematics of the disk, so that a viscous accreting disk forms. The radius of the centrifugally supported disk is easily estimated as the radius where the rotational velocity becomes equal to the infalling velocity, and is ~30 AU. Recent interferometric observations which actually resolve the compact disk structure around HL Tau indicate a size of ~60 AU (Lay et al. 1994). The case of HL Tau is, therefore, an excellent example of the formation of a centrifugally supported viscous accretion disk through dynamical accretion from a protostellar envelope.

Acknowledgement. We thank J. Najita for helpful comments on the original manuscript. N.O. was supported by a Smithsonian Postdoctoral Fellowship.

References

Adams, F.C., Lada, C.J, Shu, F.H. 1987: ApJ 312, 788
Beckwith, S.V.W., Sargent, A.I., Chini, R., Güsten,R. 1990: AJ 99, 924
Benson, P.J., Myers, P.C. 1989: ApJS 71, 89
Galli, D., Shu, F.H. 1993a] ApJ 417, 220
Galli, D., Shu, F.H. 1993b: ApJ 417, 243
Hayashi, M., Ohashi, N., Miyama, S.M. 1993: ApJL 418, L71
Kaifu, N. Hasegawa, T., Morimoto, M., Inatini, J., Nagane, K., Miyazawa, T., Akabane, K., Suzuki, S. 1984: A&A 134, 7
Kawabe, R., Ishiguro, M., Omodaka, T., Kitamura, Y., Miyama, S. 1993: ApJL 404, L63
Koerner, D.W, Sargent, A.I., Beckwith, S.V.W. 1993: ApJL 408, L93
Keene, J., Masson, C.R. 1990: ApJL 355, 635

Ladd, E.F., Adams, F.C., Casey, S., Davidson, J.A., Fullar, G.A. Harper, D.A., Myers, P.C., Padman, R. 1991: ApJ 366, 203
Lay, O.P., Carlstrom, J.E., Hills, R.E., Phillips, T.G. 1994: ApJL 434, L75
Lin, D.N.C., Hayashi, M. Bell, K.R., Ohashi, N. 1994 ApJ 435, 821
Moriarty-Schieven, G.H., Wannier, P.G., Keene, J., Tamura, M. 1994: ApJ 436, 800
Mundt, R., Ray, T.P., Burke, T. 1988: ApJL 333, L69
Mundy, L.G., Wotten, H.A., Wilking, B.A., Blake, G.A., Sargent, A. 1992: ApJL 385, 306
Myers, P.C., Fuller, G.A., Mathieu, R.D., Beichman, C.A., Benson, P.J., Schild, R.E., Emerson, J.P. 1987: ApJ 319, 340
Myers, P.C., Heyer, M., Snell, R.L., Goldsmith, P.F. 1988: ApJ 324, 907
Nakamoto, T., Nakagawa, Y. 1994: ApJ 421, 640
Ohashi, N., Kawabe, R., Hayashi, M., Ishiguro, M. 1991: AJ 102, 2054
Ohashi, N., Hayashi, M., 1995: Ap&SS 224, 130
Osterloh, M., Beckwith, S.V.W. 1995: ApJ 439, 288
Rodríguez, L.F., Cantó, J., Torrelles, J.M., Ho, P.T.P 1986: ApJL 301, L25
Sargent, A.I., Beckwith, S. 1987: ApJ 323, 294
Sargent, A.I., Beckwith, S. 1991: ApJL 382, L31
Sargent, A.I., Beckwith, S., Keene, J., Masson, C.R. 1988: ApJ 333, 936
Snell, R.L., Loren, R.B., Plambeck, R.L. 1980: ApJL 239, L17
Shu, F.H., Adams, F.C., Lizano, S. 1987: ARAA 25, 23
Strom, K.M., Strom, S.E., Edwards, S., Cabrit, S., Skrutzkie, M.F. 1989: AJ 97, 1451
Tamura, M., Gatley, I., Waller, W., Werner, M. 1991: ApJL 374, L25
Tamura, M., Ohashi, N., Moriarty-Schieven, G., Hayashi, M., Hirano, N. 1994: in Astronomy with Millimeter and Submillimeter Wave Interferometory, IAU Colloquium No. 140, edited by M. Ishiguro & W.J. Welch (Astronomical Society of the Pacific, San Francisco) 268–269
Tamura, M., Ohashi, N., Hirano, N., Itoh, Y., Moriarty-Schieven, G.H. 1996: in preparation
Terebey, S., Vogel, S.N., Myers, P.C. 1992: ApJ 390, 181

Spatially Resolved Spectroscopy of Infrared Companions to T Tauri Stars

Evidence for Ongoing Accretion

T.M. Herbst[1], C.D. Koresko[2], and Christoph Leinert[1]

[1] Max-Planck-Institut für Astronomie, Königstuhl 17, 69117 Heidelberg, Germany
[2] Department of Astronomy, University of Texas, Austin, Texas, USA 78712–1083

Summary. We describe the technique of spatially resolved spectroscopy of close binary stars and present K band spectra of the Haro 6–10 and UY Aur infrared companion (IRC) systems. The IRC's are characterized by very low infrared colour temperature and high luminosity. Our observations address the debate between those who suggest that the IRC's are at an earlier evolutionary stage than their primary stars, and those who believe that these objects are relatively normal pre-main-sequence stars with substantial local extinction and accreting matter.

In both Haro 6–10 and UY Aur, we detect 2 μm molecular hydrogen emission confined to the infrared companions. The line ratios are consistent with shock heating, strengthening the view that the IRC's are experiencing ongoing accretion. In the case of UY Aur, non-spherical geometry, such as accretion onto a circumstellar disk, can explain the presence of molecular hydrogen emission, low effective temperature, and high luminosity while allowing moderate extinction to the photosphere at near infrared wavelengths.

1. Introduction

The infrared companions (IRC) are a small but important class of pre-main-sequence objects that have very low infrared colour temperatures ($<$ 1500 K) and high infrared luminosities. This colour temperature is lower than that of any plausible star, yet they are found in binary systems containing a relatively normal primary. Although there are relatively few known infrared companion systems, several well-studied young stellar objects are in fact IRC binaries. These include T Tau, Haro 6–10, Z CMa, and XZ Tau, which is the bright source to the east of HL Tau on the poster for this conference. More infrared companion systems will almost certainly be discovered in the coming years as instrumentation and observing techniques improve.

The IRC's are usually too faint to detect at visible wavelengths, yet they are frequently more luminous than their primaries. For example, Figure 1 shows J, H, and K band images of the infrared companion system Do-Ar 24E. The IRC is barely detected at 1.2 μm, whereas at 2 μm, the companion begins to rival the primary. At longer wavelengths, the IRC dominates the system flux (Chelli *et al.* 1988).

High interstellar extinction cannot explain this behaviour. With the coolest and lowest-mass stars having T$\sim$2500 K, approximately 10 magnitudes of extinction would be required to redden such an object to give the J – K colours

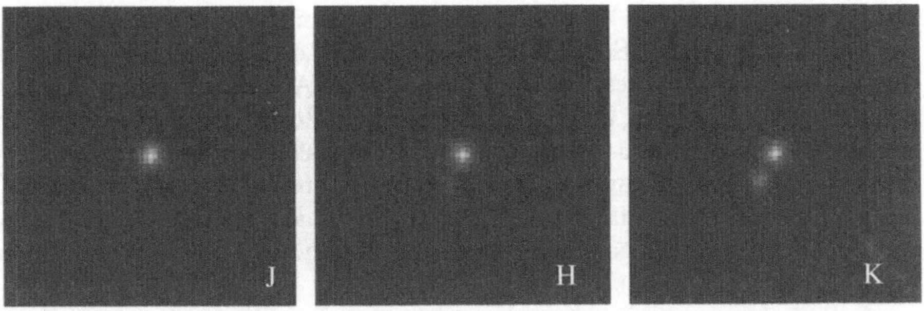

Fig. 1. The Do–Ar 24E Infrared Companion System in the J ($1.13 - 1.37\mu$m), H ($1.5 - 1.8\mu$m), and K ($2.01 - 2.42\mu$m) photometric bands. The binary separation is 2.2 arcseconds.

observed in the IRC's. If due to interstellar matter, this extinction would imply an underlying object several times more luminous than observed, and such a luminosity far exceeds that of any possible 2500 K star (Koresko, 1993). Also, the presence of an apparently normal object 0.1–1.0 arcseconds away would require implausible gradients in the interstellar extinction. This suggests that any extinction must be confined to within 100–200 AU of the IRCs.

The recognition that the IRC phenomenon is restricted to the immediate vicinity of the binary system has led some investigators to suggest that the infrared companions are at an earlier evolutionary stage than the primary star (Zinnecker, 1989). Others believe that these objects are relatively normal, albeit heavily extincted pre–main–sequence stars with matter accreting directly onto the photosphere or onto a circumstellar disk, (Bertout 1983, Ghez *et al.* 1991). In the latter case, the large infrared luminosity arises from the release of potential energy as matter accretes onto the IRC.

The accretion scenario is strengthened by the apparent correlation between the presence of an infrared companion in a binary system and spectral signatures of shocks, in particular the $v = 1 - 0$ S(1) line of molecular hydrogen at 2.12 μm. Beckwith *et al.* (1978) detected this line in emission in only one of four young stars they observed. This star, T Tau, was later discovered to have a very red binary companion, and it became the prototypical IRC system, (Dyck, Simon, and Zuckerman, 1982). In the last ten years, approximately half a dozen additional IRC systems have been discovered, and most show molecular hydrogen in emission.

If the accretion picture is correct, a binary system containing a relatively normal primary and an actively accreting infrared companion should produce spatially unresolved molecular hydrogen emission centered on the IRC with flux ratios between the various transitions characteristic of warm, shocked gas. However, with relatively faint H_2 lines, small projected separations and up to three magnitudes in brightness difference, these binaries present a formidable observational challenge.

We have begun the first program of infrared spectroscopy of IRC binaries with sufficient sensitivity and spatial resolution to identify the emission region.

Our observations of the Haro 6–10 and UY Aur infrared companion systems demonstrate that the H_2 S(1) line is indeed confined to the IRC's in these sources. The next section introduces the Haro 6–10 and UY Aur binary systems. Section 3 describes the observing and data reduction procedures, and presents spatially resolved spectra of Haro 6–10 and UY Aur. A discussion of possible origins of the molecular hydrogen emission appears in section 4, and the final section contains conclusions and some initial results from other binary systems. A substantial portion of this work appears in an *Astrophysical Journal Letters* article, now in press (Herbst *et al.* 1995).

2. Haro 6–10 and UY Aur

Haro 6–10 is a bright pre–main–sequence object embedded in a 10″ conical nebula in the L 1524 cloud in Taurus. Leinert and Haas (1989) discovered the infrared companion 1″.2 north of the primary using one–dimensional speckle interferometry. Haro 6–10 varies at short infrared wavelengths, but appears constant at 10 μm. The presence of a T Tauri primary and an infrared companion dominating the mid infrared flux naturally explains this observation. Carr (1990) detected the 2.12 μm molecular hydrogen emission line toward Haro 6–10 in a 3″.7 aperture. Herbst and Koresko (hereafter HK91) measured a molecular hydrogen flux similar to that of Carr using the CGS4 spectrometer on UKIRT (also spatially unresolved).

UY Aur is a 0″.9 separation IRC system that is unique in that the infrared companion was apparently a visible star within the last 50 years. The separation and position angle reported by Joy and van Biesbroeck (1944) are consistent with recent infrared observations (Leinert *et al.* 1993, Ghez *et al.* 1994). Any interpretation of the origin and evolutionary status of infrared companions must encompass such extraordinary changes. HK91 detected hydrogen Brγ and relatively weak H_2 emission in a $\sim$ 3 arcsecond region centered on the UY Aur system.

The UY Aur companion is currently not a visible star. An R band CCD frame taken at the Calar Alto 3.5m telescope in 1992 February shows no emission at the location of the IRC. Simple false star addition tests indicate that the companion must be at least five magnitudes fainter than the primary at this wavelength. Joy and van Biesbroeck reported a $\sim$ 0.5 mag difference in 1944.

3. Observing Technique, Data Reduction, and Spectra

We observed all the close binary systems in the K photometric band (2.0–2.4 μm), using the MAGIC infrared camera / spectrometer on the 3.5 m telescope on Calar Alto, Spain. MAGIC uses a 256×256 NICMOS3 HgCdTe array and contains optics and drive electronics that permit a wide variety of observing modes. A direct–ruled ZnSe grism gives complete K band coverage at a spectral resolution R$\sim$ 360 ($\frac{\lambda}{\Delta\lambda}$), and a resin replica grism provides *simultaneous* K and H band (1.5–1.8 μm) coverage at R$\sim$ 260. The observer can switch rapidly ($<$ 30 seconds) between direct imaging and spectroscopic modes, since the slits and grisms are on motorized wheels. Herbst *et al.* (1993, 1994a, 1994b) describe the infrared camera and grisms.

We observed Haro 6–10 on 25 September 1993 using the direct–ruled ZnSe grism and the imaging configuration in MAGIC, giving spectral resolution R$\sim$360 and 0$''$.33 pixel^{-1}. The UY Aur measurements of 28 January 1994 used the resin replica grism and the speckle interferometry configuration, giving R$\sim$260 and 0$''$.07 pixel^{-1}. Figure 2 shows an averaged "shift and add" spectrum of the UY Aur binary system. This frame represents the best 30 of 128 exposures, each of 1 second duration. The two components of this 0$''$.88 binary are clearly separated with the 0.07 arcsecond pixel scale. The individual traces are 0.22 arcseconds wide (full width at half maximum), close to the diffraction limit of the telescope at this wavelength.

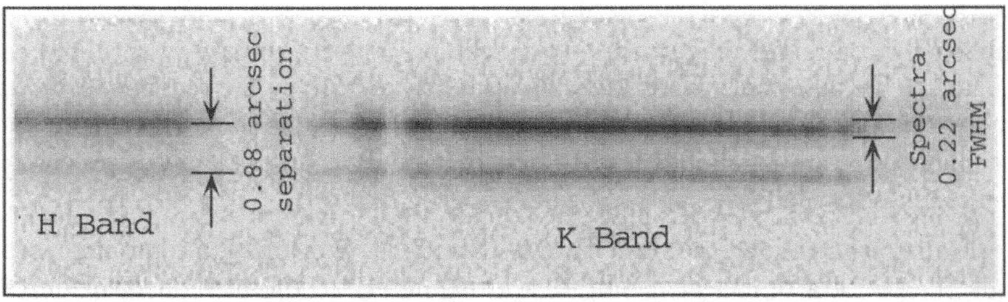

Fig. 2. Spectrum of the UY Aur infrared companion system. The two traces are completely distinct, despite the 0.88 arcescond separation of the binary. The lower star is the infrared companion.

Figure 3 shows the observing procedure schematically. After finding the binary system, we rotate the Cassegrain flange of the telescope to align the two stars along the slit. Inserting the grism produces a pair of close but distinct spectra for the two components of the binary. Taking multiple exposures with the stars shifted up and down the slit allows cancellation of thermal background, airglow lines, and bad pixels.

The data reduction proceeds as follows: After subtracting an equal–duration exposure with the stars shifted and dividing by a dome flat, we use the IRAF *twodspec* package to extract the individual traces of the two components (Figure 4). Dividing the resulting spectra by that of a spectroscopic standard, then multiplying by a blackbody at the same effective temperature (as the standard), eliminates telluric features and yields the correct spectral slope. The final spectra are the average of many such extracted traces.

The wavelength calibration comes from a fit to the atmospheric OH emission lines, extracted at the same location as the stars from the average of all frames with no stellar emission at that position (see Figure 3). The narrow slit makes absolute photometric calibration impossible. Comparison with the

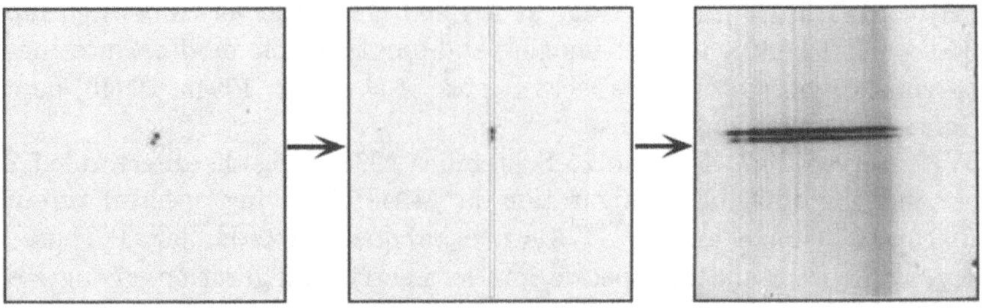

Fig. 3. Observing procedure for close binary stars. Note the airglow lines above and below the stellar traces and the thermal emission at the long wavelength (right) side of the K band spectrum.

published broadband magnitudes of the primary stars (Leinert *et al.* 1993) establishes the flux scale. The known variability of pre–main–sequence objects makes this procedure less than ideal, but the careful alignment of the slit and the small separations should lead to accurate relative brightnesses between the components.

Figure 5 shows spectra of the Haro 6–10 and UY Aur infrared companion systems. These data represent a total of approximately 10 minutes on each source. Note the vertical scales on the plots, since the infrared companions are fainter than their primaries at these wavelengths. Table 1 lists the line fluxes for the hydrogen Brγ and H_2 S(1) lines in each object, based on the photometric scaling described in the previous section. The table includes previous measurements for comparison.

Table 1. H_2 and Brγ line fluxes for the Haro 6–10 and UY Aur infrared companion systems. The earlier measurements of Carr (1990) and Herbst and Koresko (1991) did not have sufficient spatial resolution to separate the primaries from the companions.

Object	Line[a]	Primary	IRC	Total	HK91[b]	Carr (1990)[c]
Haro 6–10	H_2	–	2.14	2.14	11.0	6.8
	Brγ	15.4	–	15.4	12.6	< 3.1
UY Aur	H_2	–	7.2	7.2	8.9	not observed
	Brγ	–	12.6	12.6	52.0	not observed

[a] All fluxes in 10^{-14} erg cm^{-2} s^{-1}
[b] Herbst and Koresko unpublished data. Observed in 1991 November.
[c] Observed in 1986.

The following paragraphs summarize the results for each system.

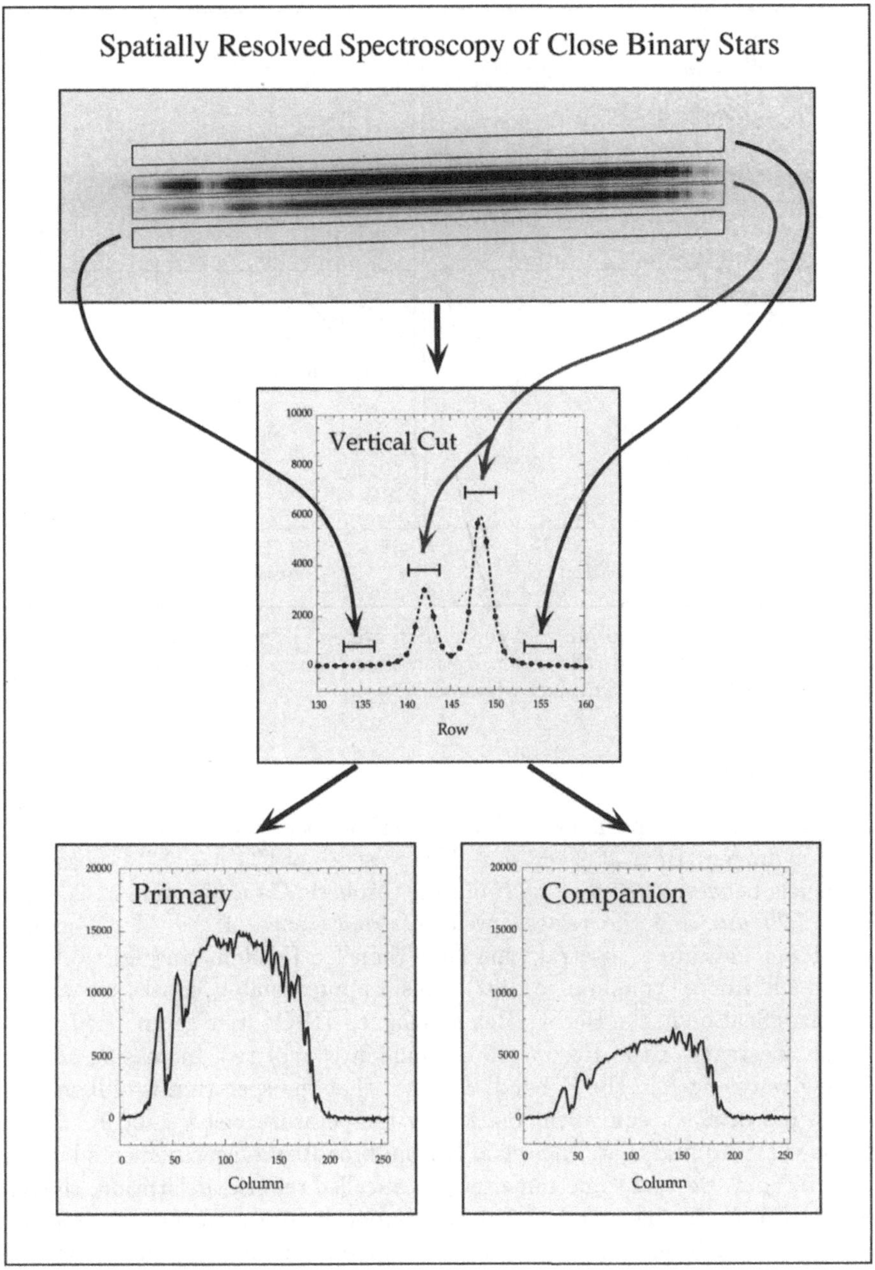

Fig. 4. Procedure for extracting the close binary spectra. The IRAF *twodspec* package subtracts the signal in two adjacent background locations from each of the stellar spectra. The paths of the four traces across the detector are not necessarily horizontal or linear. In the case of Haro 6–10, the IRC is considerably fainter, so we used the (shifted) trace of the primary.

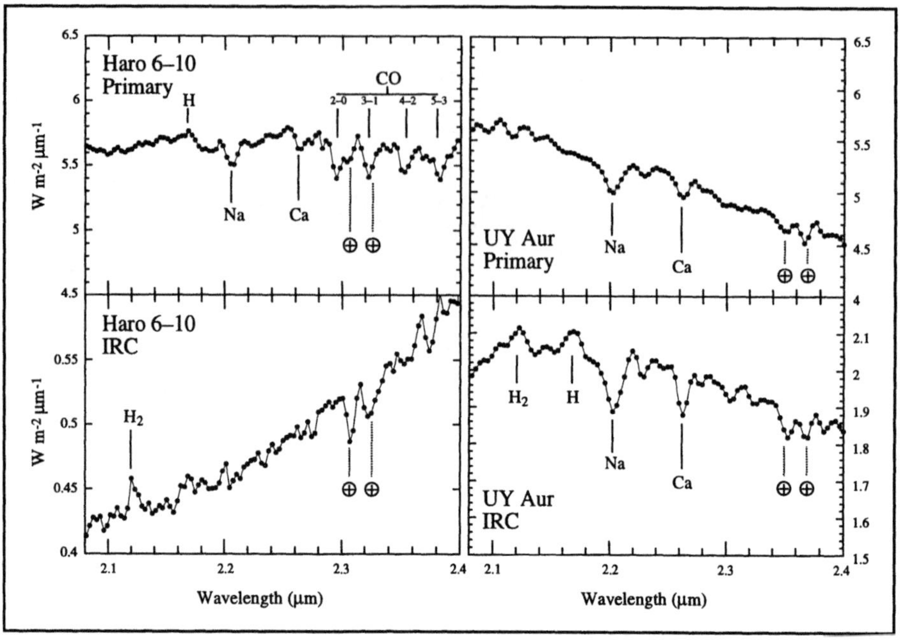

Fig. 5. Spectra of the two infrared companion systems. The features labelled with ⊕ are imperfectly subtracted telluric absorption lines. The spectrum of the UY Aur IRC has been smoothed by a two pixel window.

3.1 Haro 6–10

The spectrum of the primary of Haro 6–10 shows the $n = 7 - 4$ recombination line of hydrogen (Brγ) at 2.166 μm. The presence of CO $v = 2 - 0$ absorption bandheads between 2.29 and 2.38 μm, photospheric Ca absorption at 2.26 μm, Na at 2.20 μm, and the relative weakness or absence of the Al / Mg blend at 2.11 μm indicate a spectral type later than K5 (Kleinmann and Hall 1986, Lançon and Rocca–Volmerange 1992). This is approximately consistent with the K3,5 classification in the Herbig–Bell catalogue, (HBC – Herbig and Bell, 1988).

The spectrum of the Haro 6–10 companion is very red, increasing with increasing wavelength in the K band. The fact that the spectrum is still rising at 2.4 μm indicates an equivalent blackbody temperature below 1200 K. Leinert and Haas (1989) find that $T_{eff} \sim 650$ K from broadband measurements between 1 and 100 μm. Beyond some imperfectly–cancelled telluric absorption, the only spectral feature is the $v = 1 - 0$ S(1) emission line of molecular hydrogen at 2.12 μm. No molecular hydrogen emission appears in the spectrum of the primary.

3.2 UY Aur

The UY Aur primary spectrum shows photospheric absorption features due to Al / Mg (2.11 μm), Na (2.20 μm), and Ca (2.26 μm). This resembles the spectrum of a K5 or later dwarf, consistent with the K7V classification in the HBC (Kleinmann and Hall 1986, Herbig and Bell 1988). No hydrogen Brγ or H_2 emission appears in this object.

The spectrum of the UY Aur infrared companion is quite different from that of Haro 6–10. Photospheric absorption features appear, clearly indicating that it is a star. The overall spectrum is similar to that of the primary, although the stronger Ca (2.26 μm) feature and the weak or absent Al / Mg blend (2.11 μm) argue for a later spectral type. The UY Aur IRC shows the $v = 1 - 0$ S(1) line of H_2 and the $n = 7 - 4$ recombination line of hydrogen (Brγ) in emission.

3.3 Line Strengths and Variability

The H_2 and Brγ line fluxes in Table 1 show considerable variability. In 1986, Carr (1990) detected molecular hydrogen in Haro 6–10 but found no measurable Brγ. HK91 saw the H_2 S(1) line at approximately the same strength five years later. They also detected Brγ emission stronger than that of the molecular hydrogen. The Haro 6–10 observations presented here show the Brγ line strength unchanged and the S(1) flux considerably reduced in the intervening two years. Segregation of the Brγ to the primary and the molecular hydrogen to the IRC provides a natural explanation of the uncorrelated line strengths of these two transitions.

In the case of UY Aur, HK91 saw strong Brγ and comparatively weak S(1) line emission. Carr did not observe this object. The spectrum presented here shows essentially the same H_2 line strength and reduced Brγ. Unlike Haro 6–10, it appears that both lines originate from the UY Aur infrared companion. The primary may in fact produce Brγ emission that was strong at the time of the spatially unresolved measurements of HK91, and is weak in the current observations. This behaviour would be similar to the Haro 6–10 primary (see Table 1).

4. Origin of the H_2 Emission in Haro 6–10 and UY Aur

The $v = 1 - 0$ S(1) line of molecular hydrogen is confined to the infrared companions of both Haro 6–10 and UY Aur. A number of physical processes can produce H_2 emission, the most common of which are shock excitation and ultraviolet fluorescence (Shull and Beckwith 1982, and references therein). The ratios of the various K band molecular hydrogen lines can in principle distinguish between shock heating and UV fluorescence, although extinction and other factors can complicate the analysis. Shock models predict that the $v = 2 - 1$ S(1) transition at 2.248 μm will be approximately five times weaker than the $1 - 0$ line, while UV fluorescence calculations give a $2 - 1/1 - 0$ ratio close to 0.5. The nondetection the 2.248 μm line in both Haro 6–10 and UY Aur points to a $2 - 1/1 - 0$ ratio < 0.3 and hence to a shock origin for the H_2 emission in these objects.

4.1 Accretion Shocks

If the molecular hydrogen is excited in shocks as matter accretes onto the infrared companions, the line fluxes in Table 1 and a simple model of the infall geometry allow an estimate of the mass flux needed to power the observed H_2 emission. Here we assume steady, spherical freefall accretion with a shock front surrounding the star. This front lies at a radius such that the freefall speed (~ 20 km s^{-1}) produces shocks that are efficient at generating H_2 emission. For a one solar mass star, this shock front would be $\sim$10 A.U. in diameter.

Smith (1994) has calculated the ratio of the $v = 1-0$ S(1) line emission to the total energy in the shock. This ratio is approximately 0.02 for C shocks and $v \sim 20$ km s^{-1}. At a distance of 140 pc to the Taurus–Auriga cloud, the S(1) line fluxes correspond to total shock luminosities of 6×10^{-4} $L_\odot$ and 2×10^{-3} $L_\odot$ for the Haro 6–10 and UY Aur companions, respectively. Equating this amount of energy to $\frac{1}{2}\dot{\mathrm{M}}v^2$ gives the mass accretion rate. For a shock speed of 20 km s^{-1}, the rates are 2×10^{-8} $M_\odot$ yr^{-1} for Haro 6–10 and 6×10^{-8} $M_\odot$ yr^{-1} for UY Aur.

For spherical accretion, the mass flux through a shell of radius r is $\dot{\mathrm{M}} = 4\pi r^2 n(r)\overline{\mu}v(r)$, where $\overline{\mu}$ is the mean mass of the particles, and $n(r)$ and $v(r)$ are the number density and velocity, respectively. Setting $v(r) = \sqrt{2GM/r}$ for infall from a large distance gives:

$$n(r) = \frac{\dot{\mathrm{M}}/\overline{\mu}}{\sqrt{32\pi^2 GM}} \cdot r^{-\frac{3}{2}}, \tag{1}$$

where G is the gravitational constant and M is the mass of the accreting star. The column density of infalling matter down to a radius r is the integral of $n(r)$ from r to infinity. Bohlin *et al.* (1978) measured 5.8×10^{21} particles cm^{-2} per magnitude of selective extinction ($\mathrm{E_{B-V}}$). With $\mathrm{A_V} / \mathrm{E_{B-V}} = 3.09$ and $\mathrm{A_V} / \mathrm{A_K} = 0.112$ (Rieke and Lebofsky 1985), the K band extinction is then:

$$\mathrm{A_K} = 2.2 \times 10^5 \cdot r^{-\frac{1}{2}} \cdot \frac{\left(\frac{\dot{\mathrm{M}}}{10^{-8}\mathrm{M_\odot yr^{-1}}}\right)}{(\overline{\mu}_{\mathrm{amu}}) \cdot \left(\sqrt{\frac{\mathrm{M}}{1\mathrm{M_\odot}}}\right)}, \tag{2}$$

where $\overline{\mu}_{\mathrm{amu}}$ is the average particle mass in atomic mass units.

Koresko (1993) has modelled the spectral energy distributions of both components of the Haro 6–10 and UY Aur binary systems. Assuming that the infrared companions are the same age as their primaries, and that all the luminosity of the IRC's is reprocessed starlight, the evolutionary tracks of Cohen and Kuhi (1979) give mass $M \sim 1.75$ $M_\odot$ and stellar radius $R \sim 3R_\odot$ for Haro 6–10, and $M \sim 0.5$ $M_\odot$, $R \sim 1.9R_\odot$ for UY Aur.

Taking $\overline{\mu}_{\mathrm{amu}} = 2$, the mass and radius estimates, along with the mass accretion rates derived from the H_2 line flux, give K band extinctions of 0.74 mag and 2.5 mag to the Haro 6–10 and UY Aur companions, respectively. In both instances, the $\mathrm{A_K}$ to the $v(r) = 20$ km s^{-1} point is less than 0.1 mag. This amount of extinction centered on the IRCs will have a profound effect on the observed magnitudes and colours. The lack of photospheric absorption lines makes spectral classification of the Haro 6–10 infrared companion impossible. In the case of UY Aur, however, the 2.5 magnitudes of K band extinction derived from the H_2 flux contradicts the observed brightness and spectral characteristics.

The spectrum of the UY Aur IRC (Figure 5) indicates a spectral type between M0 and M4, and the measured brightness is approximately 1 mag fainter than the primary at 2.2 μm. An M0–M4 companion will be ~0.2–1 mag fainter than a coeval K7 primary at this wavelength, given the same interstellar extinction to both components of the binary (Schmidt-Kaler 1982, Bessel and Brett 1988). An additional 2.5 mag of local K band extinction makes the IRC intrinsically brighter than the primary, a situation incompatible with their spectral types.

A non-spherically symmetric configuration of infalling matter such as a circumstellar disk would allow a large accretion rate without high extinction. The extinction to the companion was certainly not high at the time of the Joy and van Biesbroeck (1944) visible wavelength measurements.

Alternatively, the extinguished photospheric flux could be compensated by the release of accretion energy with the appropriate wavelength distribution. Modelling this process is complex, but we note that the photospheric Na absorption features in the UY Aur primary and IRC have similar equivalent width. This is consistent with their respective spectral types and argues against a significant (1–2 mag) underlying thermal continuum from warm dust near the infrared companion.

Further evidence of accretion in the UY Aur system comes from the observation that the $v = 2 - 0$ CO overtone absorption band heads (2.29–2.38 μm) are absent or faint in the otherwise normal K–type spectra in Figure 5 and in HK91. These lines can be filled in by CO emission, a diagnostic of accretion disks and warm neutral winds (Carr, 1989).

Finally, the mass infall rate allows an estimate of the total available accretion luminosity $\frac{GM\dot{M}}{R}$. This luminosity is 0.33 $L_\odot$ for the Haro 6–10 companion, small in comparison to the total measured luminosity of 4.5 $L_\odot$ (Koresko 1993). In the case of UY Aur, however, the accretion luminosity is 0.5 $L_\odot$ which is a significant fraction of the total luminosity of 0.7 $L_\odot$ derived by Koresko. This estimate may be less reliable for UY Aur, since the mass derived from the evolutionary tracks assumes that all the luminosity is reprocessed starlight. Nevertheless, the mass of the UY Aur IRC, and hence the accretion luminosity, cannot be much smaller than the value assumed, since this would imply a significantly different spectral type from that seen in Figure 5.

4.2 Outflow or Wind Shocks

Winds from the star or circumstellar material can also produce shocks. The observations argue against a simple, spherical outflow scenario, however. For such a wind with constant velocity v_w, the radial density profile (equation 1) is:

$$n(r) = \frac{\dot{\mathrm{M}}/\overline{\mu}}{4\pi v_w} \cdot r^{-2}. \tag{3}$$

Integrating $n(r)$ as before allows an estimate of the K band extinction such winds would produce. The ratio of A_K from a wind to that from accretion is approximately 240 / v_w (km s^{-1}) for Haro 6–10 and 160 / v_w for UY Aur. For the modest velocities ($\sim$ 20 km s^{-1}) that are most efficient in producing H_2 emission, a constant velocity spherical outflow would produce high extinctions inconsistent with the K band spectra and the broadband photometric magnitudes. Higher velocity winds will have lower A_K but also lower efficiency. A well–collimated outflow cannot produce significant extinction except in the rare instance of alignment with the line of sight. Oblique shocks generated when winds from the primary strike the companion remain a possibility. This would allow the relatively low shock speeds most efficient at producing H_2 while keeping the emission confined to the companion. In any event, the wind must strike circumstellar molecular material to produce shocks. The molecular hydrogen emission is spatially confined to the IRCs in these systems, again pointing to material bound to and presumably accreting onto the companions.

5. Conclusions

Spatially–resolved spectroscopy of the Haro 6–10 and UY Aur infrared companion systems shows that the $v = 1 - 0$ S(1) emission line of molecular hydrogen is centered on the companions. The nondetection of higher vibrational lines points to shocks as the source of the H_2 emission, strengthening the hypothesis that infrared companions are currently undergoing accretion.

With reasonable assumptions about the shock geometry and speeds, the S(1) line fluxes correspond to accretion rates of 1.8 and 6.0×10^{-8} $M_\odot$ yr^{-1} for Haro 6–10, and UY Aur, respectively. In the case of UY Aur, this amount of matter in spherical infall will produce 2.5 mag of K band extinction. The spectrum of the UY Aur IRC is undoubtedly stellar, and the inferred spectral type coupled with the measured brightness argues for moderate extinction at 2 μm. This view is supported by the visible wavelength detection of the UY Aur IRC in 1944. Non–spherical geometry, such as accretion onto a circumstellar disk, can explain the presence of S(1) emission, low effective temperature, and high luminosity while allowing moderate extinction to the photosphere at near infrared wavelengths.

5.1 The Future

We are continuing and expanding our program of spatially resolved spectroscopy of young stellar objects. Figure 6 shows an image and spectra of the two components of Serpens SVS–20 (separation 1.5 arcseconds). This binary system is unusual in that the two stars have very similar colours despite having significantly different brightness. The spectrum in Figure 6 shows Brγ and CO bandhead emission in the companion, with no corresponding lines in the brighter star. The CO emission points to a warm neutral wind or an accretion disk. Figure 7 shows the two–dimensional distribution of molecular hydrogen in the close vicinity of

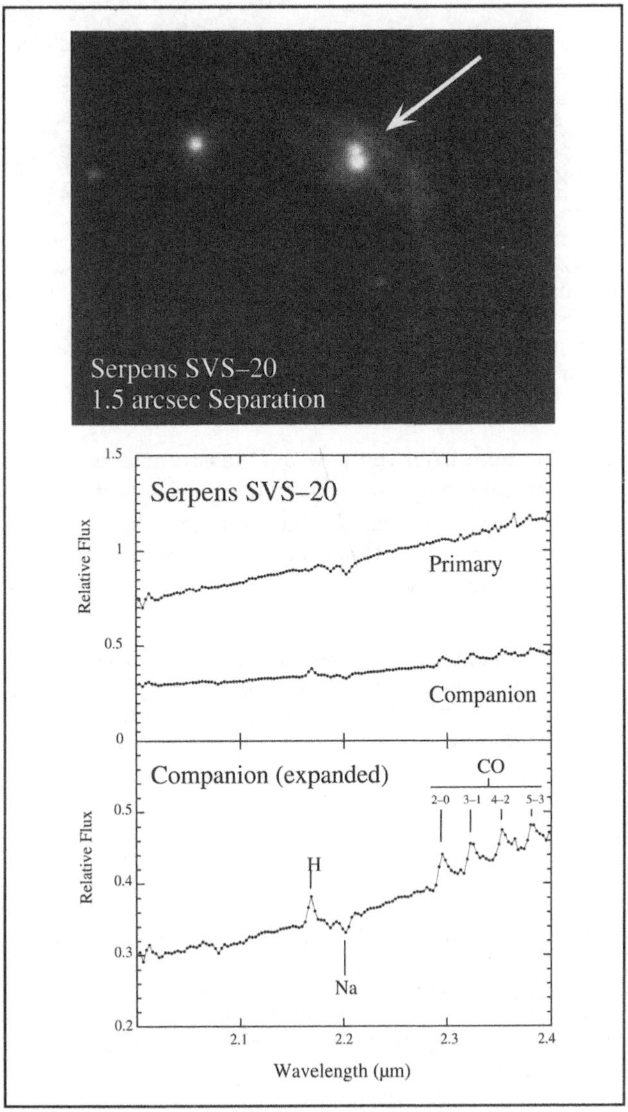

Fig. 6. Image and spectra of the Serpens SVS–20 binary system. The upper frame shows a combined J–H–K image of this object. The lower panel contains spatially resolved spectra showing hydrogen Brγ and CO $v = 2 - 0$ bandhead emission confined to the fainter companion.

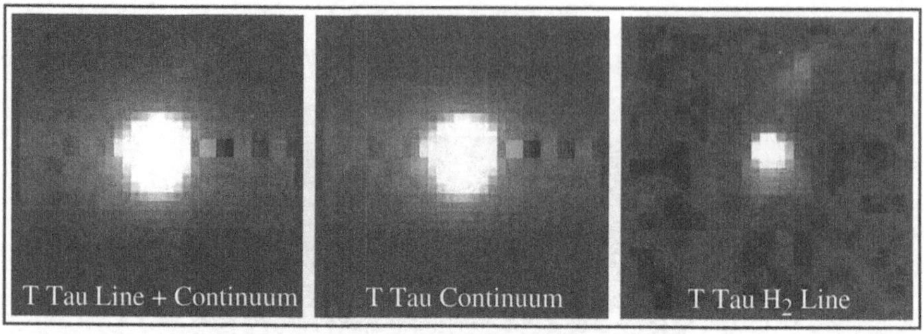

Fig. 7. Spectral imaging of T Tau with the 3-D instrument. The leftmost panel shows T Tau in the 2.12 μm line of molecular hydrogen. The center frame is an average of the adjacent continuum, and the right image shows the difference. There is substantial, extended H_2 emission in the near vicinity of this 0.7 arcsecond separation IRC system, which is spatially unresolved in these data. The banded structure in the left two frames is an artifact of processing and should be ignored.

T Tau. These data come from the new 3–D instrument developed at the MPE in Garching. The complete spatial coverage of such spectrometers is necessary to disentangle the emission regions in this complex object. Fitted with smaller pixel scales and adaptive optics, this type of instrument represents the future of spatially resolved spectroscopy of close binary stars.

Acknowledgement. T. M. H. thanks Tom Megeath for sharing observing time. Peter Bizenberger helped with the design and testing of MAGIC's grisms, and M. Haas provided observational support. This research has made use of the Simbad database, operated at CDS, Strasbourg, France.

References

Beckwith, S., Gatley, I., Matthews, K., and Neugebauer, G. 1978, *Ap. J. (Letters)*, **223**, L41.
Bertout, C. 1983, *Astr. Ap.*, **126**, L1.
Bessel, M.S. and Brett, J.M. 1988, *P. A. S. P.*, **100**, 1134.
Bohlin, R.C., Savage, B.D., and Drake, J.F. 1978, *Ap. J.*, **224**, 132.
Carr, J.S. 1989, *Ap. J.*, **345**, 522.
Carr, J.S. 1990, *Astron. J.*, **100**, 1244.
Chelli, A., Zinnecker, H., Carrasco, L., Cruz–Gonzales, I., and Perrier, C. 1988, *Astr. Ap.*, **207**, 46.
Cohen, M. and Kuhi, L.V. 1979, *Ap. J. Suppl.*, **41**, 743.
Dyck, H.M., Simon, T., and Zuckerman, B. 1982. *Ap. J. (Letters)*, **255**, L103.
Ghez, A.M., Neugebauer, G., and Matthews, K. 1994. *Astron. J.*, **106**, 2005.
Ghez, A.M., Neugebauer, G., Gorham, P.W., Haniff, C.A., Kulkarni, S.R., Matthews, K., Koresko, C., and Beckwith, S. 1991, *Astron. J.*, **102**, 2066.
Herbig, G.H. and Bell, K.R. 1988, Lick Obs. Bull. No. 1111.
Herbst, T.M., and Koresko, C.D. 1991 (HK91). Unpublished data.

Herbst, T.M., Beckwith, S.V.W., Birk, Ch., Hippler, S., McCaughrean, M.J., Mannucci, F., and Wolf, J. 1993, in: *Infrared Detectors and Instrumentation* SPIE Conference 1946.
Herbst, T.M. and Rayner, J.T. 1994a, in: *Infrared Arrays, The Next Generation.* Ed.: McLean, J.S., Kluwer Acad. Publ., Dordrecht, p. 515
Herbst, T.M., Pitz, E., and Reuther, C.M. 1994b, in *Tridimensional Optical Spectroscopic Techniques in Astrophysics*, I.A.U. Colloquium 149, in press
Herbst, T.M., Koresko, C.D., and Leinert, Ch. 1995, *Ap. J. (Letters)*, **444**, L93.
Joy, A.H. and van Biesbroeck, G. 1944, *P. A. S. P.*, **56**, 123.
Kleinmann, S.G. and Hall, D.N.B. 1986, *Ap. J. Suppl.*, **62**, 501.
Koresko, C.D. 1993, PhD. Thesis, Cornell University.
Lançon, A. and Rocca–Volmerange, B. 1992, *Astr. Ap. Suppl.*, **96**, 593.
Leinert, Ch., and Haas, M. 1989, *Ap. J. (Letters)*, **342**, L39.
Leinert, Ch., Zinnecker, H., Weitzel, N., Christou, J., Ridgway, S.T., Jameson, R., Haas, M., and Lenzen, R. 1993, *Astr. Ap.*, **278**, 129.
Oliva, E., and Origlia, L. 1992, *Astr. Ap.*, **254**, 466.
Rieke, G.H., and Lebofsky, M.J. 1985, *Ap. J.*, **288**, 618.
Schmidt-Kaler, Th. 1982, in: *Landolt-Börnstein Numerical Data and Functional Relationships in Science and Technology*, Group 4, Vol. 2, Subvol. b, eds. K. Schaffers and H.H. Voigt, p. 451.
Shull, J.M. and Beckwith, S. 1982, *Ann. Rev. Astr. Ap.*, **20**, 163.
Smith, M.D. 1995, A&A **296**, 789
Zinnecker, H. 1989, in *Low-Mass Star Formation and PMS Objects*, ed. B. Reipurth.

Physical Processes in Circumstellar Disks

Fred C. Adams

Physics Department, University of Michigan, Ann Arbor, MI 48109

Summary. We review the basic physical processes which occur in circumstellar disks. We discuss the possible role of gravitational instabilities in producing both disk accretion and secondary bodies within the disk. Next, we consider the viscous evolution of disks and the possible sources of the viscosity. We briefly consider magnetic instabilities, tidal effects, and vortex motions. We also discuss the process of planet formation.

1. Introduction

The purpose of this review is to describe the physical processes which occur in the disks associated with young stellar objects (YSOs). These processes, in turn, determine the structure and subsequent evolution of the disks (see also the reviews of Pringle 1981; Adams & Lin 1993). Circumstellar disks play an important role in the process of star formation. In recent years, compelling observational evidence has established the presence of YSO disks and has begun to determine their physical properties (cf. the reviews of Beckwith & Sargent 1993; Appenzeller & Mundt 1989; Bertout 1989; Shu, Adams, & Lizano 1987). The available observational evidence indicates that these YSO disks may produce significant luminosity ($L_D \sim L_* \sim 1\ L_\odot$) and may have masses which are a substantial fraction of the stellar mass (see, e.g., Kenyon & Hartmann 1987; Bertout, Basri, & Bouvier 1988; Adams, Lada, & Shu 1988; Beckwith et al. 1990; Adams, Emerson, & Fuller 1990; see also the large amount of recent work represented by the posters at this conference). The stability, structure, and evolution of these disks are determined by mass, angular momentum, and radiation transport processes.

Much of our discussion will focus on the physical processes which can lead to disk accretion. Before discussing any particular disk accretion mechanism, however, we consider the following general argument: A given star/disk system always tends to evolve toward a configuration of lower total energy; this trend holds true for *any* kind of dissipative mechanism. On the other hand, the system must also conserve its total angular momentum. Next we note that the lowest energy state accessible to the system is to have essentially all of the mass in the central star (so that the material is as far down the gravitational potential well as possible), with one point particle in a large radius orbit to carry all of the angular momentum. We expect any star/disk system to evolve toward this minimum energy configuration. And we note that this state of affairs is very nearly met within our solar system: almost all of the mass resides in the Sun itself, and almost all of the angular momentum is carried by the orbital motion of the giant planets (at large radii).

In this review, we discuss many different possible physical processes that can occur in these circumstellar disks. We first review the observational evidence and related theory that determine the basic disk properties (see Sect. 2). We then consider the growth of gravitational instabilities in star/disk systems (Sect. 3). These instabilities may lead to accretion through the disk and hence to the observed disk luminosities; in the limit of runaway growth of a single mode, these instabilities may also lead to the formation of a binary companion. Next, we discuss viscous accretion disks (in Sect. 4) and the role of convective instabilities in determining the viscosity. We then discuss magnetic instabilities (Sect. 5.5), tidal effects (Sect. 6), and the possibility of vortices in circumstellar disks (Sect. 7). In Sect. 8, we briefly discuss the process of planet formation in circumstellar disks. We conclude, in Sect. 9, with a summary and discussion of our results.

2. Basic Disk Properties

The basic characteristics of circumstellar disks vary with the stage of evolution. For purposes of this review, we will conceptually divide the evolution of disks into two regimes. The first regime corresponds to the protostellar phase of evolution; here, the central star/disk system is still deeply embedded within an envelope of infalling gas and dust. The second regime corresponds to the T Tauri phase of evolution where the star/disk system is optically revealed; in this case, the system is essentially isolated from its environment, and the disk no longer has a mechanism for gaining mass.

2.1 Protostellar Disks

In the protostellar phase of evolution, the central star/disk system is deeply embedded in an infalling envelope of gas and dust. Essentially all of the intrinsic radiation from the star/disk system is absorbed and re-radiated by dust grains in the envelope. As a result, no direct signature can be measured for these disks. However, we can establish the existence of these disks and estimate their properties through indirect methods.

One way to view protostellar evolution is through the $L{-}A_V$ diagram (Adams 1990), which is roughly analogous to the H–R diagram for ordinary stars. In this diagram (see Figure 1), protostars evolve toward the upper right as they continually gain mass from the interstellar medium. As the mass increases, the luminosity increases, and the column density (and hence the visual extinction) decreases (see Shu 1977; Terebey, Shu, & Cassen 1984). Any successful theory of star formation must produce $L - A_V$ diagram tracks which are consistent with observations of protostellar candidates. As shown by the dashed curve in Figure 1, a purely spherical theory (with no rotation and hence no disk) is overluminous; we can thus conclude that disks must be present in protostellar systems. As shown by the lower dotted curve in Figure 1, a theory with rotation and no disk accretion is severely underluminous; we can thus conclude that some disk accretion must occur during the protostellar phase. The solid curves in Figure 1 are calculated on the basis of the current picture of star formation (which includes disk accretion). Thus, disk accretion is necessary to account for the observed properties of protostellar objects. In particular, the intrinsic disk luminosity L_D arising from accretion is expected to be comparable to (or larger than) the stellar luminosity.

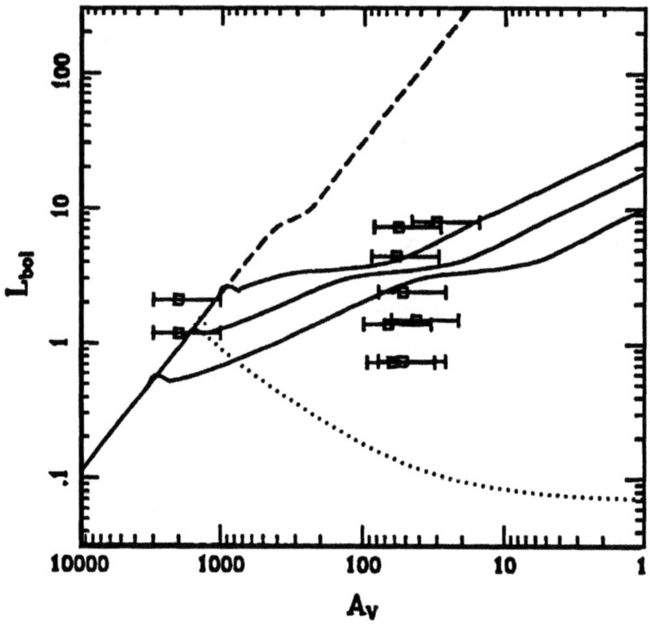

Fig. 1. The $L - A_V$ Diagram for protostars in the Taurus molecular cloud (adapted from Adams 1990). The solid lines in the diagram are the set of tracks calculated for varying core rotation rates Ω; the open symbols represent observed sources.

The comparison of theoretical tracks in the $L - A_V$ diagram with observed sources provides a *necessary* but not *sufficient* condition on the theory. The theory must also be able to account for both the spectral energy distribution and the spatial distribution of emission. In the protostellar (embedded) phase, circumstellar disks are required to produce the correct spectral energy distributions (cf. Adams, Lada, & Shu 1987; Ladd et al. 1991; Calvet et al. 1994) and to understand the observed emission distribution (see Keene & Masson 1990).

The disk masses during the protostellar phase are likely to be substantial because much of the infalling material falls directly onto the disk rather than onto the star. However, as we discuss below, the disk masses cannot be arbitrarily large, because such disks would become highly unstable. In addition, if the disk did not somehow give mass to the star, the resulting stellar masses would be too small. Thus, the disk mass must be roughly comparable to the stellar mass during this phase of evolution.

The physical size of protostellar disks is determined by the angular momentum of the initial state and by conservation of angular momentum during the subsequent collapse. We can measure the initial core rotation rates (e.g., Goldsmith & Arquilla 1985), and we can determine the rotation rates necessary to understand the observed protostellar spectra (Adams et al. 1987). These considerations suggest that the disk radius is in the range $\sim 10 - 100$ AU for much of the protostellar phase of evolution.

The properties of protostellar disks can be summarized as follows. These disks must undergo disk accretion and produce intrinsic disk luminosity. The protostellar disk mass, M_D, is expected to be a significant fraction of the total mass $M_{total} \sim 1 M_\odot$. Finally, the disks are expected to have radii of 10 – 100 AU, comparable in size to our solar system.

2.2 T Tauri Disks

For T Tauri disk systems, the intrinsic spectral energy distribution is directly measurable and we can more directly study the properties of the circumstellar disks. One of the most striking properties of these systems is the large infrared excess (e.g., Rucinski 1985; Rydgren & Zak 1987) which indicates the presence of circumstellar disks.

We can understand the spectral signature of circumstellar disks as follows. The basic idea is that the disk radiates at a wide range of temperatures, and hence its spectral energy distribution is made up of a superposition of blackbodies. As a first approximation, the disk can be assumed to be both spatially thin (and hence two-dimensional) and optically thick. The flux density emitted by the disk is then given by

$$\nu F_\nu \sim \int 2\pi r dr \, \nu B_\nu[T(r)] \sim r^2 \nu B_\nu , \tag{2.1}$$

where the integral is over the surface of the disk and where we have assumed that the disk radiates locally like a blackbody (B_ν is the Planck function). Now we consider the radial temperature profile in the disk to be a simple power-law,

$$T(r) \sim r^{-q} . \tag{2.2}$$

If we use the fact that each temperature in the disk radiates at a characteristic frequency ($T \sim \nu$), we find that the spectral energy distribution of the disk has the form

$$\nu F_\nu \sim \nu^n \qquad \text{where} \qquad n \equiv 4 - 2/q . \tag{2.3}$$

Since the observational properties of T Tauri disk systems have been extensively discussed elsewhere (cf. the poster papers of this conference), here we only summarize the basic properties of these disks. The characteristic radial size is ~100 AU; this size can be deduced from the characteristics of the spectral energy distributions (Adams et al. 1988; Adams et al. 1990), from studies of spectral lines in outflows (Edwards et al. 1987), or from our own solar system (the mean distance of Neptune from the Sun is ~30 AU). The disks in T Tauri systems can be either *passive* and merely reprocess stellar radiation (see, e.g., Figure 2), or *active* and possess additional intrinsic luminosity (see, e.g., Figure 3).

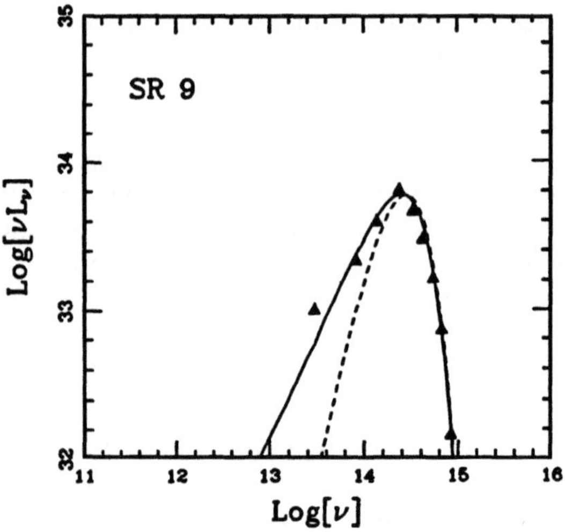

Fig. 2. Observed and theoretical spectral energy distribution of the passive disk system of SR 9. Dashed curve shows a blackbody spectrum; solid curve shows the effects of including a passive disk (adapted from Adams, Lada, & Shu 1987).

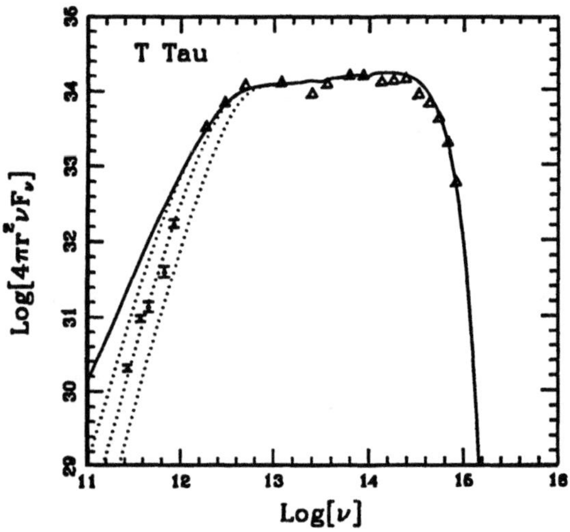

Fig. 3. Observed and theoretical spectral energy distribution of the active disk system T Tauri. Solid curve shows the spectrum in the limit of an optically thick disk ($M_D \to \infty$); dotted curves show spectra with finite disk masses of 0.01, 0.1, & 1.0 $M_\odot$ (adapted Adams, Emerson, & Fuller 1990).

The reprocessing of stellar photons by the disk is generally an important process. As long as the disk is optically thick to the stellar radiation, the disk will intercept and re-emit 1/4 of the stellar luminosity (Adams & Shu 1986). If the disks are flared (Kenyon & Hartmann 1987), or if scattering from circumstellar envelopes is important (Natta 1993), the fraction of the absorbed luminosity can be much higher. Finally, we note that for high mass stars, the stellar radiation field is sufficiently strong so that the disk becomes photoevaporated (Hollenbach et al. 1994).

For sources with active disks, the intrinsic disk luminosity L_D can be comparable to that of the star, i.e., L_D can be a significant fraction of the total luminosity. In addition, the spectral energy distributions of active disk sources indicate that the disk effective temperature distribution often appears to be flatter than the expected form $T_D \sim r^{-3/4}$ for a classical steady Keplerian accretion disk (Lynden-Bell & Pringle 1974). This claim follows from measuring the index n of the spectral energy distribution at infrared wavelengths and using equation (2.3) to determine the index q of the temperature profile. Keep in mind, however, that residual infalling envelopes can alter this interpretation.

The disk masses can be estimated by making observations at submillimeter wavelengths where the disk emission becomes optically thin. In this regime, the amount of observed emission is directly proportional to the mass of the disk. Current estimates suggest that the disk masses lie in the range

$$0.01 M_\odot \leq M_D \leq 1.0 M_\odot \tag{2.4}$$

(see, e.g., Weintraub, Sandell, & Duncan 1989; Adams et al. 1990; Beckwith et al. 1990). However, these mass estimates depend directly on the assumed dust opacity at submillimeter wavelengths, and this quantity remains uncertain at present. An additional uncertainty in these mass estimates arises because of the presence of circumstellar envelopes (see Calvet et al. 1994), which can contribute substantially to the observed emission at far-infrared wavelengths ($\lambda \sim 100$ μm).

Notice that the disk masses and radii (and hence the angular momenta) for the T Tauri disks are consistent with expectations from protostellar theory. Notice also that both the protostellar disks and the T Tauri disks must be capable of disk accretion, but no definitive theory of disk accretion exists at this time. In the following sections, we will discuss several physical mechanisms which may lead to disk accretion and angular momentum transport. The first mechanism is self-gravitating disk instabilities, which require a fairly large disk mass (i.e., M_D comparable to M_*) and are expected to be important during the protostellar phase of evolution. The second mechanism is convective instabilities, which are expected to arise during the T Tauri phase of evolution (when M_D is of order 10^{-2} $M_\odot$). We also consider magnetic instabilities, tidal effects, and vortices.

3. Gravitational Instabilities

In this section, we discuss gravitational instabilities in star/disk systems. During the early phase of protostellar evolution, the mass of the disk can be comparable to that of the forming star itself. In this case, gravitational instabilities may be particularly important for producing angular momentum transport and an accretion flow through the disk. These gravitational instabilities may also play a role in producing binary companions and/or planets in the disk.

Two general classes of gravitational instabilities can arise: *global gravitational instabilities*, in which extended regions of the disk participate in promoting the growth of some initial disturbance, and *local gravitational instabilities*, in which the characteristic wavelengths are much shorter than the physical dimension of the disk. Global gravitational instabilities may occur when the mass of the disk is comparable to that of the central star. Local gravitational instabilities may occur in disks which are considerably less massive than the star, if the disks are sufficiently cool. In this present discussion, we concentrate on the global gravitational instabilities, which may be responsible for inducing energy dissipation and global energy transport in disks with relatively large masses.

3.1 The Initial State of the Star/Disk System

For the sake of definiteness, we begin by specifying the basic unperturbed system, which consists of a star and an accompanying gaseous disk. The growth of spiral modes is mainly determined by three elements: self-gravity, pressure, and differential rotation. The gravitational forces are determined by the potential of the star and by the disk's surface density distribution $\sigma_0(r)$, which we can consider as a simple power-law in radius, r, from the central star (the disk is assumed to be in centrifugal equilibrium). The pressure is determined by the temperature distribution, or equivalently, the distribution of sound speed in the disk; we can also consider the sound speed $a(r)$ to be a power-law in radius.

The rotation curve $\Omega(r)$ in the disk is then determined by the potential of the star, the potential of the disk, and the pressure gradients. However, since the potential well of the star dominates that of the disk everywhere except near the disk's outer edge, the rotation curve is nearly Keplerian throughout most of the disk's radial extent. Recall that for thin gaseous disks, by definition, the pressure gradients are small compared to the gravitational forces and do not significantly affect the rotation curve.

Finally, we must specify the radial size of the disk. As discussed in the previous section, both observations and theory indicate that the characteristic disk size is approximately 100 AU. In other words, the disk radius R_D can be up to $\sim 10^4$ times the radius R_* of the star.

3.2 Spiral Modes with Azimuthal Wavenumber $m = 1$

Much of our discussion concentrates on modes with azimuthal wave number $m = 1$ (spiral patterns with one arm), since these modes can be global in extent and may also be the most difficult modes to suppress in unstable gaseous disks. Modes with $m = 1$ correspond to elliptic streamlines (eccentric particle orbits), a special characteristic of Keplerian potentials. Thus, for a disk with an exact Keplerian rotation curve and no interactions between particles, $m = 1$ disturbances correspond to purely kinematic modes of the system (see Figure 4); for realistic disks (with pressure), a relatively 'small' amount of self-gravity is required to overcome the pressure (and shear) and sustain modal growth. In addition, modes with $m = 1$ are the most global, i.e., these modes can affect a larger radial extent in the disk than higher m modes (see, e.g., Adams, Ruden, & Shu 1989).

One unique and important aspect of $m = 1$ modes is that the center of mass of the perturbation in the disk does *not* lie at the geometrical center of the system; hence, the frame of reference centered on the star is *not* an inertial reference frame. The star is actually in orbit about the center of mass (in other

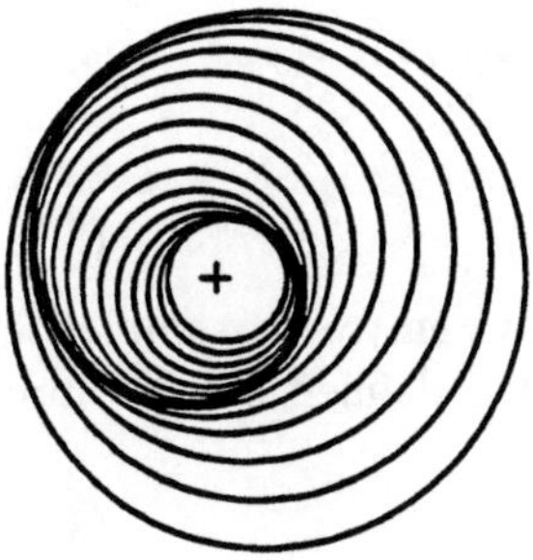

Fig. 4. Schematic diagram of $m = 1$ elliptical streamlines oriented to produce a one-armed spiral. The cross denotes the position of the star. (Adapted from Adams, Ruden, & Shu 1989)

words, the star is accelerating) and creates an effective forcing potential. This extra term is denoted as the "indirect potential" and has the form

$$\psi_{ind} = \omega^2 R_0 r \exp[i(\omega t - \theta)], \tag{3.1}$$

where ω is the (complex) eigenvalue of the mode. The quantity R_0 measures the strength of the indirect term and is given by

$$R_0 = \frac{\pi}{M_D + M_*} \int_{R_*}^{R_D} r^2 dr\, \sigma_1(r), \tag{3.2}$$

where σ_1 is the perturbation of the surface density of the disk. Under some circumstances, the indirect potential can have additional contributions from the displacement of the disk edge. Recent results show that this indirect potential is essential for the growth and maintenance of spiral modes with azimuthal wavenumber $m = 1$. In fact, the interaction of this indirect potential with the outer Lindblad resonance in the disk (see below) can be the dominant amplification mechanism for these modes.

3.3 Wave Physics and Spiral Instabilities in Gaseous Disks

In the simplest description of spiral instabilities, self-excited spiral disturbances can grow through the feedback and amplification of spiral density *waves*. In this section, we quickly review the theory of spiral density waves. In the following sections, we describe the feedback cycle and the amplification mechanism for spiral modes.

In the asymptotic (WKBJ) limit, the dispersion relation for spiral density waves in a gaseous disk has the form

$$k^2 a^2 - 2\pi G \sigma_0 |k| + \kappa^2 = (\omega - m\Omega)^2, \tag{3.3}$$

where k is the radial wavenumber, κ is the epicyclic frequency, and where ω is the complex eigenvalue of the system (Lin & Lau 1979). Since the gravitational term is proportional to $|k|$, this dispersion relation has four branches,

$$k = \pm(k_0 \pm k_1), \tag{3.4}$$

where

$$k_0 \equiv \frac{\pi G \sigma_0}{a^2}, \qquad k_1 \equiv \frac{\pi G \sigma_0}{a^2}[1 - Q^2(1 - {\nu_m}^2)]^{1/2},$$

and

$$\nu_m \equiv (\omega - m\Omega)/\kappa. \tag{3.5}$$

Here, the quantity Q determines the stability of the system to *axisymmetric* disturbances ($Q > 1$ is the condition for stability) and is defined by (Toomre 1964):

$$Q \equiv \frac{\kappa a}{\pi G \sigma_0}. \tag{3.6}$$

The overall sign of k determines whether the waves are *leading* ($|k| > 0$) or *trailing* ($|k| < 0$); the inner sign determines whether the waves are *short* [$k \propto (k_0 + k_1)$] or *long* [$k \propto (k_0 - k_1)$].

The quantity ν_m is a dimensionless frequency of the spiral density waves. The radius in the disk where $\nu_m = 0$ (i.e., where $\Re(\omega) = m\Omega$) is known as the *corotation resonance* (CR); the energy and angular momentum of the perturbation (and the action) are positive outside the corotation radius and negative inside. Notice that for $Q > 1$, the wavenumber k_1 becomes imaginary for radii sufficiently close to the CR; in other words, a classical turning point exists for the density waves. The resulting "forbidden" region surrounding the CR is known as the "Q-barrier". Notice also that for long waves, $k \to 0$ at any radius where $|\nu_m| = 1$. The radius in the disk where $\nu_m = +1$ is known as the *outer Lindblad resonance* (OLR) and plays an important role in the physics of $m = 1$ modes. In particular, long waves have a classical turning point at the Lindblad resonances and are thus confined to radii in the disk between the Q-barrier and the Lindblad resonances.

3.4 Feedback Loop: The Four-Wave Cycle

We now describe the feedback cycle for $m = 1$ modes in gaseous disks (see Shu et al. 1990). One unique aspect of this feedback cycle is that all four types of waves are utilized (see Figure 5):

[1] Begin (somewhat arbitrarily) with the excitation of a long trailing (LT) spiral density wave at the outer Lindblad resonance (OLR) by the indirect term. The LT wave propagates inward (its group velocity is negative) until it encounters the outer edge of the Q-barrier.

[2] At the Q-barrier, the LT wave refracts into a short trailing (ST) spiral density wave that propagates back outward, through the OLR to the outer disk edge.

[3] The ST waves that propagate to the outer disk edge reflect there to become short leading (SL) waves. The SL waves then propagate back to the interior, through the OLR, until they encounter the outer edge of the Q-barrier, where they refract into long leading (LL) spiral density waves that propagate back toward the OLR.

[4] At the OLR, the LL waves reflect to become LT waves. If the reflected LT wave possesses the correct phase relative to the LT wave launched from OLR by the indirect term in step A above, then we have constructive reinforcement

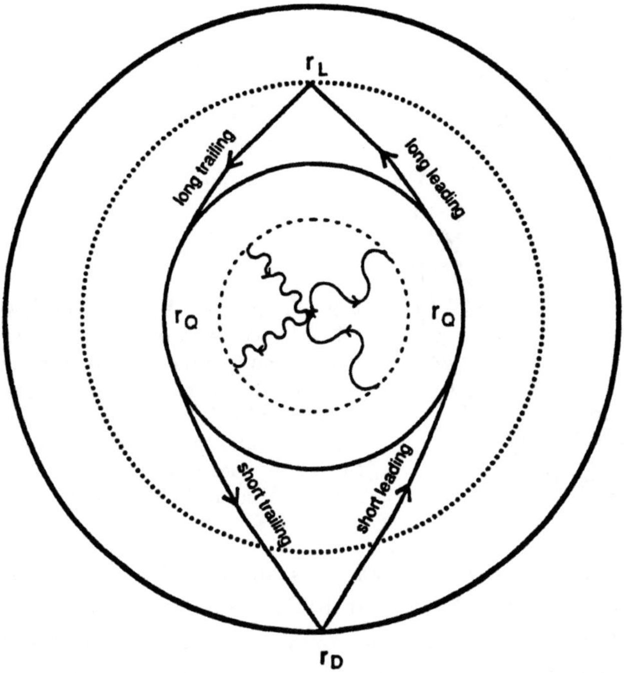

Fig. 5. The Four-Wave Feedback Cycle for $m = 1$ modes. The propagation of each type of wave is shown by the solid lines with arrows. (See text for further discussion; adapted from Shu et al. 1990)

of the entire wave cycle, and the basis for the establishment of a *resonant wave cavity*.

Using a WKBJ analysis, this four-wave cycle can be used to derive a quantum condition for $m = 1$ spiral modes:

$$\oint k dr + \sum \Delta\phi = 2\pi n, \tag{3.7}$$

where the wavenumber k is integrated over the entire four wave cycle described above, the changes in phase $\Delta\phi$ occur due to the reflection and refractions of the waves at their turning points, and where n is an integer. This quantum condition accurately predicts the pattern speeds (the real part of the eigenfrequencies) for these modes; for strongly growing modes, the analytical results (Shu et al. 1990) and numerical results (see Figure 6; Adams et al. 1989) agree to within a few percent.

3.5 SLING Amplification

In this section, we describe the amplification mechanism for eccentric modes. Detailed analysis indicates that the dominant mechanism for amplification arises from the indirect potential, which provides an effective forcing term. The indirect term varies slowly with radius in the disk; since a slowly varying force can only couple to oscillatory disturbances at the disk edges or at the Lindblad resonances (Goldreich & Tremaine 1979), the main coupling occurs at the outer Lindblad

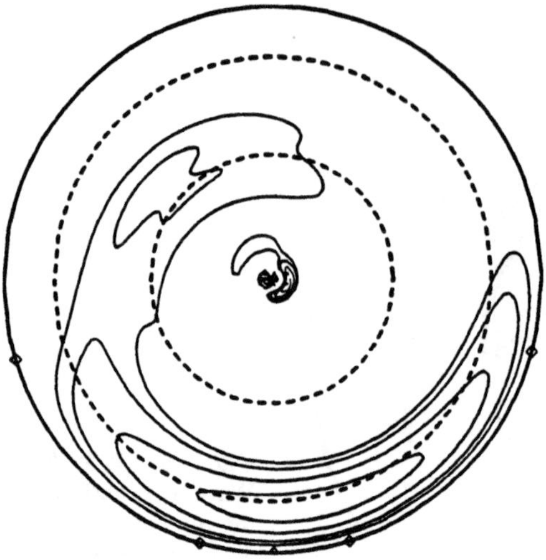

Fig. 6. Contour plot for the lowest order ($n = 0$) one-armed spiral mode. The two dashed circles show the location of the corotation and outer Lindblad resonances. The spiral arm trails in the sense of rotation of the material. (Adapted from Adams, Ruden, & Shu 1989)

resonance for the modes considered here. Thus, this amplification mechanism differs substantially from the previously studied mechanisms, which utilize the process of *super-reflection* across the corotation resonance (super-reflection can still occur in these disks and is implicitly included in the numerical treatment, but it does not dominate the amplification). In an analytic treatment, the growth rates can be determined for the modes under the assumption that *all* of the amplification arises from this coupling of the indirect term to the outer Lindblad resonance in the disk (and that the indirect term arises mostly near the outer disk edge). In other words, the indirect potential is considered as an external forcing term acting on the disk and the torque exerted on the disk at the OLR is calculated accordingly. Since the long-range coupling of the star to the outer disk provides the essential forcing, this new instability mechanism is called SLING: **S**timulation by the **L**ong-range **I**nteraction of **N**ewtonian **G**ravity.

The combined numerical and analytical treatments indicate the dependence of the growth rates (i.e., the imaginary part of the eigenfrequencies) on the parameters of the problem. Most importantly, a finite threshold exists for the SLING Amplification mechanism. When all other properties of the star/disk system are held fixed, this effect corresponds to a threshold in the ratio of disk mass M_D to the total mass $M_* + M_D$. We find that the growth rates are largest for the case of equal masses $M_D = M_*$ and decrease rapidly with decreasing relative disk mass. In the optimal case, $M_D = M_*$, the growth rates can be comparable to the orbital frequency at the outer disk edge, i.e., the modes can grow on nearly a dynamical time scale. On the other hand, the presence of the finite threshold implies a critical value of the relative disk mass, i.e., the maximum value of $M_D/(M_* + M_D)$ that is stable to these $m = 1$ disturbances. For the simplest case of a perfectly Keplerian disk and $Q(R_D) = 1$, this critical ratio has the value

$$\frac{M_D}{(M_* + M_D)} = \frac{3}{4\pi} \approx 0.24. \tag{3.8}$$

This critical value of the disk mass is known as the "maximum solar nebula" (Shu et al. 1990). Thus, when the disk mass is greater than about 1/3 of the stellar mass, gravitational instabilities grow strongly and the disk is highly unstable. When the disk mass becomes less than about 1/3 of the stellar mass, instabilities can still grow, but at a much slower rate.

3.6 Shutting off Gravitational Instabilities

The gravitational instabilities described above can be turned off via several mechanisms. If these instabilities lead to mass accretion through the disk, the fraction of the total mass that resides in the disk must eventually decrease below the critical fraction (see above) and the growth rates for the instabilities will decrease substantially. These instabilities can also be stabilized through excessive heating of the disk. If the Q parameter becomes sufficiently large everywhere in the disk, the growth rates for these instabilities will again decrease. Finally, these modes can be stabilized through the behavior of the outer disk edge. If the disk edge is very efficient at absorbing the (outgoing) short leading waves, the feedback cycle will be broken. However, the SLING amplification mechanism produces a *strong* amplifier, and relatively little feedback is required to sustain the growth of these modes. The disk edge can also distort itself in such a manner as to cancel the effects of the indirect potential (see Shu et al. 1990). However, a fairly wide distribution of disk edge configurations will allow the growth of these modes.

3.7 Nonlinear Simulations and Other Unstable Modes

In addition to the case of $m = 1$ modes discussed above, other gravitationally unstable modes can arise in circumstellar disks. For example, the condition for stability against the growth of axisymmetric disturbances in a gaseous disk does not require the disk mass to be comparable to the central star. In the cool outer regions of protoplanetary disks, the temperature can be sufficiently cool so that $Q < 1$, even when $M_D/(M_* + M_D) < 3/4\pi$. In this limit, local gravitational instabilities may be excited; in this case, the characteristic wavelength derived from the dispersion relation can be considerably less than the disk radius. Gravitational instability against non-axisymmetric perturbations (when they can be excited) may also provide an effective mechanism for angular momentum transfer (Paczynski 1977; Lin & Pringle 1987).

Nonlinear simulations of self gravitating disks have been performed using a sticky particle hydrodynamics scheme (Anthony & Carlberg 1988), traditional hydrodynamics codes (Yang et al. 1991; Woodward, Tohline, & Hashisu 1994; Laughlin & Bodenheimer 1994), and SPH simulations (Adams & Benz 1992). These calculations indicate that non-axisymmetric perturbations can grow strongly into the nonlinear regime and can indeed transfer angular momentum through their tidal torques (see also Larson 1984). In addition, these simulations show that spiral modes with higher azimuthal wavenumbers ($m = 2, 3, 4$) can also grow strongly.

In many cases, these hydrodynamic simulations show that the amplitude of the perturbations eventually saturates at some finite value (e.g., Savonije & Papaloizou 1989; Laughlin & Bodenheimer 1994). When this saturation occurs at a fairly low amplitude (very roughly, $\Delta\rho/\rho \sim 1$), the angular momentum transfer leads to an accretion flow through the disk. In other words, the effective angular momentum transport induced by these growing modes affects the *secular* evolution of the disks (Lin & Pringle 1990). Furthermore, this accretion flow can often be described by an effective viscosity prescription (see Laughlin & Bodenheimer 1994 for further details; see the following section for a discussion of viscous evolution of disks).

For relatively massive disks, the gravitational instabilities grow more strongly. In some cases, a spiral arm can collapse to form a gravitationally bound "clump" of gas when the equation of state of the disk is isothermal (Figure 7; Adams & Benz 1992; G. Laughlin, private communication). These collapsed "clumps" typically have masses of $\sim 0.01\ M_D$ and travel on elliptical orbits. The possibility that these clumps survive to form either binary companions or giant planets is especially interesting. Notice, however, that these clumps are formed as a result of the assumption of isothermal evolution of the disk. This assumption limits the local pressure support available. The density enhancements produced by the unstable modes are sufficient for pushing small regions over the local Jeans mass and thus triggering the collapse. This collapse to gravitationally bound objects does not occur in all cases. In particular, the formation of these bound clumps of gas can occur in real physical systems only when these systems can efficiently radiate away their energy. Finally, we note that the break up of a system into two parts can also occur during the collapse itself, i.e., before a well-defined star/disk system has developed (see the contribution of Bonnell in this volume).

3.8 Summary

The gravitational instabilities described here may have important applications in circumstellar disks. In the earliest stage of star formation – the protostellar phase – the mass of the disk is likely to be comparable to that of the star. Global gravitational instabilities are thus likely to grow and may lead to mass accretion through the disk and may produce the observed disk luminosities. Perturbations with one spiral arm prove especially interesting because they force the star to move from the center of mass and thereby transfer angular momentum to the stellar orbit. These modes may thus lead to the formation of a binary companion within the disk or perhaps to the formation of giant planets (see Figure 7).

4. Viscous Accretion Disks

In the later stages of disk evolution, the disk mass becomes small enough that gravitational instabilities will shut off, or at least become less important. In the realm of intermediate mass disks ($M_D \sim 10^{-1} - 10^{-2}\ M_\odot$), circumstellar disks

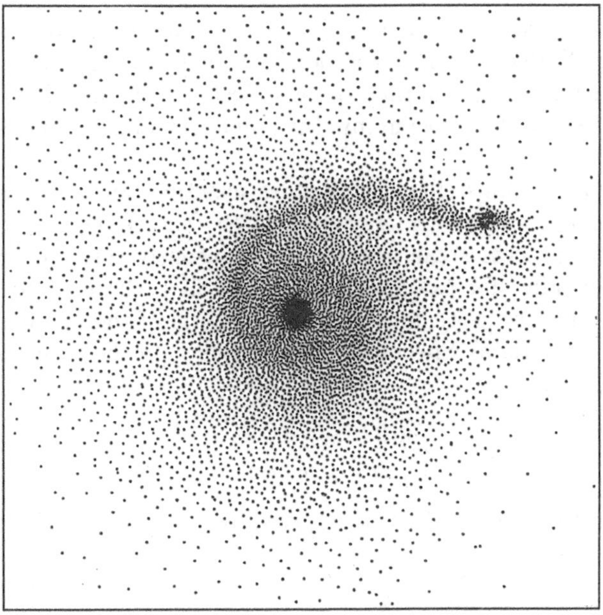

Fig. 7. Result of an SPH simulation of a star / disk system. The main spiral arm has collapsed to form a gravitationally bound "clump" of gas. (Adapted from Adams and Benz 1992)

evolve through the process of viscous diffusion. We expect young stellar objects to have disk masses in this range after the infall has been terminated, i.e., after the disk no longer has a source for gaining mass.

4.1 Viscous Evolution of Circumstellar Disks

The evolution of circumstellar disks is governed by the laws of fluid dynamics. In the limit that viscous forces drive the evolution of the disk, the behavior is described by a time-dependent diffusion equation, which can be written in the form

$$\frac{\partial \sigma}{\partial t} - \frac{3}{r}\frac{\partial}{\partial r}\left[r^{1/2}\frac{\partial}{\partial r}(\sigma \nu r^{1/2}) - \frac{2S_\sigma(r,t)J(r,t)}{\Omega}\right] - S_\sigma(r,t) = 0, \qquad (4.1)$$

where Ω, σ, and ν are the angular frequency, surface density, and viscosity of the disk material respectively, $S_\sigma(r,t)$ and $J(r,t)$ are the mass flux and excess angular momentum (respectively) of the infalling material (see Lüst 1952; Lynden-Bell & Pringle 1974; Lin & Papaloizou 1985). In other words, disk evolution occurs through a viscous diffusion process. As in any diffusion process, the net result is for the system (here, the disk) to spread out. In this case, the inner parts of the disk move farther inward and some disk material is accreted onto the star in the center. On the other hand, the outer parts of the disk gain angular momentum and spread farther outward. As this process continues, the disk mass becomes smaller relative to the star, the surface density of the disk decreases, and the outer radius of the disk increases. Thus, this process makes a star/disk system evolve toward the current state of our own solar system, with most of the mass in the central star and most of the angular momentum carried by objects at large radii (the planets). Notice that the evolutionary time scale depends on the magnitude of the viscosity. Notice also that viscous effects depend on the *local* disk properties, in contrast to the *global* nature the gravitational instabilities discussed earlier.

In a differentially rotating circumstellar disk, a mass accretion flow requires both angular momentum transport and energy dissipation (e.g., Lynden-Bell & Pringle 1974). These two processes can operate simultaneously through an effective viscous stress. Consider, for example, an annulus in a circumstellar disk. The difference in viscous torque on the inside and outside of the annulus leads to a net inward flow of material. Specifically, the torque τ on the annulus is determined by the viscosity ν,

$$\tau = 2\pi r \sigma \nu r^2 \frac{d\Omega}{dr} . \tag{4.2}$$

The corresponding mass accretion rate $\dot{M}_D$ is then given by

$$-\dot{M}_D \frac{dJ}{dr} = \frac{d\tau}{dr} , \tag{4.3}$$

where J is the specific angular momentum of the gas at the radial position r. Thus, both the torque τ and the mass accretion rate $\dot{M}_D$ are linearly proportional to the viscosity ν.

The energy release from the net inward flow of mass produces a characteristic spectral signature. If we equate the loss of energy from the accretion process with the energy radiated through the upper and lower surfaces of the disk, we obtain the relation

$$\frac{\tau}{2\pi r} \frac{d\Omega}{dr} = 2\sigma_B T^4 , \tag{4.4}$$

where T is the surface temperature of the disk. For the particular case of a Keplerian rotation curve (where $\Omega \sim r^{-3/2}$), the temperature profile takes the form $T \sim r^{-3/4}$ and hence the spectral energy distribution has the slope $n = 4/3$ at infrared wavelengths (see equation [2.3]).

The above discussion illustrates the importance of viscosity in disks. However, molecular viscosity in accretion disks is generally too small to be of astrophysical interest. In a variety of accretion disks, an anomalous "turbulent" viscosity is often assumed to be responsible for both angular momentum transport and viscous dissipation, despite the lack of rigorous proof that turbulence may occur intrinsically (Shakura & Sunyaev 1973; Lynden-Bell & Pringle 1974). In this prescription, an effective viscosity is introduced through the ansatz

$$\nu_{\rm eff} = \frac{2}{3} \alpha_s a_s H, \tag{4.5}$$

where a_s is the sound speed, H is the pressure scale height in the disk, and α_s is a dimensionless efficiency factor. The evolution of the disk is specified once the viscosity (or, equivalently, the α_s parameter) is given. The goal of any study of viscous evolution in disks is thus to calculate the parameter α_s in an *a priori* manner.

In the case of YSO disks, several different processes can produce an effective viscosity. In the previous section, we discussed how gravitational instabilities can drive an accretion flow which can be modeled with an α parameter (Laughlin 1994). In this section, we discuss how thermal convection can occur in the direction normal to the plane of the disk and can produce turbulent flow and hence an effective viscosity (Lin & Papaloizou 1980, 1985; Lin 1981ab). Finally, in the following section (Sect. 5.5), we discuss how magnetic effects can produce an effective viscosity.

4.2 Intrinsic Nature of Convective Instability

In order to show that YSO disks are intrinsically unstable against thermal convection, we analyze the structure of a turbulence-free disk in which rotation prevents gas from migrating in the radial direction. However, gas can contract in the vertical direction. If the disk is not in hydrostatic equilibrium initially, it will rapidly evolve towards such a state. Using a one-dimensional numerical hydrodynamic scheme, Ruden (1986) showed that a disk of cold gas contracts towards the midplane. After the disk has settled into a quasi-hydrostatic equilibrium, slow contraction continues as thermal energy is lost from the disk surface. In the absence of an energy source, surface radiation leads to heat diffusion from the midplane to the surface region. The associated reduction in pressure support leads to a readjustment towards a new hydrostatic equilibrium. In the typical temperature range ($10 - 10^3$ K), such adjustment can lead to a *superadiabatic* structure in the vertical direction (see Ruden 1986). This structure is produced because the opacity (which is primarily due to dust grains) is a sufficiently strongly increasing function of temperature. In the case of YSO disks, the cooler surface region has a lower opacity and cools less efficiently (Lin & Papaloizou 1980, 1985; Lin 1981ab; Ruden & Lin 1986). The standard criterion for convection can be written in the form

$$\frac{d\log T}{d\log P} \geq \frac{\gamma - 1}{\gamma} \equiv \nabla_{\rm ad}, \tag{4.6}$$

where γ is the adiabatic exponent of the gas (e.g., Schwartzschild 1958). When the temperature gradient in the disk becomes superadiabatic, the disk becomes convectively unstable in the vertical direction.

4.3 Global Convective Pattern and Angular Momentum Transport

Once convection has become established in a circumstellar disk, a self-consistent cycle of disk accretion can occur. In this cycle, the convective motions lead to turbulence. This turbulence leads to the production of eddies on a wide range of size scales, i.e., a *cascade of eddies*. On the smallest eddy scale, energy dissipation occurs through an effective viscosity. This viscosity, in turn, leads to energy generation in the disk and hence to disk accretion. The energy generation leads to steep vertical temperature gradients, which can be superadiabatic and therefore can support further convection. Thus, a self-consistent cycle of convective instability can be maintained (see Figure 8).

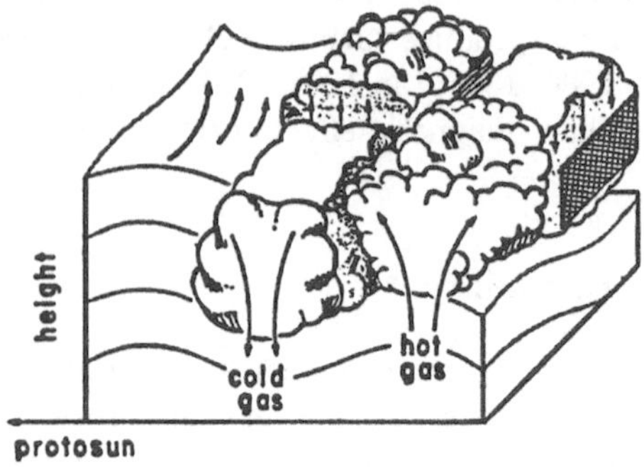

Fig. 8. Schematic diagram of convective cell in a circumstellar disk (see text).

The role of convection in a disk is not limited to heat transport in the vertical direction. Convective eddies can also induce mixing over a radial extent comparable to their own size. Through this mixing process, angular momentum is transferred. Convection also generates turbulence which causes dissipation of energy stored in differential rotation. Perhaps the simplest treatment for convection is to use the mixing length prescription in which the eddy viscosity is assumed to be the product of the convective speed and an effective mixing length which is comparable to the size of the eddies (Lin & Papaloizou 1980). From such a treatment, we can build self-consistent models in which convection is responsible for energy dissipation, vertical heat transport, and the radial transport of mass and angular momentum. Thus, energy dissipation is distributed from the midplane to the surface throughout the disk.

Notice, however, that the mixing length model is based on an *ad hoc* prescription of eddy viscosity. In a convective disk, eddies with a variety of scales are generated. The largest eddies often provide the dominant momentum transfer, whereas the smallest eddies provide the dominant energy dissipation in the disk. The scale of the largest convective eddies is usually comparable to the vertical scale height of the disk. On these large scales, global effects such as rotation and radiative losses are important. Thus, convection must be examined with a global analysis. In an attempt to carry out such a global analysis, Cabot et al. (1987a,b) computed a vertically-averaged effective viscosity which is derived from integrating the linear growth rate through various distances above the midplane of the disk. This growth rate varies greatly in the vertical direction and therefore cannot be attributed to any given eddy.

A more appropriate global treatment is to determine a unique growth rate (which is an eigenvalue) and its associated eigenfunction for each characteristic convective mode (Ruden, Papaloizou, & Lin 1988; see Figure 9). The resulting convection cells extend over finite radial distances. These global linear stability analyses indicate the following results:

1. Rotation and compressibility tend to reduce the growth rate of the disturbances. However, they cannot suppress the onset of convective instability in protostellar disks.
2. The growth rate is proportional to the square root of the radial wave number and is bounded by the maximum value of the Brunt-Vaisala frequency.
3. The maximum radial size of the eddies scales as the square root of the superadiabaticity times the size of the convective region.
4. Due to radiative losses, the short wavelength modes become overstable and only the fundamental and the first harmonic modes can grow effectively (in these modes, the wavelength is comparable to the disk thickness).
5. Both even and odd modes exist; in other words, a single eddy may either be confined to one side of the disk or may thread through the midplane and have a characteristic scale comparable to the thickness of the entire disk.

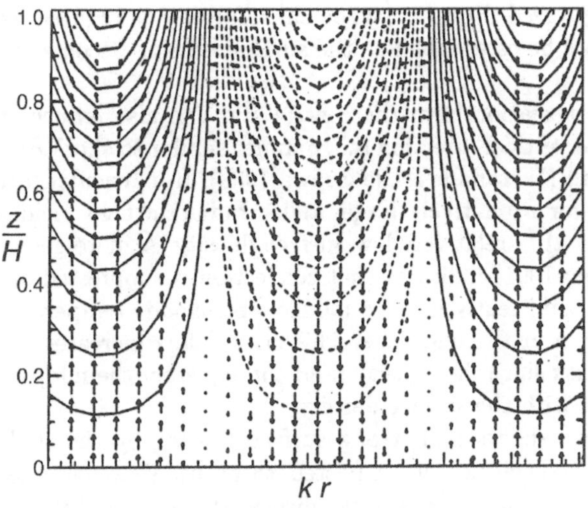

Fig. 9. Convective cell of eddies for an unstable convective mode in the protosolar nebula. (Adapted from Ruden, Papaloizou, & Lin 1988)

Convective eddies can thus provide a relatively effective coupling between different parts of the disk and can induce both heat dissipation and angular momentum transport. The magnitude of the effective viscosity can be derived under the assumption that all of the gas within a radial wavelength mixes efficiently during the characteristic growth time scale. This calculation implies an effective viscosity with $\alpha_s \sim 10^{-2}$. The corresponding evolutionary time scale for the disk is $\sim 10^6$ yr and the mass accretion rate is $10^{-7} - 10^{-8}\ M_\odot\ \mathrm{yr}^{-1}$. The resulting intrinsic luminosity of the disk is fairly *low*, $L_D \leq 1\ L_\odot$, and compares well with estimates of L_D from various observations.

4.4 Suppression of Convection: Infall and Surface Heating

In order for convection to occur, the vertical gradient of temperature in the disk must be sufficiently large (superadiabatic); in particular, the central disk temperature must be sufficiently hotter than the disk surface. This temperature structure can be significantly modified by the boundary condition at the surface of the disk. In fact, any heating of the disk surface will tend to flatten the vertical temperature gradient and thereby suppress convection. In this section, we discuss two mechanisms which can heat the surface of the disk and can suppress convection: disk heating through surface shocks from infalling material and disk heating by the stellar radiation field.

In the protostellar phase of evolution, infalling material falls onto the disk surface and dissipates its kinetic energy; this dissipation is achieved through a shock on the disk surface (e.g., Adams & Shu 1986). This shock heats the disk from its surface and thereby decreases the vertical temperature gradient; if the shock is sufficiently energetic, this heating from above can stabilize convection. If the infall rate onto the disk remains constant, shock dissipation near the surface can dominate the energy loss due to contraction towards the midplane and convection can be suppressed. However, when the infall rate is reduced, the superadiabatic gradient may be established as the surface heating becomes less important (Ruden 1986).

Surface heating due to radiation from the central star can also reduce the temperature gradient in the vertical direction in a manner similar to shock dissipation associated with infall (Watanabe & Nakazawa 1988). In particular, convection may be stabilized for a relatively large stellar radiation field incident onto the surface of a disk with a relatively low surface density. A more general analysis (Bell, Lin, & Ruden 1991) indicates that convection is suppressed only when the surface heating is sufficiently large to induce a blackbody temperature comparable to temperature near the midplane. This critical flux increases with the surface density of the disk since the opacity (and consequently the midplane temperature) also increases with the surface density.

In the absence of surface shocks and strong surface heating by the stellar radiation field, convective instabilities are expected to occur. However, these instabilities lead to accretion through the disk (as discussed above) and thereby reduce the disk mass and hence the disk surface density. When the surface density in the disk becomes sufficiently small, the disk becomes optically thin and can no longer maintain vertical gradients of temperature. The disk thus becomes nearly isothermal in the vertical direction, and convection again becomes suppressed.

4.5 Magnetic Effects and Instabilities

In addition to the mechanisms for angular momentum transport outlined above, other possibilities exist. For example, if the magnetic field strength is large enough, magnetic torques can lead to the transport of angular momentum (see, e.g., Hoyle 1960; Hayashi 1981; Stepinsky & Levy 1990). One way to view magnetic effects is to define an effective magnetic viscosity through the relation

$$\nu_B \equiv \frac{B_r^2 \tau_B}{4\pi\rho}, \tag{5.1}$$

where B_r is the radial component of the magnetic field, τ_B is the decay time of the field, and ρ is the density (Hayashi 1981). This magnetic viscosity drives disk evolution in a manner which is analogous to the usual viscous evolution described above. Here, the magnetic field strength determines the size of the viscosity and hence the evolutionary time scale of the disk.

On the other hand, if the magnetic field is relatively weak (i.e., if the energy density of the field is less than the thermal energy density), then a local axisymmetric shear instability can arise (Balbus & Hawley 1991; Hawley & Balbus 1991). Very roughly, we can describe the instability as follows: Begin with vertical magnetic field lines threading a differentially rotating disk. Now consider a field line to become displaced radially outward. The subsequent evolution is determined by two competing effects. The first effect is due to magnetic field tension, which tends to straighten the field line and hence provides a stabilizing influence. The second effect is that the displaced field line tends to enforce rigid rotation in the gas at its new radial position. Hence, the parcel of gas tied to the field line acquires too large of a velocity for its radial position and hence tends to move out even farther in the disk. This second effect is thus destabilizing. A detailed mathematical analysis shows that this second effect is dominant, and hence the system is unstable.

This magnetic instability can lead to turbulence in the disk. The turbulence, in turn, can produce a cascade of eddies and thereby provide a source of viscosity. As discussed above, once viscosity is present, the disk will evolve through a viscous diffusion process and can undergo both disk accretion and angular momentum transport. Numerical simulations of magnetic accretion disks are now being done (e.g., Stone & Norman 1994; see also the contribution by Stone in this volume). These simulations show that when the magnetic field strength is sufficiently strong, magnetic braking effects dominate. When the field is weaker, the weak-field magnetic instability occurs. In nearly all cases, these time-dependent simulations show that the global evolution of magnetic disks is intrinsically *unsteady.* However, further work is necessary to adequately assess all of the possible evolutionary paths.

Finally, we note that magnetic fields may also play an important role in the generation of winds from young stellar objects (e.g., Shu et al. 1988; Ruden & Königl 1993). In many of the current wind models, the outflow is coupled to mass accretion through the disk; in addition, the associated magnetic fields tend to brake the rotation of the star and may expand the disk. The star/disk/wind interface is thus another important issue concerning the evolution of circumstellar disks.

5. Interaction of Secondary Bodies with Circumstellar Disks

The interaction of disks with secondary bodies can also be important in the disks of young stellar objects. We expect secondary bodies to be present in most star/disk systems. These secondary bodies can either be giant planets or binary companions. If planets form in the disk, they can clear out gaps in much the same way as moons of Saturn clear gaps in the rings (see, e.g., Lin & Papaloizou 1986ab). On the other hand, most stars are found in binary systems, and hence binary companions should also be present in most star/disk systems; furthermore, more than half of the binary pairs have orbit separations less than the expected disk radius of ~100 AU (Duquennoy & Mayor 1992). Thus, companions are often present and can transport angular momentum through the disk.

Secondary bodies in star/disk systems can clear out gaps, drive angular momentum transport, and "shepherd" the disk edges to keep them sharp and well-defined. We can conceptually divide the interaction of secondaries and disks into two mechanisms (see the contribution of Artymowicz and Lubow in this volume). In the first case, denoted as *tidal interactions*, the secondary perturbs the gas in the disk through its non-axisymmetric gravitational effects. These interactions are *not* in resonance and can affect parcels of gas with any orbits within the disk. This interaction essentially amounts to gas parcels scattering in the gravitational potential well of the secondary (see the discussion below). In the second case, denoted as *resonant interactions*, the parcels of gas in the disk are forced to respond to the periodic forcing of the secondary. These resonant interactions occur only at the appropriate positions in the disk where the orbits of the gas parcel are in resonance with the orbit of the secondary (see Artymowicz and Lubow for further discussion).

Although a detailed treatment of tidal effects is beyond the scope of this review, we can summarize the basic physical mechanism as follows. We consider a secondary body (either a planet or a binary companion) in orbit about the primary star. This secondary will interact with a given parcel of gas in the disk due to its gravitational force, which is strongest at the point of closest approach of the secondary and the parcel of gas. In order to understand the basic principles, we consider the simplest case, in which all of the interaction takes place at the point of closest approach; this is known as the *impulse approximation.* We thus have a simple scattering problem in a $1/r$ potential: The parcel of gas "scatters" off of the secondary with the scattering angle θ given by

$$\cot^2[\theta/2] = \frac{u^4 b^2}{G^2 M_S^2} , \qquad (6.1)$$

where u is the relative velocity between the gas and the secondary, b is the impact parameter, and M_S is the mass of the secondary.

This scattering process transfers angular momentum. For the sake of definiteness, consider the case in which the orbit of the gas parcel is inside that of the secondary. It can easily be shown that, in the rest frame of the secondary, the parcel of gas is deflected with the kinetic energy of the parcel conserved. As a result, the velocity of the parcel of gas obtains a component perpendicular to its initial direction and the speed in the parallel direction decreases. This result, in turn, implies that the parcel of gas *loses* angular momentum and hence the secondary gains angular momentum. The change in angular momentum is given by

$$\Delta J = uR_S(1 - \cos\theta), \tag{6.2}$$

where R_S is the radius of the orbit of the secondary. Thus, the net result of this interaction is to move the parcel of gas farther in towards the star. For the opposite case, where the parcel of gas has an orbit outside that of the secondary, the net result of the interaction is to move the parcel of gas farther out.

We can estimate the rate of angular momentum transfer as follows: We consider the motion of both the secondary and the gas parcel to be nearly circular with orbital frequencies Ω and ω, respectively. The relative velocity is then $u = (\Omega - \omega)\, R_S$. Next, we note that the time between successive encounters is simply $2\pi/(\Omega - \omega)$. The rate of angular momentum transfer (per unit mass) is then given by

$$\frac{dJ}{dt} = \frac{G^2 M_S^2}{\pi R_S^2 b^2 (\Omega - \omega)^2} \,. \tag{6.3}$$

To fix ideas, consider the case in which $b \sim R_S$. In this case, we find $\dot{J} \sim J\Omega$; in other words, the specific angular momentum is transferred on nearly a dynamical time scale. Thus, this mechanism of angular momentum transfer is fairly efficient for parcels of gas near the secondary. For a more detailed treatment of this subject, see, e.g., Lin & Papaloizou (1985).

6. Vortices in Circumstellar Disks

In this section, we discuss possible effects that vortices have on disk evolution and dynamics. These vortices are essentially long-lived storm systems. These vortices can enhance the formation of giant planets within the mechanism of gravitational instability; vortices can also enhance energy dissipation and thereby affect disk accretion.

The most well known and well studied example of a fluid vortex in an astrophysical setting is the Great Red Spot of Jupiter. This stable swirling storm system has lived for many dynamical time scales. Although the exact nature of the Great Red Spot remains somewhat controversial (see, e.g., Marcus 1993 and Petviashvili & Pokhotelov 1992 for different perspectives), the basic physics are known and provide a starting point for this present discussion. Similar vortices arise in most rotating fluid systems, such as the atmosphere and oceans of the Earth (e.g., Ghil & Childress 1987) and the disk of the galaxy (Korchagin & Petviashvili 1985; Korchagin & Ryabtsev 1991). These previous studies have not included the self-gravity of the system and have not considered nearly Keplerian rotation curves such as those found in circumstellar disks (see, however, von Weizsäcker 1944). In addition, two recent papers (Tanga et al. 1995; Barge & Sommeria 1995) also discuss planet formation and dust settling in vortices.

To begin, we outline the derivation of the basic equation of motion for vortices in circumstellar disks and describe the relevant approximations. We consider disturbances which are small compared to the radial extent of the disk. As a general trend, the vortex size is roughly comparable to the scale height H of the disk and for thin disks $H/r \ll 1$. We thus work in terms of the local variables x and y centered on a point (r_0, ϕ_0) rotating with the disk.

As the next simplification, we separate out the mean flow from the disturbance. In equilibrium, the disk is azimuthally symmetric and has no radial flow, i.e., $u = 0$ and $v = v_0 = r\Omega(r)$. Here, we want to separate the perturbed flow of the vortex from the unperturbed motion; we therefore write the azimuthal velocity v as the sum of two parts:

$$v = v_0 + v_1 = r\Omega(r) + v_1 \,, \tag{7.1}$$

where $\Omega(r)$ is the unperturbed rotational speed as a function of radius. The usual force equation for the gas can then be written in the form

$$[\partial_t + r(\Omega - \Omega_0)\partial_y]\mathbf{v} + (\mathbf{v}\cdot\nabla)\mathbf{v} + 2\Omega(\hat{z}\times\mathbf{v}) + \nabla h_1 + \nabla\psi_1 + ur(\partial_x\Omega)\hat{y} = 0 \,, \tag{7.2}$$

where $\mathbf{v} = (u, v_1)$ is the perturbed velocity field. Notice the presence of the Coriolis term, $2\Omega(\hat{z}\times\mathbf{v})$, which arises because we are working in a rotating frame of reference.

We want to write the equations of motion in terms of vorticity, instead of velocity. This approach makes sense because we are interested in vortex solutions and vorticity is thus the relevant physical quantity. The vorticity ω is defined by

$$\omega \equiv \hat{z}\cdot(\nabla\times\mathbf{v}) \,, \tag{7.3}$$

i.e., we consider only the $\hat{z}$ component. We now take the curl of the force equation and combine it with the continuity equation. After some rearrangement, the resulting equation of motion becomes

$$\mathbf{D}\left[\frac{\omega + 2\Omega + r\partial_x\Omega}{\sigma}\right] = \mathbf{D}\Upsilon = 0 \,, \tag{7.4}$$

where we have defined the operator $\mathbf{D}$ to be the total advective derivative operator,

$$\mathbf{D} \equiv \partial_t + r(\Omega - \Omega_0)\partial_y + \mathbf{v}\cdot\nabla \,, \tag{7.5}$$

and where we have defined Υ to be the total vorticity per unit surface density. This quantity Υ, which we denote as the *vortensity*, has two contributions: the perturbative part and that of the mean flow. The equation of motion (7.4) implies that the *total* vortensity is advectively conserved. However, the vortensity in the mean flow can be freely exchanged with the vortensity in the perturbations.

For the vortices in circumstellar disks, gravity can be important and hence the fluid is nearly in a generalized type of geostrophic balance. Exact geostrophic balance occurs when pressure forces are balanced by Coriolis forces, so that the

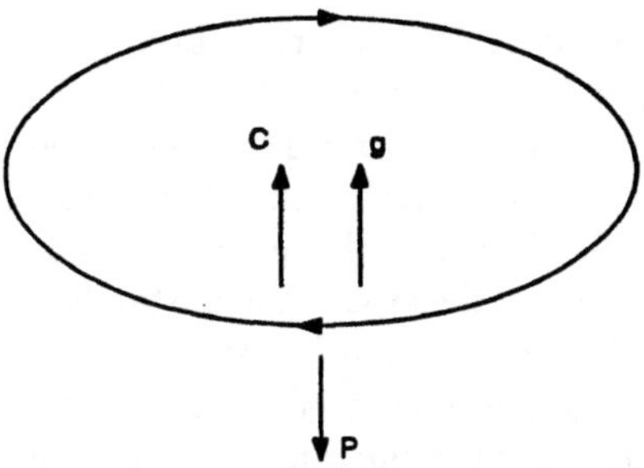

Fig. 10. Schematic diagram illustrating generalized geostrophic balance. For the case shown, both the Coriolis force (C) and the gravitational force (g) point inward, whereas the pressure force (P) is directed outward. (Adapted from Adams & Watkins 1995)

inertial forces are negligible. In this context, we generalize this concept to include gravity, so that both the Coriolis and gravitational forces are balanced by the pressure force (see Figure 10). In particular, we write the velocity $\mathbf{v}$ in the form

$$\mathbf{v} = \frac{1}{2\Omega_0} \left[\hat{z} \times (\nabla h_1 + \nabla \psi_1) \right] , \tag{7.6}$$

where Ω_0 is the value of the rotation rate in the disk at the origin of the local (x, y) coordinate system. In the geostrophic approximation, the vorticity can be written

$$\omega = \frac{1}{2\Omega_0} \left(\nabla^2 h_1 + \nabla^2 \psi_1 \right) . \tag{7.7}$$

Thus, our equation of motion in the *quasi-geostrophic approximation* now becomes

$$\mathbf{D}\Upsilon = 0 \qquad \text{where} \qquad \Upsilon = \frac{1}{2\Omega_0 \sigma} \left(\nabla^2 h_1 + 4\pi G \rho_1 \right) + \frac{2\Omega + r\partial_x \Omega}{\sigma} , \tag{7.8}$$

where we have used the Poisson equation to eliminate the gravitational potential in favor of the density. Equation (7.8) represents the fundamental equation of motion for vortices and implies that the total vortensity Υ is advectively conserved. Furthermore, equation (7.8) defines the relationship between the total vortensity and the density perturbation.

The equation of motion (7.8) has many different types of solutions (Adams & Watkins 1995) and can be used to elucidate the basic physics of vortices in circumstellar disks. These vortices are basically storm systems in a generalized geostrophic balance, i.e., a balance between pressure forces, gravitational forces, and the Coriolis force. We have obtained a number of basic results concerning the physical properties of vortices in circumstellar disks. These results are summarized below:

1. Many different types of vortex solutions are possible in circumstellar disks. In other words, circumstellar disks can exhibit many different types of vortical motions. Point vortices are the simplest type of vortex. Their properties can be found analytically and hence these vortices provide a prototype for understanding the basic physics of these systems.

2. In the absence of self-gravity, the size scale Λ for this type of vortex is determined by the inverse of the Rossby wavenumber, i.e., $\Lambda = a/\Omega$, where a is the local sound speed and Ω is the local rotation rate. For circumstellar disks, this size scale is comparable to the thermal scale height H in the disk.
3. The leading order effect of self-gravity is to make the effective Rossby wavenumber smaller. This effect, in turn, makes the size of the vortex larger than in the case without gravity. When the surface density of the disk becomes large (when the Toomre Q stability parameter becomes less than about 2), the effective Rossby wavenumber becomes imaginary. In this case, the vortex motion ceases and becomes a gravitational instability.
4. Nonlinear effects can lead to qualitatively new behavior. In particular, under special circumstances, solitary wave solutions can arise.
5. Magnetic fields can have two different effects on the evolution of vortices. The fields exert a force on the fluid and hence add an additional force term to the equation of motion. For the simplest case of a magnetic field in the vertical ($\hat{z}$) direction, this force corresponds to a magnetic pressure only. The second effect is to allow for ohmic dissipation; for expected parameters in circumstellar disks, however, this effect is small.
6. Circumstellar disks can, in principle, generate vorticity (or vortensity) through baroclinic effects. These effects arise whenever the surfaces of constant density do not line up with the surfaces of constant temperature. Disks can naturally obtain nearly axisymmetric temperature profiles (from reprocessing of stellar photons) and non-axisymmetric density profiles (from self-gravitating spiral modes). The estimated time scale for the vortensity to change is $\sim$500 yr.
7. Vortices can enhance the formation of giant planets through the mechanism of gravitational instability. In particular, dust grains settle to the center of these vortices. In order for the time scale for dust settling to be interesting for planet formation, either the dust grains must be quite large ($\langle b \rangle \gg 1$ μm) and/or the vortex circulation must be very strong ($|\Gamma| \gg aH$). Vortices also enhance the formation of giant planets by gravitational instability by increasing the local surface density of the disk. Planets formed in vortices should generally rotate in a prograde sense with respect to their orbits. In order to produce a planet with retrograde rotation, the circulation of the vortex must be sufficiently strong that the approximations used in this model break down.
8. Vortices can enhance the dissipation of energy in circumstellar disks and can thereby help disk accretion. For a given fluid viscosity, vortices dissipate energy faster than the disk as a whole. The physical reason for this enhancement is that the differential rotation in the vortex is strong and the size scale is short compared to the entire disk. Thus, vortices can enhance disk accretion, provided that some mechanism exists to transfer energy from the mean circumstellar flow into the vortices.
9. Finally, we have derived a general argument which illustrates the conditions under which secondary bodies can form in circumstellar disks. For moder-

ately massive disks associated with young stellar objects, a moderately large perturbation amplitude $\beta = \Delta\sigma/\sigma \sim 3$–$5$ is required to form a secondary.

7. Planet Formation

Another important physical process that occurs within circumstellar disks is the process of planet formation, which we briefly discuss in this section. The hypothesis that planets form within circumstellar disks – *the nebular hypothesis* – was first stated two centuries ago (Kant 1755; Laplace 1796) and continues to be the most viable scenario for planet formation. The disk properties thus provide the initial conditions for the process of planet formation.

Very roughly, we can identify two conceptually different ways for planets to form within a circumstellar disk: The first of these processes, *accumulation of planetesimals*, assumes that planets form by the gradual accretion of small rock-like bodies. In this case, the planets form "from the bottom up". Alternatively, planets could also form through a gravitational instability in the disk. In this case, the circumstellar disk becomes gravitationally unstable and breaks up into secondary bodies which become the planets. In this second case, the planets form "from the top down". Although both of these planet formation scenarios have some difficulties, the accretion of planetesimals is generally thought to be the leading theory. We discuss both of these mechanisms below. We note that the problem of planet formation is on a somewhat different footing than the overall problem of star formation. As discussed throughout this volume, we now have many examples of observed objects which we think are forming or newly formed stars. However, we cannot directly observe planets near other stars; we only have direct information concerning the planets associated with our Sun. Thus, any discussion of planets and planet formation is necessarily more specific to our own solar system.

Our solar system contains two different types of planets. The inner four planets (the *terrestrial planets*) are composed primarily of heavy elements. The next four planets (the *giant planets* or the *Jovian planets*) are much more massive than the terrestrial planets and contain substantial amounts of hydrogen and helium. They do not, however, contain as much hydrogen and helium as the Sun; they are still somewhat enriched in heavy elements. The giant planets also differ from stars (and the Sun) in that they contain solid rocky cores in their centers. The elemental abundances of the different planets place important constraints on how they must have formed. Finally, for completeness, we note that the solar system contains many other bodies, including asteroids, the planet Pluto, and comets.

7.1 Formation of Planets by Accumulation

In this section, we consider planet formation through the accumulation of small solid bodies (see, e.g., Cameron 1973; Cameron 1988; Hayashi, Nakazawa, &

Nakagawa 1985). In the disks which form planets, most of the heavy elements are found initially in the form of dust grains. In the interstellar medium, the dust grains are very small with typical radii $b \sim 0.1\ \mu m \sim 10^{-5}$ cm. In the denser environment of a circumstellar disk, the grains are expected to become somewhat larger, but remain microscopic in size (small compared to 1 cm). Thus, the process of planet formation begins with tiny dust grains and somehow produces huge objects with characteristic sizes of $\sim 10^9$ cm.

Dust grains accumulate on a fairly short time scale into large rock-like bodies called *planetesimals*. In the absence of turbulence, the dust grains settle to the midplane of the disk because they do not feel the same pressure force as the gas. In this case, the resulting thin layer of dust becomes gravitationally unstable and breaks up into planetesimals (Goldreich & Ward 1973). These planetesimals are basically large rocks with sizes of a $\sim$ 1 km, i.e., about the size of an asteroid. In other words, these planetesimals are much larger than the original dust grains, but are still much smaller than a planet. In actuality, this picture of planetesimal production is complicated by the presence of turbulence, which prevents the dust grains from settling all the way to disk midplane and inhibits gravitational instability (e.g., Weidenschilling 1984; Weidenschilling & Cuzzi 1993; Hayashi et al. 1985). In this case, the sticking together of dust grains becomes important. However, the net result is similar in that large rocks of roughly asteroid size and mass can be quickly produced. Here, the time scale for the production of planetesimals is about 10^4 yr, which is short compared to the expected lifetime of the disk (roughly 10^7 yr).

The next stage of planet formation involves the accumulation of planetesimals into the planets themselves. In the preceding phase, dust grains accumulate into planetesimals; during this second phase, these planetesimals accumulate into planets. The terrestrial planets are composed mostly of the rocky material that makes up the planetesimals, so their formation can be understood entirely in terms of planetesimal accumulation. However, the giant planets also contain substantial amounts of the lighter gases. For this latter type of composition to arise, the planetesimals must accumulate into a massive rocky core which subsequently accretes gas from the circumstellar disk. Giant planets thus require an additional stage of formation.

The accumulation of planetesimals takes considerably longer than the initial buildup of the planetesimals. One relevant time scale for this process is the orbit time at the radial position of the forming planet. This natural "clock" runs considerably slower in the outer solar system than in the inner parts. Another important effect is that the density of the disk (and hence the number density of the planetesimals) decreases with increasing radius. This effect makes the likelihood of planetesimal collisions smaller in the outer solar system. As a result, it takes a long time for planetesimals to accumulate in the outer solar system.

The time scales to build up planet sized bodies from planetesimals remain poorly understood. Under the most favorable circumstances, the cores of the giant planets can be produced in $\sim 10^6$ yr. Since the gas in the disk has a lifetime at least this long, these cores can accrete enough gaseous material to account for the observed compositions of the giant planets. Thus, this scenario for planet formation basically works. We should mention, however, that other calculations suggest that the time scale for building the cores of the outer planets could be as long as 10^8 yr (e.g., Hayashi et al. 1985), which is *longer* than the expected lifetime of the gaseous disk. These issues are still currently being studied.

7.2 Formation of Planets by Gravitational Instability

We now consider the possibility of forming planets through a gravitational instability in the disk (e.g., DeCampli & Cameron 1979; Cameron 1978). We know that gravitational condensation alone cannot account for the composition of the planets. Gravity does not distinguish between gas and rocky particles. Thus, any objects formed through the action of gravity alone are expected to have the same composition as the Sun. However, it is possible for the outer giant planets to have formed from a gravitational condensation in the disk, provided that some other mechanism can produce the observed enrichment in heavy elements.

Circumstellar disks can be gravitationally unstable provided that their temperatures are low enough and their masses are large enough. Furthermore, many observed star/disk systems seem to live near this point of gravitational instability. The natural mass scale for forming a gravitationally bound object within a disk is roughly 0.01 $M_\odot$ for a star/disk system with nearly equal masses (for solar-type stars – see Adams & Benz 1992; Adams & Watkins 1995); this mass is more than enough to form a giant planet. The relevant time scale for gravitational condensation is comparable to the orbital time scales of $10 - 10^3$ yr, which is much shorter than the accumulation time scale discussed above. Thus, the gravitational instability mechanism has both a time scale and a mass scale that are appropriate for the formation of giant planets.

The difficulty in this scenario of planet formation is to understand how the heavy elements become enriched and how most of the heavy elements end up in the cores of the planets (Stevenson 1982; Cameron 1988). Dust grains (which contain the heavy elements) will naturally settle toward the center of a self-gravitating protoplanet, because the grains feel less force from pressure than the surrounding gas. However, the time scale for this dust settling process is generally thought to be too long (more than 10^8 yr) to account for the observed composition and structure of the giant planets. Other mechanisms to provide dust separation and core formation have also been proposed, including dust settling in vortices (Sect. 7; Adams & Watkins 1995; Barge & Sommeria 1995; Tanga et al. 1995), gas drag capture of planetesimals (Pollack, Burns, & Tabuer 1979), and diffusive redistribution of water vapor (Stevenson & Lunine 1988). However, this issue remains unresolved.

8. Summary and Discussion

In this review, we have outlined the basic physical processes which occur in circumstellar disks. We first considered the possible role of gravitational instabilities in determining disk evolution. Gravitational instabilities can grow strongly (on nearly the dynamical time scale) when the disk mass is comparable to that of the star. Thus, we expect these effects to be important early in the evolution of the star/disk system when the disk mass is relatively large. Numerical simulations suggest that these instabilities can grow well into the nonlinear regime. Gravitational instabilities can be important for producing both disk accretion and secondary bodies within the disk.

Next, we considered the viscous evolution of disks. In this regime, the disk evolves according to a viscous diffusion process, i.e., the disk basically spreads out. The evolutionary time scale is determined by the size of the viscosity, which remains somewhat controversial. Several different possible sources of the viscosity have been proposed. Convective instabilities may give rise to a turbulent cascade of eddies and eventually lead to an effective viscosity. Magnetic fields can lead to magnetic braking in the strong field limit and magnetic instabilities in the weak field limit. When these latter instabilities occur, they can lead to turbulence and an effective viscosity.

We have briefly discussed the interaction of secondary bodies (either binary companions or giant planets) with circumstellar disks. These interactions can either be resonant interactions or non-resonant (tidal) interactions. In any case, secondary bodies can gain angular momentum from gas in inner orbits and will give angular momentum to gas in outer orbits. These interactions can dynamically clear gaps in disks, drive angular momentum transport, and shepherd the edges of disks.

Next, we considered the possible role played by vortices, which can exist in many different forms in circumstellar disks. Vortices may enhance giant planet formation via gravitational instability by allowing dust grains (heavy elements) to settle to the center on a short time scale; the gravitational instability itself is also enhanced, because the vortices create a larger local surface density in the disk. Vortices can also enhance energy dissipation in disks and thereby affect disk accretion.

Finally, we have discussed the process of planet formation in circumstellar disks. Planet formation can occur either by accumulation of planetesimals or through a gravitational instability. Although both options have some difficulty, the accumulation scenario is currently the leading theory.

In this review, we have suggested that gravitational instabilities in disks (and vortices) may be important for both the formation of binary companions (and/or planets) and for disk accretion. It might seem paradoxical that one physical process can produce two very different results – namely secondary bodies and disk accretion. However, this apparent contradiction can be resolved for both mechanisms as follows: Energy dissipation and hence disk accretion arises for "failed" structures, whereas secondary bodies can form only within "successful" structures.

We first consider energy dissipation and disk accretion. Self-gravitating spiral modes lead to disk accretion when they saturate at a fairly *low* amplitudes (Laughlin & Bodenheimer 1994), i.e., when the instabilities "fail" to achieve strong nonlinear growth. Similarly, vortices can lead to energy dissipation and hence help disk accretion when they dissipate their energy faster than they can grow, i.e., when they also "fail" to achieve strong nonlinear growth.

On the other hand, secondary bodies can be produced by either mechanism for the case of "successful" structures. Self-gravitating spiral modes can collapse to form secondary bodies when they successfully grow well into the nonlinear regime; a density contrast $\beta = \Delta\sigma/\sigma \sim 3 - 5$ is required. Vortices can lead to giant planet formation if they are sufficiently long-lived (so that dust grains collect at the vortex center) and/or highly nonlinear (so that they must collapse gravitationally), i.e., if the vortices are "successful".

References

Adams, F. C.: 1990, ApJ, **363**, 578
Adams, F. C., & Benz, W.: 1992, in *Complementary Approaches to Double and Multiple Star Research*, ed. H. McAlister, (Provo: Pub. A S P), p. 185
Adams, F. C., Emerson, J. E., & Fuller, G. A.: 1990, ApJ, **357**, 606
Adams, F. C., Lada, C. J., & Shu, F. H.: 1987, ApJ, **213**, 788
Adams, F. C., Lada, C. J., & Shu, F. H.: 1988, ApJ, **326**, 865
Adams, F. C., & Lin, D.N.C.: 1993, in *Protostars and Planets III*, eds. E. H. Levy and J. I. Lunine (Tucson: Univ. Arizona Press), p. 721
Adams, F. C., Ruden, S. P., & Shu, F. H.: 1989, ApJ, **347**, 959
Adams, F. C., & Shu, F. H.: 1986, ApJ, **308**, 836
Adams, F. C., & Watkins, R.: 1995, ApJ, **451**, 314
Anthony, D. M. and Carlberg, R. G.: 1988, ApJ, **332**, 637
Appenzeller, I., & Mundt, R.: 1989, Astron. Astrophys. Rev., **1**, 291
Balbus, S. A., & Hawley, J. F.: 1991, ApJ, **376**, 214
Barge, P., & Sommeria, J.: 1995, A & A, **295**, L1
Beckwith, S.V.W., Sargent, A. I., Chini, R. S., & Gusten, R.: 1990, AJ, **99**, 924
Beckwith, S.V.W., Sargent, A. I.: 1993, in *Protostars and Planets III*, eds. E. H. Levy and J. I. Lunine (Tucson: Univ. Arizona Press), p. 521
Bell, K. R., Lin, D.N.C., & Ruden, S. P.: 1991, ApJ, **372**, 633
Bertout, C.: 1989, A R A & A, **27**, 351
Bertout, C., Basri, G., & Bouvier, J.: 1988, ApJ, **330**, 350
Cabot, W., Canuto, V. M., Hubickyj, O., & Pollack, J. B.: 1987a, Icarus, **69**, 387
Cabot, W., Canuto, V. M., Hubickyj, O., & Pollack, J. B.: 1987b, Icarus, **69**, 423
Calvet, N., Hartmann, L. W., Kenyon S. J., & Whitney, B. A.: 1994, ApJ, **434**, 330
Cameron, A.G.W.: 1973, Icarus, **18**, 407
Cameron, A.G.W.: 1978, Moon and Planets, **18**, 5
Cameron, A.G.W.: 1988, A R A & A, **26**, 441
DeCampli, W. M., & Cameron, A.G.W.: 1979, Icarus, **38**, 367
Duquennoy, A., & Mayor, M.: 1992, A & A, **248**, 485
Edwards, S., Cabrit, S., Strom, S. E., Heyer, I., Strom, K. M., and Anderson, E.: 1987, ApJ, **321**, 473
Ghil, M., & Childress, S.: 1987, *Topics in Geophysical Fluid Dynamics* Applied Math. Sci. Ser. **60** (New York: Springer-Verlag)
Goldreich, P., & Tremaine, S.: 1979, ApJ, **222**, 850
Goldreich, P. and Ward, W. R.: 1973, ApJ, **183**, 1051

Goldsmith, P.F. and Arquilla, R.: 1985, in *Protostars and Planets II*, eds. D. Black and M.S. Matthews: (University of Arizona Press: Tucson), p. 137
Hawley, J. F., & Balbus, S. A.: 1991, ApJ, **376**, 223
Hayashi, C.: 1981, *Prog. Theo. Phys. Suppl.*, **70**, 35
Hayashi, C., Nakazawa, K., & Nakagawa, Y.: 1985, in *Protostars and Planets II*, eds. D. Black and M.S. Matthews, (University of Arizona Press: Tucson), p. 1100
Hoyle, F.: 1960, QJRAS, **1**, 28
Hollenbach, D., Johnstone, D., Lizano, S., & Shu, F. H.: 1994, ApJ, **428**, 654
Kant, I.: 1755, *Allgemeine Naturgeschichte und Theorie des Himmels*
Keene, J., & Masson, & C.: 1990, ApJ, **355**, 635
Kenyon, S. J. and Hartmann, L.: 1987, ApJ, **323**, 714
Korchagin, V. I., & Petviashvili, V. I.: 1985, Sov. Astron. Lett., **11**, 121
Korchagin, V. I., & Ryabtsev, A. D.: 1991, A & A, **246**, 368
Ladd, E. F., Adams, F. C., Casey, S., Davidson, J. A., Fuller, G. A., Harper, D. A., Myers, P. C., & Padman, R.: 1991, ApJ, **366**, 203
Laplace, P. S.: 1796, *Exposition du systeme du monde*, (Paris)
Larson, R.: 1984, MNRAS, **206**, 197
Laughlin, G. P.: 1994, PhD Thesis, U. C. Santa Cruz
Laughlin, G. P., & Bodenheimer, P.: 1994, ApJ, **436**, 335
Lin, C. C., & Lau, Y. Y.: 1979, Studies in Applied Math, **60**, 97
Lin, D.N.C.: 1981a, ApJ, **242**, 780
Lin, D.N.C.: 1981b, ApJ, **246**, 972
Lin, D.N.C., & Papaloizou, J.C.B.: 1980, MNRAS, **191**, 37
Lin, D.N.C., & Papaloizou, J.C.P.: 1985, in *Protostars and Planets II*, eds. D. C. Black and M. S. Mathews (University of Arizona Press: Tucson), p. 981
Lin, D.N.C., & Papaloizou, J.C.B.: 1986a, ApJ, **307**, 395
Lin, D.N.C., & Papaloizou, J.C.B.: 1986b, ApJ, **309**, 846
Lin, D.N.C., & Pringle, J. E.: 1987, MNRAS, **225**, 607
Lin, D.N.C., & Pringle, J. E.: 1990, ApJ, **358**, 515
Lüst, R.: 1952, Z. Naturforschung, **7a**, 87
Lynden-Bell, D., & Pringle, J. E.: 1974, MNRAS, **168**, 603
Marcus, P. S.: 1993, A R A & A, **31**, 523
Natta, A.: 1993, ApJ, **412**, 761
Paczynski, B.: 1977, Acta Astr., **28**, 91
Petviashvili, V. I., & Pokhotelov, O. A.: 1992, *Solitary Waves in Plasmas and in the Atmosphere* (Philadelphia: Gordon and Breach)
Pollack, J. B., Burns, J. A., & Tauber M. E.: 1979, Icarus, **37**, 587
Pringle, J. E.: 1981, A R A & A, **19**, 137
Rucinski, S. M.: 1985, AJ, **90**, 2321
Ruden, S. P.: 1986, PhD Thesis, University of California, Santa Cruz
Ruden, S. P., & Königl, A.: 1993, in *Protostars and Planets III*, eds. E. H. Levy and J. I. Lunine: (Tucson: Univ. Arizona Press), p. 641
Ruden, S. P., & Lin, D.N.C.: 1986, ApJ, **308**, 883
Ruden, S. P., Papaloizou, J.C.B., & Lin, D.N.C.: 1988, ApJ, **329**, 739
Rydgren, A. E., & Zak, D. S.: 1987, P A S P, **99**, 141
Savonije, G. J., & Papaloizou, J.C.B.: 1989, in *Dynamics of Astrophysical Discs*, ed. J. A. Sellwood, (Cambridge University Press: Cambridge), p. 115
Schwartzschild, M.: 1958, *Structure and Evolution of the Stars* (Princeton University Press: Princeton)
Shakura, N. I., & Sunyaev, R. A.: 1973, A & A, **24**, 337
Shu, F. H.: 1977, ApJ, **214**, 488
Shu, F. H., Adams, F. C., & Lizano, S.: 1987, A R A & A, **25**, 23
Shu, F. H., Lizano, S., Ruden, S. P., & Nagita, J.: 1988, ApJ, **328**, L19
Shu, F. H., Tremaine, S., Adams, F. C., Ruden, S. P.: 1990, ApJ, **358**, 495

Stepinsky, T. F., & Levy, E. H.: 1990, ApJ, **350**, 819
Stevenson, D. J.: 1982, Planet. Space Sci., **30**, 755
Stevenson, D. J., & Lunine, J.: 1988, Icarus, **75**, 146
Stone, J. M., & Norman, M. L.: 1994, ApJ, **433**, 746
Tanga, P., Babiano, A., Dubrulle, B., & Provenzale, A.: 1995, submitted to Icarus
Terebey, S., Shu, F. H., & Cassen, P.: 1984, ApJ, **286**, 529
Toomre, A.: 1964, ApJ, **139**, 1217
Watanabe, S., & Nakazawa, K.: 1988, Prog. Theo. Phys. Suppl., **96**, 130
Weidenschilling, S. J.: 1984, Icarus, **60**, 555
Weidenschilling, S. J., & Cuzzi, J. N.: 1993, in *Protostars and Planets III*, eds. E. H. Levy and J. I. Lunine (Tucson: Univ. Arizona Press), p. 1031
Weintraub, D. A., Sandell, G., & Duncan, W. D.: 1989, ApJ, **340**, L69
Weizsäcker, C. F. von: 1944, Zeit für Astrophys., **22**, 319
Woodward, J. W., Tohline, J. E., & Hashisu, I.: 1994, ApJ, **420**, 247
Yang, S., Durisen, R. H., Cohl, H. S., Imamura, J. N., & Toman, J.: 1991, Icarus , **91**, 14

The FU Orionis Outburst as a Thermal Accretion Event: Theoretical and Observational Implications

K.R. Bell

Space Sciences Division, NASA Ames Research Center, MS 245-3 Moffett Field, CA, 94035 USA, bell@cosmic.arc.nasa.gov

Summary. Detailed calculations of viscous accretion disks suggest that FU Ori outbursts signify the existence of protostellar disks transporting mass at a time-averaged rate of $(1-10) \times 10^{-6}$ $M_\odot$/yr around some fraction ($\approx 10\%$) of young stellar objects. Accretion through the inner parts of these disks is self regulated by the thermal ionization instability such that long periods (≈ 1000 yr) of low mass flux: $(1-10) \times 10^{-8}$ $M_\odot$/yr, are punctuated by short periods (≈ 100 yr) of high mass flux: $(1-10) \times 10^{-5}$ $M_\odot$/yr. The unstable region of the disk extends only to $\approx$ 1/4 AU. Beyond this region matter is transported stably through the disk at the infall rate, $\dot{M}_{in}$.

In systems for which $M_* = 1$ $M_\odot$ with inner disk edges of 3 $R_\odot$, the critical rate for outburst is 5×10^{-7} $M_\odot$/yr *independent of the magnitude of the viscosity.* The magnitude of the alpha parameter may be constrained by global timescale fitting to be 10^{-4}, where hydrogen is neutral, and 10^{-3} where ionized. This low viscosity has implications for self gravity as a possible mass transport mechanism through much of the disk. Light curves of V1515 Cyg, FU Ori, and V1057 Cyg are reproduced, and model results are compared to observations.

1. Introduction

In this contribution are discussed some theoretical and observational consequences of the Bell & Lin (1994) disk accretion model for FU Orionis outbursts. In Section 2., the importance of these outbursts is briefly discussed. In Section 3., some useful accretion disk formulae are outlined, and in Section 4., the Bell & Lin outburst model is summarized. To match global timescales of FU Orionis statistics, a very small viscosity is required. In Section 5., theoretical implications of the small viscosity for self gravity as a mass transport mechanism are discussed. In Section 6., model fits to the light curves of FU Ori, V1515 Cyg, and V1057 Cyg are shown. Model features are used to make detailed reply to objections raised for the accretion disk model for outbursts, and predictions of future line-width evolution are made.

2. Background

FU Orionis outbursts are temporary large increases in luminosity by factors of 40 – 250, thought to occur repeatedly in all low mass, young stellar systems (Herbig 1977). Time to peak light ranges from one year in the case of prototype FU Ori to 20 years in the case of V1515 Cyg. Outbursts are estimated to last

≈ 100 yr and are thought to recur on a timescale of $10^3 - 10^4$ yr (Herbig 1989). For a discussion of FU Ori timescales, see the review by S.J. Kenyon in this volume. Although FU Ori objects are optically revealed, all show long wavelength excesses consistent with slightly flattened envelopes which extend in to several AU (Kenyon & Hartmann 1991).

Broad band photometry of FU Ori objects (hereafter Fuors) show a characteristic broader than black body spectral shape which can be well fitted under the assumption that the system radiates simultaneously at a range of temperatures. Spectra are consistent with the model that the system is a flattened nebula with a temperature distribution: $T_s \sim r^{-3/4}$, characteristic of protostellar disks (Kenyon, Hartmann, & Hewett 1988). Spectral analysis of systems in outburst show absorption features which decrease in line width as a function of wavelength consistent with emission from a Keplerian disk in which rotation rate and surface temperature both decrease with increasing radius (Hartmann & Kenyon 1987). Estimates of accretion rates through FU Orionis disks are in the range of $(1 - 10) \times 10^{-5}$ $M_\odot$/yr (Kenyon, Hartmann, & Hewett 1988).

From a preoutburst optical spectrum of V1057 Cyg (Herbig 1977) which shows typical T Tauri emission features, it seems that the precursors of FU Ori outbursts are not dramatically different from the normal T Tauri or Class II objects. The suggestion that the preoutburst object was "rather red" together with a surprisingly small estimate of R–I, however, suggest that the precursor may have had a rising spectral energy distribution more typical of the younger Class I objects (Bell et al. 1995).

Accretion disks are thought to play a role in the evolution of most protostellar systems. Observations of broader than black body infrared excesses in young T Tauri objects are widely accepted to be indicative of the presence of flattened solar nebula type disks. Observed correlations of infrared and ultraviolet excesses in visible T Tauri systems suggest that the disks actively transport mass inward onto the central object and do not merely passively reradiate light from the central object (Bertout, Basri, & Bouvier 1988). Estimates of the highest accretion rates in T Tauri systems rarely exceed several times 10^{-7} $M_\odot$/yr, two to three orders of magnitude lower than the mass fluxes estimated for Fuors. How, then do these systems make the transition from low to high mass flux state, and why are there so few observations of intermediate mass fluxes?

3. Accretion Disk Formulae

In this section, some useful formulae are briefly derived. Viscosity (responsible for local energy generation and mass transport) is the classic unknown in disk calculations. Molecular viscosity is too small to transport mass through the nebula on reasonable timescales by many orders of magnitude. Shakura & Sunyaev (1973) therefore assume that mass transport occurs due to the turbulent mixing of adjacent annuli over the local pressure scale height, H, at the the local sound speed, c_s, with some unknown efficiency, α; thus arose the standard "alpha law" viscosity: $\nu = \alpha c_s H$. The source of the underlying turbulence remains unknown, although proposed processes include convection, magnetic instability, and surface accretion shocks.

Many global properties of accretion disks such as mass flux and surface temperature can be derived independent of the viscosity. While it is convenient that much can be learned about disks without specifying the underlying mass transport mechanism, this necessarily means that the form of the viscosity cannot be derived from observations alone; it must be derived from numerical fits. In the standard disk model (Shakura & Sunyaev 1973; Lynden-Bell & Pringle 1974), optically thick ($\tau >> 1$), geometrically thin ($H/r << 1$) disks are in vertical thermal balance; all locally generated and absorbed energy (Q^+) is locally radiated (Q^-). In such systems, the vertical and radial evolution is decoupled and can be calculated independently. With $\dot{M}$ and Σ as full plane mass flux and surface density respectively, the vertically averaged equation of motion is given by

$$3\pi\Sigma\nu_c = \dot{M}(r)(1-\beta\sqrt{R_*/r}). \tag{1}$$

The surface flux, $\sigma T_d^4(r)$, can be shown to depend only on the local mass flux $\dot{M}$:

$$\sigma T_d^4(r) = \frac{3GM_*\dot{M}(r)}{8\pi r^3}\left(1-\beta\sqrt{\frac{R_*}{r}}\right), \tag{2}$$

where R_* and M_* are the stellar radius and mass flux respectively. The factor β depends on the inner boundary condition and is generally taken to be one (eg. Lynden-Bell & Pringle 1974).

The radial temperature dependence of the disk can also be derived assuming that the disk has no internal energy sources but merely reradiates light absorbed from the central object. For a strictly flat disk, one quarter of the luminosity from the central object is intercepted by the disk (Adams & Shu 1986). With the projection factor of R_*/r this "reprocessing" flux depends on the stellar flux σT_*^4:

$$\sigma T_{rp}^4(r) = \frac{\sigma T_*^4}{4}\left(\frac{R_*}{r}\right)^3. \tag{3}$$

For flat disks, at large radius, accretion and reprocessing therefore result in the same radial surface temperature gradient: $T_s \sim r^{-q}$, $q = 0.75$. If there is significant flaring of the disk, q will be smaller. From fitting of T Tauri spectra, estimates of temperature profiles range from 0.5 to 0.75 (Beckwith et al. 1990) consistent with the presence of flat or slightly flared disks.

4. Bell & Lin Thermal Outburst Models

Observations confirm the presence of disks and can even provide estimates of disk mass flux, but tell nothing about the physical processes within. Just as in stellar atmospheres where interior properties are derived from model fits to surface observations, so in disk theory intrinsic disk properties such as surface density and central temperature can only be derived from numerical models.

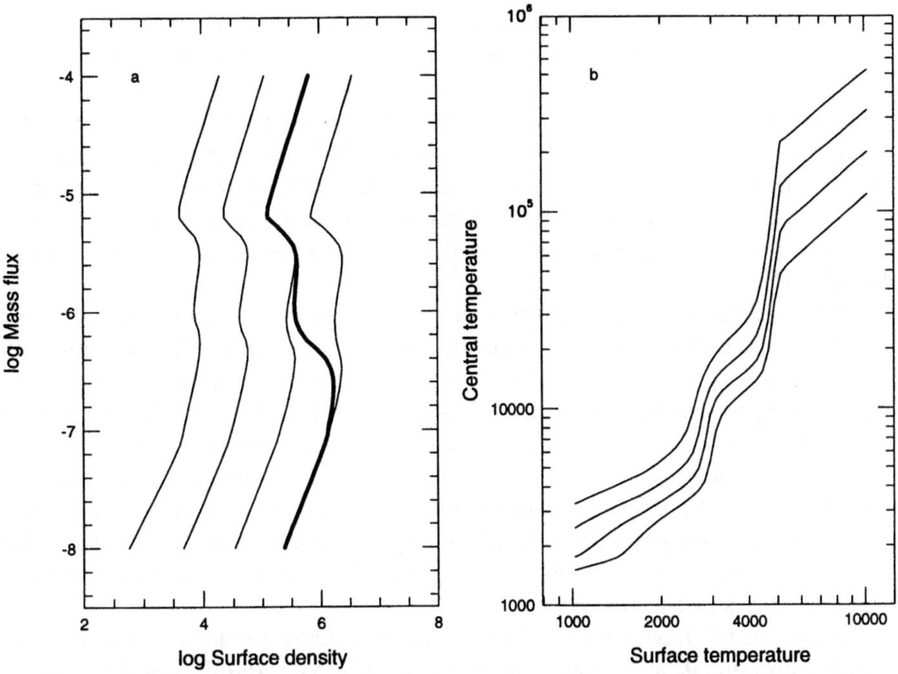

Fig. 1. Stability at 4.1 $R_\odot$ for $\alpha = 10^{-1}, 10^{-2}, 10^{-3}$, and 10^{-4} (a: left to right, b: top to bottom). (a) $\dot{M}$ ($M_\odot$/yr) vs. $\Sigma(gm/cm^2)$. Regimes with positive slope are thermally stable; regimes with negative slope are unstable. Heavy line is for $\alpha_c = 10^{-4}$ and $\alpha_h = 10^{-3}$. (b) $T_c(K)$ vs. $T_d(K)$.

The details of the thermal instability and of the numerical calculations used to investigate this instability are discussed fully in Bell & Lin (1994, hereafter BL94) and Bell et al. (1995, hereafter BLHK95). The essence of the thermal instability is the temperature dependence of the opacity ($\kappa \sim T^{10}$) and the rapid changes in the equation of state variables (C_V, μ, ∇_{ad} etc.) during the ionization of hydrogen. Any system with a temperature range which spans the region of hydrogen ionization is subject to some aspect of the thermal instability. An example of a similar mechanism in a very different context is the κ mechanism responsible for the well known "instability strip" in the HR diagram inhabited by Cepheids and RR Lyrae variables.

Under the assumption of vertical thermal balance ($Q^+ = Q^-$), the vertical structure of the disk may be calculated given a radius and mass flux. Examples of such calculations for four different values of α are shown in Figure 1. In Figure 1a where the slope is positive, the system is stable. If, for example, the disk is perturbed off equilibrium upward, local cooling becomes stronger than local heating, and the disk cools and returns to vertical thermal balance. Where the slope is negative, however, an upward perturbation results in thermal runaway and the disk heats until reaching the stable upper branch.

From the vertical structure models, it can be seen that instability occurs in the range of $(1-10) \times 10^{-6}$ $M_\odot$/yr. This mass flux is in agreement with theoretical and observational estimates of molecular cloud core collapse rates. The maximum stable quiescent mass flux: $\dot{M}_{crit} \approx 5 \times 10^{-7}$ $M_\odot$/yr, is consistent with estimates of boundary layer mass flux in T Tauri stars. The critical mass flux can be demonstrated with the simple hydrostatic, vertical structure models shown in Figure 1 and is clearly *independent of the magnitude of the viscous* α *parameter*. Critical mass fluxes depend to some degree on the assumed opacity law (BL94).

All derivations so far performed in steady state do not constrain value of α. *Changes* in the disk, however, occur on the local viscous or thermal timescales and, therefore, provide a rare opportunity to probe the details of the underlying mass transport mechanism. Time dependent calculations are therefore fit to timescales of FU Orionis outbursts (BL94). To produce systems in which long periods of low mass flux are punctuated by well-separated episodes of high mass flux, the value of α must be smaller in the quiescent or cold state (α_c) than in the outburst or hot state (α_h). In BL94, FU Ori timescales are reproduced when $\alpha_c = 10^{-4}$ and $\alpha_h = 10^{-3}$ (heavy line Figure 1a). Larger values of α result in unacceptably short timescales.

These models have very few free parameters. Once the form of the viscosity, opacity, and equation of state are specified, $\dot{M}_{in}$, α_h, and α_c are the principal variables. In addition, the stellar mass is chosen to be 1 $M_\odot$, and the inner disk radius to be 3 $R_\odot$. Critical mass flux is dependent on the latter two parameters as follows. Because the initiation of outburst requires potential energy sufficient to ionize hydrogen, $\dot{M}_{crit}$ is inversely related to the central object stellar mass. Larger mass systems show smaller critical mass flux ($\dot{M}_{crit} \approx 10^{-7}$ $M_\odot$/yr for $M_* = 3$ $M_\odot$; Bell 1994). Also, there is some stellar mass below which outbursts are not expected to occur; the very youngest protostars are, therefore, not subject to FU Ori outbursts. The critical mass flux also depends on the radius of the inner edge of the disk which in the standard case is taken to be 3 $R_\odot$. A test case in which the inner disk radius is 6 $R_\odot$ (BL94) confirms that larger critical mass fluxes are possible for larger inner disk radius. If the disk is truncated at larger radius (for example, matter may be transported along magnetic field lines), larger stable mass fluxes may be possible. An example of such an object may be DR Tau in which an unusually high accretion rate (10^{-5} $M_\odot$/yr) is coupled with a large estimated stellar radius (4 $R_\odot$) (Hartigan, Edwards, & Ghandour 1995).

5. Implications for Mass Transport

The small value of $\alpha = 10^{-4}$ implied by model timescale fitting is consistent with calculations in which turbulent convection provides the local viscosity (Ruden, Papaloizou, & Lin 1988). Quiescent models have typical surface temperatures of 2000 K in the inner disk (BL94) and can be shown to be strongly unstable to convection. Viscosity in disk regions expected to go into outburst is, therefore, consistent with convective turbulence as the physical mechanism for mass transport. Beyond the thermally unstable region, however, as the disk temperature drops, convection is expected to be weak. Such convection as there is has been shown to be incompatible with the presence of local viscous energy generation (Różyczka, Bodenheimer, & Bell 1994).

In a unified accretion scenario, there may be different sources of mass transport operating at different radii. If α is, nevertheless, taken to be constant with radius, disk evolution timescales and masses become uncomfortably large (BLHK95). Taking $\tau_\nu \sim r^2/\nu \sim (r/H)^2(\alpha\Omega)^{-1}$, with $\nu \approx \alpha H^2\Omega$, $H/r \approx 0.1$, and $\alpha = 10^{-4}$, it takes and estimated 5×10^6 yr for matter to viscously accrete from only 10 AU and 2×10^8 yr from 100 AU. Disk mass is inversely related to viscosity: $\Sigma \approx \dot{M}/(3\pi\nu)$, and low values of α result in uncomfortably large disk masses. By the above estimate, an only 10 AU disk accreting at 10^{-7} $M_\odot$/yr is 0.2 $M_\odot$. These timescales and masses are at odds with evidence which suggests that optically thick disks have lifetimes of less than 10^7 yr (Strom, Edwards, & Skrutskie 1993) and disk masses of no more than 0.1 M_* (Beckwith et al. 1990).

A way out of the dilemma is offered by theoretical derivations which suggest that disks more massive than 0.2 M_* are unstable to mass transport due to a global self gravity instability (Shu et al. 1990). The radius beyond which self gravity is important can be estimated from the radius where the the Toomre parameter $Q_T = c_s\Omega/\pi G\Sigma$ (Toomre 1964) drops to one:

$$\begin{aligned} R(Q_T{=}1) &= 1.5\mathrm{AU}\left(\frac{\alpha}{10^{-4}}\right)^{2/3}\left(\frac{H/r}{0.1}\right)^2 \\ &\times \left(\frac{\dot{M}}{10^{-6}\ \mathrm{M}_\odot/\mathrm{yr}}\right)^{-2/3}\left(\frac{\mathrm{M}_*}{1\ \mathrm{M}_\odot}\right)\left(\frac{Q_T}{1}\right)^{-2/3} . \end{aligned} \tag{4}$$

It can be argued that the process of self gravity acts to maintain the disk at some constant Q_T. By inverting the above equation, the process of self gravity can be expressed as an effective local viscosity for which, in a constant mass flux disk, α increases radially as $r^{3/2}$ (see BLHK95 for further discussion). In the inner disk, self gravity is, therefore, ineffective and some other form of mass transport (perhaps convective turbulence) must be present to transport mass while at "large" radius ($r > 1/2$ AU), self gravity may dominate.

6. Detailed Light Curve Fitting

Using the values of α derived from global timescale fitting in BL94, we then attempt to model specific light curves and to test the predictive power of this outburst model by comparing observations of spectral and line width evolution to model results. Applying standard extinctions taken from the literature, the dereddened light curves, $M_B(t)$, of FU Ori, V1515 Cyg, and V1057 Cyg are fitted with model outbursts (Figure 2); outburst parameters are given in Table 1.

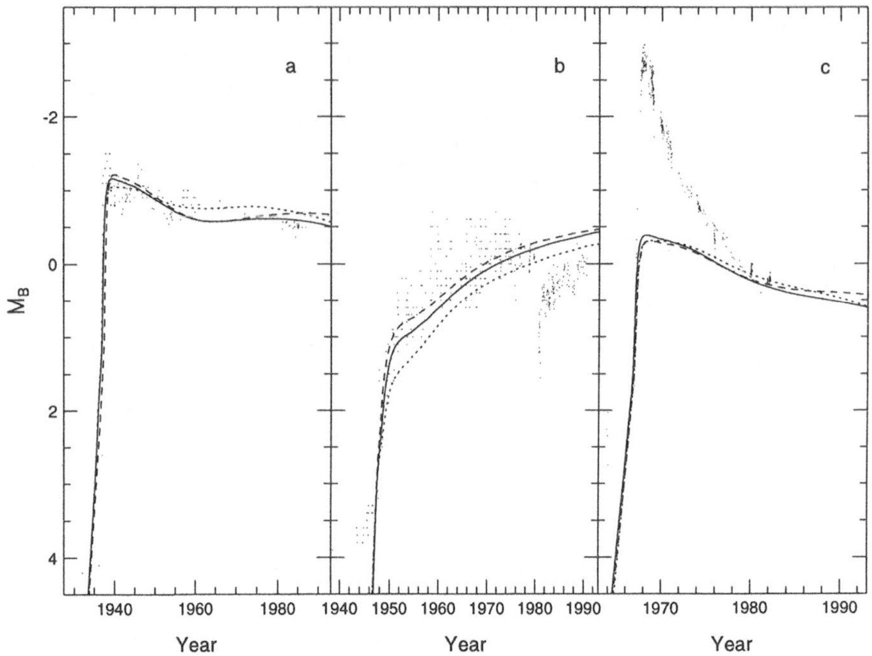

Fig. 2. Detailed fitting of FU Orionis light curves for (a) FU Ori, (b) V1515 Cyg, and (c) V1057 Cyg. Dots are extinction corrected data points; lines are models with parameters given in Table 1.

V1515 Cyg can be fitted with a self regulated outburst of the BL94 type; the other two objects require application of small perturbations in surface density to produce the observed rapid rise times. Perturbations in surface density are small (M_p less than about 0.01 $M_\odot$) and are applied at a radius (R_p) designed to force the outer edge of the thermally unstable region to go into outburst first. In a triggered model, the ionization front propagates in from the outside and has a considerably faster rise time than the self regulated model (see the discussion in BLHK95). In both triggered and self regulated models, outbursts propagate to radii (R_t) less than 1/4 AU (54 $R_\odot$). Outside-in outbursts may be triggered by the close passage of low mass binary companions, by the accretion of protoplanetary masses, or perhaps by a viscosity which increases with radius in the quiescent state.

The dip in the light curve of V1515 Cyg is thought to be the result of a dust condensation event (Kenyon, Hartmann, & Kolotilov 1991) and is not considered in light curve fitting. The poor fit to the early light curve of V1057 Cyg may be due to numerical limitations during the rapid onset of outburst. Fitting requires a strong triggering of an otherwise only marginally supercritical disk ($\dot{M}_{in} \approx \dot{M}_{crit}$). The ionization front therefore travels only a short distance through the disk (1/10 AU), and model evolution is subject to uncertainties and inaccuracies in the inner boundary condition. Because of these uncertainties, fitting of the

Table 1. Parameters and Results of Outburst Models

Model	$\dot{M}_{in}$ [M$_\odot$/yr]	M_p [10^{-2} M$_\odot$]	$\Delta\Sigma/\Sigma$	R_p [R$_\odot$]	$T^\dagger_{max}$ [K]	$\dot{M}_* M^\dagger_*$ [M$^2_\odot$/yr]	$R^\dagger_t$ [R$_\odot$]
A1	3×10^{-6}	1.0	3	13	7800	5.5×10^{-5}	35
A2	3×10^{-6}	1.0	10	13	7800	5.8×10^{-5}	35
A3	5×10^{-6}	0.7	5	20	8000	6.5×10^{-5}	35
B1	1×10^{-5}	...	...	...	7900	6.0×10^{-5}	26
B2	8×10^{-6}	...	...	...	7700	5.3×10^{-5}	25
B3	5×10^{-6}	0.1	4	34	7900	6.1×10^{-5}	26
C1	1×10^{-6}	0.2	3	10	6700	2.6×10^{-5}	24
C2	1×10^{-6}	0.2	5	10	6700	2.6×10^{-5}	24
C3	8×10^{-7}	0.3	5	10	6900	3.0×10^{-5}	24

† "Current" values at end of light curve fitting.

probably related but less powerful Exor outbursts (Herbig 1989) has not been attempted. Note that a range of outburst parameters may be used to produce acceptable fits to any given light curve. Note also that fits are made using *absolute* magnitudes as derived from one value of the extinction and may be somewhat altered with different extinctions.

7. Model Predictions

Both detailed light curve fits and the more general self regulated models may be used to make *predictions* about observable Fuor features. Solutions to several of the principal objections to the accretion disk model for FU Ori outbursts may be found in the *radially restricted* nature of model results. It has been argued (Herbig 1989) that decay from peak light in the disk outburst model ought to occur simultaneously at all wavelengths, because all disk models, regardless of mass flux, have the same temperature profile (eq. 2) and, therefore, have the same spectral slope: $\lambda F_\lambda \sim \lambda^{-4/3}$ (Lynden-Bell & Pringle 1974). Observations of V1057 Cyg, however, show that decay since peak light is greater in the optical and ultraviolet then in the near infrared (Kenyon & Hartmann 1991). Furthermore, the few existing preoutburst observations suggest that changes are also greatest at short wavelength during the rise to outburst. These features find natural explanation in a disk in which only the innermost annuli are subject to outburst.

Another major objection to the disk outburst model for FU Ori events is the observation that since peak light the optical line widths of V1057 Cyg have decreased (Herbig 1989). The observed inverse dependence of line width on wavelength in Fuors is due to the radial temperature structure of the disk. A given wavelength probes the disk at a temperature given by the Wien law. The radius which is at this temperature is indicated by the line width (Hartmann & Kenyon 1987). Observation of the *evolution* of these line widths can therefore reveal important information about the evolving radial temperature structure.

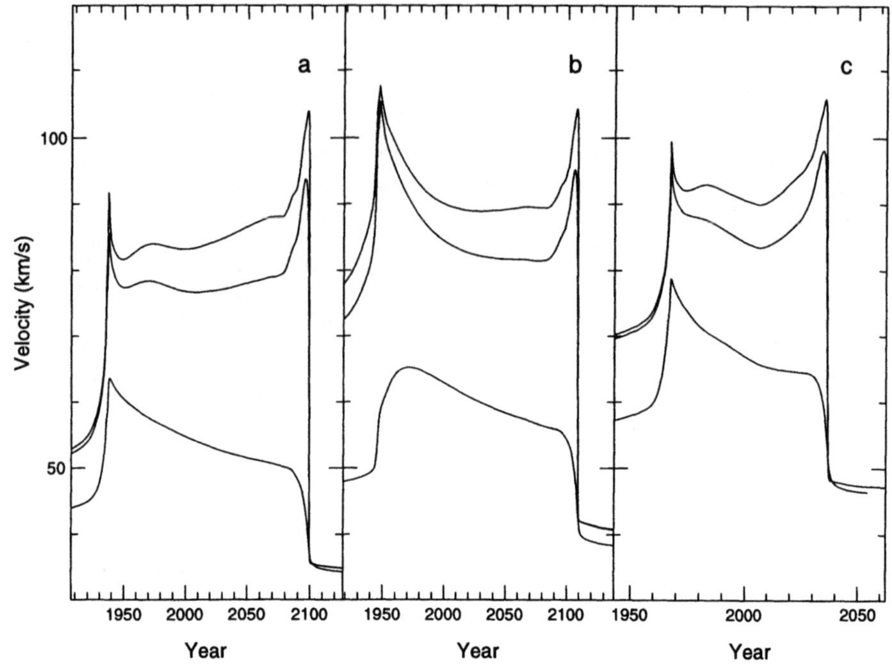

Fig. 3. Model line-width evolution for models (a) A1, (b) B1, and (c) C1 in km/s. Lines are for 4000 Å, 6000 Å, and 2.3 μm top to bottom corresponding to temperatures of 7200, 4800, and 1300 K respectively. Note the initial downward trend caused by the outward movement of ionization front. The upturn for the shorter two wavelengths at later times is due to the nearly simultaneous collapse of the entire unstable region of the disk during the decay from outburst.

In the standard disk outburst model during the decay from peak light, cooling occurs simultaneously at all radii. Line widths increase at every wavelength as a given temperature is radiated by successively smaller radii. Calculations of model line width evolution is shown in Figure 3 for three wavelengths and shows considerably more complicated behavior. The initial drop in velocity is due to the outward propagation of the ionization front; as successively larger radii go into outburst, lower velocity components are added to the line widths. Once the ionization front has propagated to its maximum radial extent, mass fluxes throughout the inner disk drop essentially simultaneously; velocities at the shorter wavelengths, therefore, rise as a given temperature is radiated by ever smaller radii. Longer wavelengths, however, ($\lambda > 1.5\ \mu$m corresponding to a surface temperature of $2000K$) probe radii which never go into outburst. In this case, velocities only increase during outburst because of contributions from the long wavelength tails of the bright inner annuli. As the disk decays after peak light, this component decreases, and the longer wavelength velocities decline monotonically (see discussion in BLHK95).

This line width evolution is a strong prediction of the thermal disk model for FU Orionis outbursts. The decline in velocities after the onset of outburst has already been observed in V1057 Cyg. It would be interesting to follow the line width evolution of known Fuors to see whether they follow the pattern shown by the models.

8. Discussion

The ubiquity of FU Orionis outbursts may have important implications for our understanding of the evolution of young stellar objects. Because the largest T Tauri stellar mass fluxes are estimated to be a little over 10^{-7} $M_\odot$/yr, even integrated over the probable lifetime of active T Tauri disks of a million years, disk mass fluxes would not contribute significantly to the final stellar mass. In this outburst model, however, a system's instantaneous stellar mass flux, $\dot{M}_*$, may be very different from its time averaged mass flux, $\dot{M}_{in}$. If the disk transports the $(1-10) \times 10^{-6}$ $M_\odot$/yr expected by this thermal outburst model even for only a hundred thousand years, the fraction of the final stellar mass which was accreted through the disk may be large.

Acknowledgement. This paper is based in part on detailed collaborations with D. Lin of the University of California, Santa Cruz and L. Hartmann and S. Kenyon of the Harvard-Smithsonian Center for Astrophysics. I would also like to acknowledge helpful discussions with P. Bodenheimer, P. Cassen, P. D'Alessio, C. Lada, S. Lizano, B. Reipurth, and S. Ruden.

References

Adams, F.C. & Shu, F.H. 1986, ApJ, 308, 836
Beckwith, S.V.W., Sargent, A.I., Chini, R.S., & Güsten, R. 1990, AJ 99, 924
Bell, K.R. 1994, in The Nature and Evolutionary Status of the Herbig Ae/Be Stars ed. P.S. Thé, M.P. Pérez, & E.P.J. van den Heuvel (ASP Conf. Series 62), 215
Bell, K.R. & Lin, D.N.C. 1994, ApJ, 427, 987 (BL94)
Bell, K.R., Lin, D.N.C., Hartmann, L.W., & Kenyon, S.J. 1995, ApJ, May 1 (BLHK95)
Bertout, C., Basri, G., & Bouvier, J. 1988, ApJ, 330, 350
Hartigan, Edwards, & Ghandour 1995 ApJ, 452, 736
Hartmann, L. & Kenyon, S.J. 1987, ApJ, 312, 243
Herbig, G.H. 1977, ApJ, 217, 693
Herbig, G.H. 1989, in ESO Workshop on Low Mass Star Formation and Pre-Main Sequence Objects, ed. B. Reipurth (ESO: Garching), No. 33, p. 233
Kenyon, S.J. & Hartmann, L.W. 1991, ApJ, 383, 664
Kenyon, S.J., Hartmann, L., & Hewett, R. 1988, ApJ, 325, 231
Kenyon, S.J., Hartmann, L.W., & Kolotilov, E.A. 1991, PASP, 103, 1069
Lynden–Bell, D. & Pringle, J.E. 1974, MNRAS, 168, 603
Różyczka, M.N., Bodenheimer, P.H., & Bell, K.R. 1994, ApJ, 423, 736
Ruden, S.P., Papaloizou, J., & Lin, D.N.C. 1988, ApJ, 329, 739

Shakura, N.I. & Sunyaev, R.A. 1973, A&A, 24, 337
Shu, F.H., Tremaine, S., Adams, F.C., & Ruden, S.P. 1990 ApJ, 358, 495
Strom, S.E., Edwards, S., & Skrutskie, M.F. 1993, in Protostars and Planets III, eds. E.H. Levy & J. Lunine (Tucson: Univ. of Arizona Press)
Toomre, A. 1964, ApJ, 139, 1217

Interaction of Young Binaries with Protostellar Disks

Pawel Artymowicz[1] and Stephen H. Lubow[2]

[1] Stockholm Observatory, Stockholm University, S-133 36 Saltsjöbaden, Sweden; pawel@astro.su.se
[2] Space Telescope Science Institute, 3700 San Martin Drive, Baltimore, MD 21218, USA; lubow@stsci.edu

Summary. We discuss the dynamical interaction between young binary stars and their protostellar accretion disks. We concentrate on circumbinary disks, which may play a major role in endowing the forming binary with its final (main-sequence) orbital characteristics. The new phenomena reported in this paper include the gap eccentricity and tidal stream formation. The latter phenomenon, effective in disks which have thickness-to-radius ratio of order 0.1, is responsible for the mass transfer from the circumbinary to the circumstellar environment, which equalizes stellar masses. In the following order, we discuss: observations of young binary systems with disks, tidal versus resonant disk perturbation, the special role played by the (m,l)=(2,1) harmonic of potential, resonant theory of gap sizes, orbital evolution of the binary, dynamically cold disks with eccentric gaps, and warm disks with tidal streams, with application to GG Tau.

1. Observations of Young Binary Systems with Disks

Gaseous disks surrounding the individual components of a binary system (the circumstellar disks or CS disks), and/or the whole binary systems (circumbinary, or CB disks, hereafter) are observed to be common among the forming and newly formed binary stars (for reviews see Mathieu 1992, 1993, 1994; Bodenheimer et al. 1993). They survive for $10^6 - 10^7$ yrs around the young stellar objects and are thought to control the early evolution of the protostellar binary orbit, as well as the later pre-main-sequence phase of dynamical evolution. The disk-binary interaction may be the dominant factor shaping the statistics of, e.g., orbital eccentricity of all main-sequence binaries. More than a half of the observed binary systems have orbital separations of < 30 AU (Duquennoy & Mayor 1992), which is smaller than a typical protostellar disk size of $\gtrsim 100$ AU. This underscores the importance of CB disks. In close binaries we can expect the CB disks to be much more massive than the CS disks, because the small inner disks may be depleted (cf. Jensen et al. 1994, Mathieu 1994, Zinnecker 1995), most likely because the gas supply from outside is hampered by the damming effect of stellar duplicity, and possibly also through active depletion by torques due to the binary (Ostriker et al. 1992).

There are at present more than 20 spectroscopic pre-main sequence binaries with estimated orbital elements, but only a few for which we have direct information on the geometry of disks. In the case of GW Ori – a classical single-lined spectroscopic T Tau star with period $P = 242$ days, strong Hα emission, and excess IR flux comparable with photospheric component (Mathieu et al. 1991; Mathieu 1994) – the presence of > 0.3 $M_\odot$ of hot and cold dust in the circumstellar environment is inferred. The location of the 'missing' material at intermediate temperatures coincides with the CB disk gap expected on dynamical grounds around the binary star. The hotter dust resides in CS disks. The observed eccentricity of binary orbit, $e = 0.04 \pm 0.06$, is atypically small among both the comparable-period young binaries, and the general population of main-sequence stars.

A strikingly beautiful and direct example of a binary with a CB disk is provided by GG Tau (Simon & Guilloteau 1992, Kawabe et al. 1993, Kitamura et al. 1993, Skrutskie et al. 1993), a classical T Tau star recently directly imaged with IRAM Plateau de Bure interferometer (Dutrey et al. 1994, Dutrey & Guilloteau 1995). The central binary system consists of two approximately equal-luminosity pre-main-sequence stars (with, by inference, masses of 0.65 and 0.5 $M_\odot$), has a semi-major axis equal to 67 AU, and a non-zero orbital eccentricity, probably of order $e \sim 0.2$. It is surrounded by a flat disk (at the current resolution resembling a doughnut) of cold, $T < 40$ K, molecular gas with an estimated mass ~ 0.17 $M_\odot$, in Keplerian rotation around the binary.

Other examples of systems with CB and the CS disks include: T Tau (Weintraub et al. 1989), Haro 6–10 (Menard et al. 1993, Monin & Menard 1995), Z CMa (Koresko et al. 1991, Malbct ct al. 1993), and UZ Tau (Ghcz ct al. 1994). Continued observations are certain to provide data on the above systems and their kin. It remains to be seen whether, as expected, the distribution of orbital properties of binary stars can be understood in terms of their interaction with disks via Lindblad resonances (Artymowicz et al. 1991, Lubow & Artymowicz 1993). In the present paper, we concentrate on the basic physics of the CB disks. (The principles of interaction are in many respect identical for CB and CS disks.)

2. Tides, Waves, and Resonances

Following the core formation in a molecular cloud, a prerequisite for both single and multiple star formation, the phase of a double protostar formation may occur either early during the dynamical collapse of a core, as in the fragmentation scenario, or alternatively after the formation of a single disk-shaped primordial nebula, as in the so-called $m = 1$ disk instability scenario (Pringle 1989; Adams et al. 1989; Boss 1991; Bonnell & Bastien 1992; Bonnell 1994). We shall be concerned with a much more prolonged (up to 10^7 yrs) and thus more easily observed phase of the pre-main-sequence evolution, where two young stars are already present but can still accrete mass from the circumstellar medium (disks). We assume coplanarity of disks with the binary orbit, neglecting the interesting complication connected with inclined subsystems (e.g., Clarke & Pringle 1991), as well as possible effects of gas selfgravity in the most massive systems. The important intrinsic parameters of the binary system are: the mass ratio $\mu = m_2/(m_1 + m_2)$, where m_1 and m_2 are the component masses ($m_2 < m_1$ by convention), and the eccentricity e. The semi-major axis a and period P can be rescaled to fit particular systems. Disks can be roughly characterized by their azimuthally averaged density distributions and half-clearing radii (disk sizes), as well as total masses. Their temperature profile may be expressed as thickness-to-radius ratio $z/r \lesssim 0.1$.

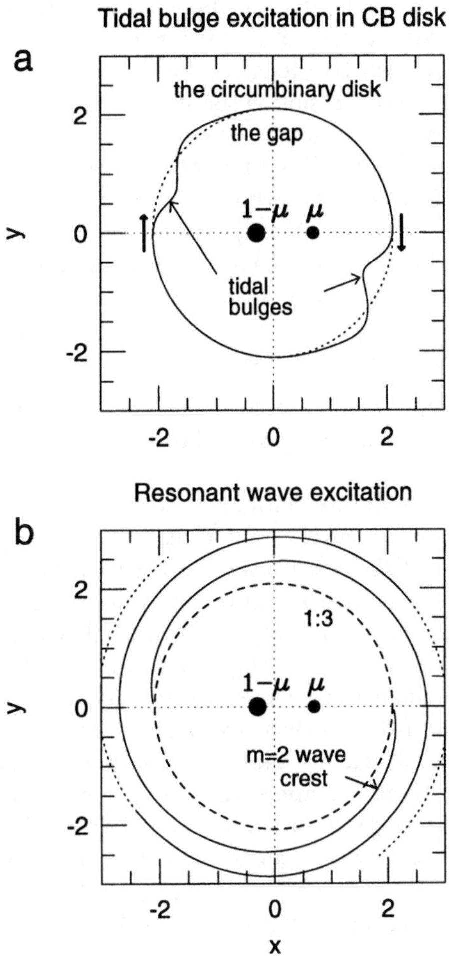

Fig. 1. Two ways to excite structure in a circumbinary disk: **a** – by streamline perturbation near the stars (tidal interaction). A circular binary with mass parameter μ is shown in its rest frame. Thick lines indicate the motion of gas in that frame; **b** – by resonant excitation of spiral density waves. The diagram is for a 2-armed wave in a nonselfgravitating disk excited at an outer Lindblad resonance (1:3 orbital period commensurability) of the eccentricity-dependent $(m, l) = (2, 1)$ harmonic of the binary gravity. Both the potential and the wave have pattern rotation speed equal to half that of the mean binary motion and rotate clockwise in the frame rotating with binary's mean motion.

The tidal and the resonant interaction, schematically shown in Figure 1, determine the evolution of the system. Under **tidal** interaction – or **non-resonant** interaction – we understand the generation of a (tidal) disturbance of the disk due to the non-axisymmetric gravitational pull of the binary (Figure 1a). Bulges, waves, or filaments are created whether or not their motion is commensurate with the binary period. Weak tidal interaction creates weak, bar-like or slightly sheared bulges – lagging somewhat behind the binary rotation because of orbital kinematics and the bulk gas viscosity. Strong tidal interaction occurs when, e.g., one of the binary stars passes close to the disk and changes the gas streamlines to such an extent that the tidal bulge becomes elongated into the form of a stream or quasi-spiral, time-varying, filament entering the otherwise rarefied gap region. In case of zero binary eccentricity ($e = 0$), the bulge(s) may be stationary in the corotating frame. In a highly elliptic case ($e \sim 1$), the perturbation is impulsive and may create a separate tidal filament for each half-period or period of the binary (depending on mass ratio). The total number of filaments seen at any time will then be proportional to the orbital period of the disk rim.

The **resonant** interaction (Figure 1b) can be thought of as forced disk response under an amplifying condition of the mean motion (period) resonance between the disk parts and the binary potential harmonics. The notion of harmonics arises, purely mathematically, because the gravitational potential $\Phi(r, \theta, z, t)$ of an eccentric binary system has period 2π in θ and orbital period $P = 2\pi/\Omega_B$ in t. By Fourier analysis, we may thus expand the binary potential in the disk plane – as well as any density variations due to wave phenomena – into an infinite double series of terms with the following azimuthal and time–dependence:

$$\phi(r, \theta, t) = \Sigma_{\{m,l\}} \phi_{ml} \exp i(m\theta - l\Omega_B t) , \tag{1}$$

where m is the azimuthal number or simply number of arms, and l is the time-harmonic number. Pattern speed of an (m, l)-harmonic is equal to $\Omega_p = (l/m)\Omega_B$ and does not vary in space or time.

The most important resonances due to potential harmonics are the Lindblad resonances (LR). They involve excitation of epicyclic vibrations of the disk at radii where the natural frequency of the disk matches the forcing frequency – felt by the moving disk element, i.e., Doppler-shifted with respect to the inertial frequency $m\Omega_p$. The resonant disturbance has a form of a spiral wave departing away from the binary in CB disk (e.g., Yuan & Cassen 1994), and toward the disk center in the CS disks – assuming the dominance of gas pressure over self-gravity. The wave originates in the vicinity of the LR, and that is where the transfer of angular momentum and energy between the binary and the *wave* occurs. The spiral wave becomes progressively more tightly wrapped as it propagates. The transfer from a wave to *disk material* takes place only after the effective disk viscosity damps the wave. The level of viscosity thought to exist in protostellar disks, described by parameter $\alpha \sim 10^{-2}$ (cf. Adams & Lin 1993), implies that the wave does not propagate far from the resonance but is damped locally. The condition for a LR at radius $r_{\rm LR}$ reads:

$$\frac{\Omega(r_{\rm LR})}{\Omega_B} = \frac{l}{m \pm 1} \, . \tag{2}$$

The plus sign in the denominator corresponds to an outer Lindblad resonance (OLR), and the lower to an inner LR (ILR). The (m,l)=(2,1) harmonic component is of special importance for slightly eccentric binaries (Artymowicz et al. 1991). From Eqn. (2), we see that its OLR lies at the 1:3 mean motion commensurability radius, or at $r = 2.08a$, where a denotes the binary semi-major axis (unit of distance in all our orbital-plane plots). At such a radius there is enough gas around a slightly eccentric binary ($e \lesssim 0.1$) to give rise to efficient wave emission. Strictly speaking, for the efficiency of the OLR finite density at the 1:3 resonance is not strictly required. It suffices that there is some gas at a distance of order one characteristic first radial wavelength of the spiral wave outside the OLR radius given by commensurability criterion. Perturbation in such a regime is called near-resonant.

A corotational resonance (CR) may also be present under certain conditions, causing wave emission from annulae corotating with the potential pattern, where angular velocity of the disk, $\Omega(r)$, satisfies $\Omega(r_{\rm CR})/\Omega_B = l/m$. In practice, we find that CR resonances usually play a minor role in determining the evolution of a young binary star system compared with LRs. For example, the (2,1) potential causes a CR at the 1:2 period commensurability in the CB disk. Neither the CR nor the yet closer lying ILR at the 1:1 commensurability are important for typical parameters of the binary and the disk (non-extreme mass ratio, thin disk), because they are nominally present in the circumbinary gap region devoid of gas.

The question about what processes are dominant in disk-binary interaction is still an open one. The disk-binary interaction is not trivial and, in general, cannot be adequately described as either tidal or resonant alone. By studying the time-dependent structure of perturbed disks we find that both modes of coupling operate at the same time. The tidal interaction is evidenced by material spiral arms moving with particle orbital velocities. The resonant interaction is manifested by directly observable, sometimes highly nonlinear, spiral density waves with appropriate radial location (close to LR), number of arms, and pattern speed (Lubow & Artymowicz 1991). Alternatively, it can be revealed by Fourier analysis of the time-dependent disk density, if the density contrast is small.

3. The (m,l)=(2,1) Harmonic and Its OLR

In order to provide an intuitive basis for the resonant theory of disk-binary interaction, we consider a generic binary of intermediate mass parameter $\mu = 0.3$ (mass ratio 7:3), and a small eccentricity equal to $e = 0.1$. We have argued above that the binary acts as a sum of its potential harmonics. If many harmonics are at play, pursuing such a description brings little if any insight to the analysis. However, Artymowicz et al. (1991) proposed that a slightly eccentric binary interacts with a CB disk mainly via one particular harmonic, namely the two-armed one, rotating at half the speed of the binary ($m = 2$, $l = 1$, $\Omega_p = \Omega_B/2$). Here, we prove their assertion graphically by comparing two smoothed particle hydrodynamics (SPH) simulations: one with the full binary potential, and one with a bare-bone potential purged of all the harmonics except the monopole term $\phi_{00} = -GM/r$ (where $M = m_1 + m_2$), i.e., the (m,l)=(0,0) harmonic, and the (2,1) harmonic[1]. At time $t = 0$, as in all our SPH calculations, the binary was at periastron. The disk was locally isothermal with radial temperature distribution

[1] Analytical approximation to the (2,1) harmonic is given by $\phi_{21} = (9GMa^2/4r^3)e\mu(1-\mu)\exp i(2\theta - \Omega_B t)$ (Artymowicz & Lubow 1994).

simulating a constant opening angle of a few degrees, corresponding to thickness-to-radius ratio equal to $z/r \simeq c/v_k = 0.025 = const$. The disk was initially confined between $r = 2.08a$ – roughly equal to the expected final gap size – and $r = 4a$. To increase spatial resolution for our adopted total starting number of particles $N = 20\,000$, we have run the model as a 2-D disk with zero vertical extent. The 3-D models yield very similar results (cf. Figure 6 below). The SPH viscosity parameters, α and β, were set to $\alpha = 0.5$ and $\beta = 0$, which resulted in Reynolds number of the disk flow $\mathcal{R} \sim 10^4$.

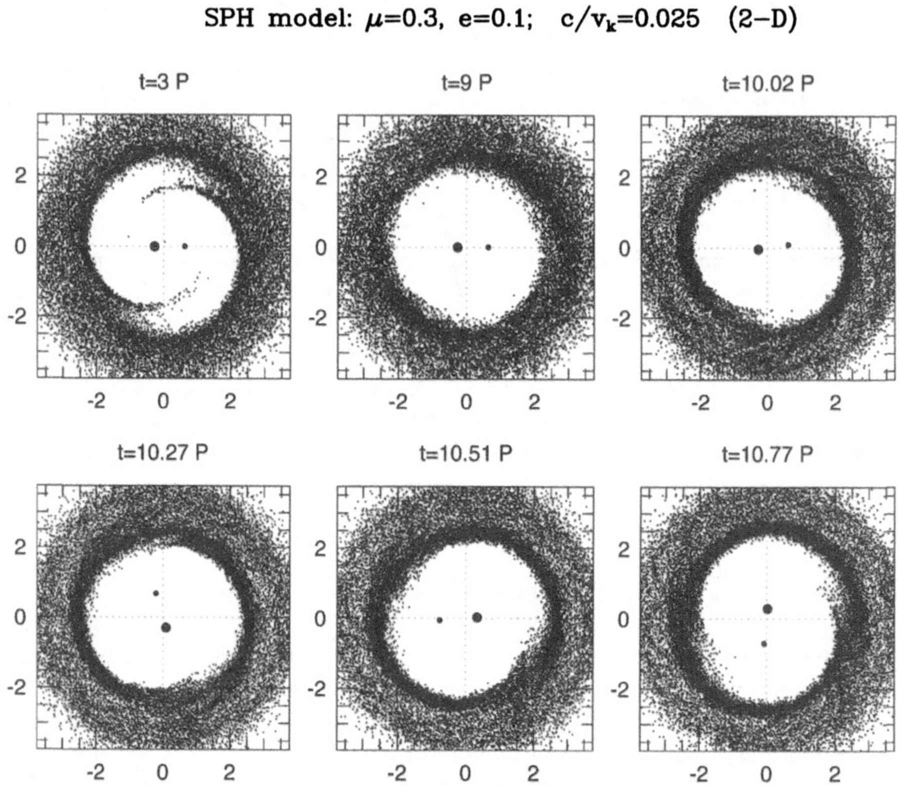

Fig. 2. SPH simulation of a binary system with $\mu = 0.3$ and eccentricity $e = 0.1$ surrounded by a 2-dimensional disk with temperature distribution corresponding to a constant soundspeed-to-circular velocity ratio equal to $c/v_k = 0.025$. Face-on views at 6 different times are shown. Axes are labeled in units of binary semi-major axis a.

Figure 2 shows several snapshots from the simulation with the full potential. Figure 3 shows the simulation with the bare-bone binary potential. It can immediately be noticed that both potentials yield essentially the same results, with some unimportant differences in the gap region, where the strong, localized forces of higher harmonics clear the immediate vicinity of the binary in Figure 2. This proves the assertion about the dominance of the (2,1) harmonic and its ability to singlehandedly withstand the tendency of the disk to viscously spread into the gap region. Next, observe how the wave patterns evolve with time. During the first 3 to 6 orbital periods, a strong transient spiral density wave is present. At later times, the wave's outer tail weakens, and the wave is very nonlinear and apparent only in a narrow region surrounding the 1:3 orbital commensurability (the OLR of the (2,1) harmonic at $r \approx 2.08a$). At all times, the density variations of the gas are a true density wave, rotating at half the binary mean motion, rather than at 1/3 the mean motion as would be the case for "material spiral arms" corotating with gas. In Figs. 2 and 3, we see thus a vivid confirmation of the resonant excitation of density wave at OLR. The wave does not propagate far out; indeed, already in its first wavelength, it is strongly shocked and damped viscously. The linear theory of OLR formally predicts the density contrast of 9!.

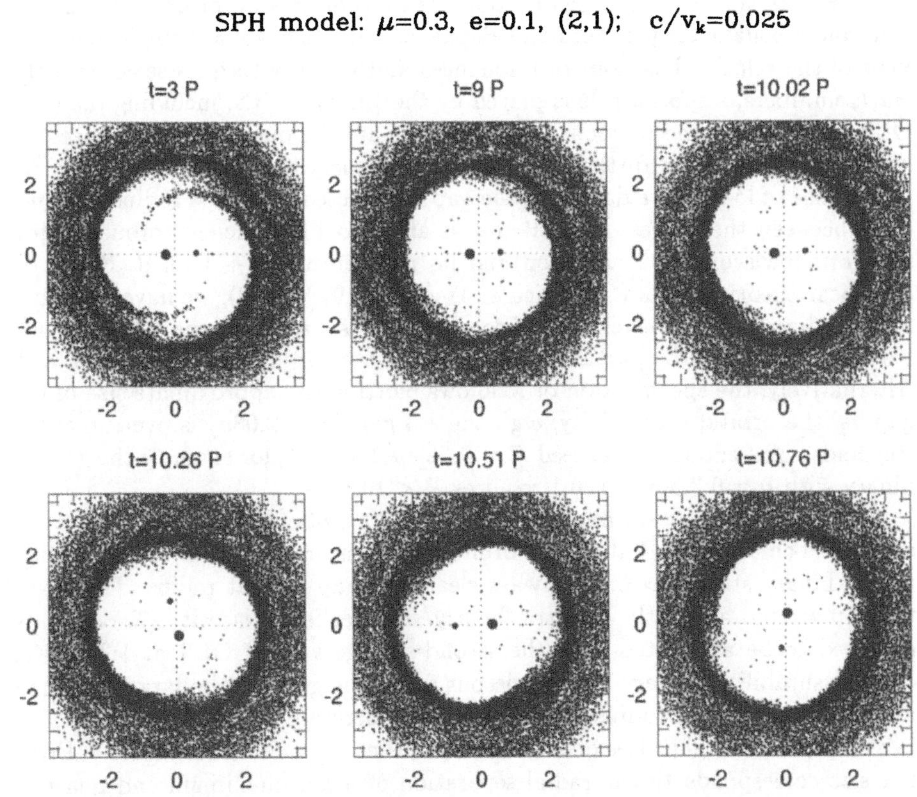

Fig. 3. SPH simulation of a binary system and a disk identical to that of Figure 2, in which the binary potential has been reduced to its monopole term and the (2,1) harmonic. Notice close correspondence of gap edge location and the pattern speed of the $m = 2$–armed density wave in gas with Figure 2 – the (2,1) harmonic dominates totally.

4. Gap Sizes: a Simple Resonant Theory

The gap region is cleared by the strong resonances surrounding the binary. How big are the CB gap and the CS disks? Are they changing in time? We have tried to answer these questions using independently an analytical theory and SPH (Artymowicz & Lubow 1994). The SPH has shown that after an initial clearing of the gap (occurring on dynamical time scale) the system is in steady state preserving the average size of the gap the outer disk is subject to viscous spreading in time inversely proportional to Reynolds number, and thus very long. This result justifies the assumptions used in the analytic theory. The two approaches give compatible results with regard to gap sizes in the quasi-stationary regime. The quantitative description of interaction in terms of resonances turns out to be successful here, despite the neglect of the tidal features arising at the very edge of the disk.

Our analytical theory is based upon the concept of equilibrium between viscous and resonant torques (Lin and Papaloizou 1979a,b, 1993; Pringle 1991). In view of the relative locations of resonances and the way torques scale with the (m, l) numbers, a special role is played by the $(m, 1)$ OLR's, including the (2,1) OLR presented in the previous section. Those are the outermost resonances for any given m, reaching furthest into the disk, namely to radii, $r = (m+1)^{2/3}a$. In the case of local wave damping, the gap is expected to have a radius intermediate between the highest-m OLR that is able provide sufficient torque to stop the normal viscous disk accretion and its neighboring $m+1$ OLR. Based on the linear theory of LRs (Goldreich & Tremaine 1979, 1980), we have derived a set of criteria for viscous closing of consecutive resonances giving the minimum viscosity[2] at which the particular resonance is overrun by gas spiraling inward. Alternatively, the specification of $\mathcal{R}$ allows plotting the approximate size of the gap vs. the orbital eccentricity, e.g., the 1:3 gap ($r = 2.08a$) is overrun if the Reynolds number does not exceed $\mathcal{R} = 0.32[e\mu(1-\mu)]^{-2}$; for example the generic binary with $\mu = 0.3$ and $e = 0.1$ requires $\mathcal{R} < 10^{3.4}$.

In Figure 4, we present a series of gap-size vs. e plots (dotted lines) for $\mu = 0.3$ systems. The $(1,1)$ OLR at the 1:2 orbital commensurability is in fact so strong, that a binary star ($\mu > 0.01$) always clears the gap at that radius. Therefore, the gap around a slightly eccentric binary with realistic viscosity ($\mathcal{R} \sim 10^4$) is expected to be associated with the second-strongest $(2,1)$ OLR at 1:3 orbital commensurability. In very weakly viscous disks, the gap may be larger, residing at the 1:4 or even 1:5 radius particularly in moderately eccentric systems.

We plotted the gap sizes obtained with SPH as open circles in Figure 4. Error bar size corresponds to the radial separation of a half-maximum and quarter-maximum surface density in the disk. A good agreement is found with the predicted trend, and, as a by-product, we verified that standard SPH codes can

[2] The viscosity is conveniently expressed as the Reynolds number $\mathcal{R}$ of the disk flow, very roughly speaking equal to the number of dynamical times needed for substantial inward spiraling of a gas parcel in a viscous accretion disk. In a thin disk with $z/r = 0.03$, the logarithms of Reynolds number and of the standard Shakhura-Sunyayev α are connected via $\log \mathcal{R} = 3 - \log \alpha$.

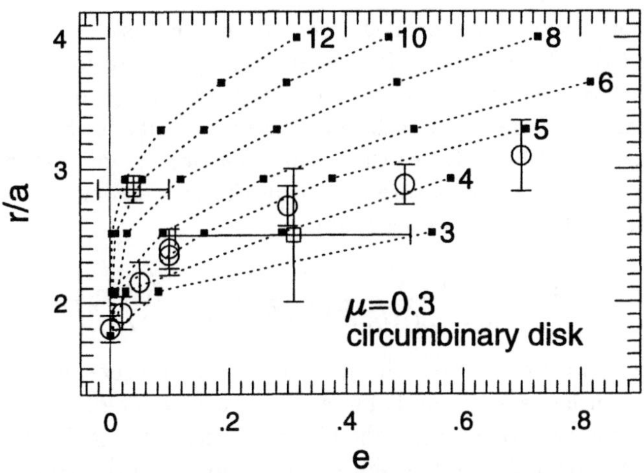

Fig. 4. Radius of the inner edge of the circumbinary disk for a binary with mass parameter $\mu = 0.3$. Dotted lines connect theoretical predictions of Artymowicz & Lubow (1994) for 7 different Reynolds numbers of the disk gas (whose logarithms label the curves). Circles with error bars are the results of SPH modeling. Two white squares are the estimated parameters for GW Ori ($e = 0.04 \pm 0.06$ and $r/a = 2.85 \pm 0.10$) and GG Tau ($e = 0.3 \pm 0.2$ and $r/a = 2.5 \pm 0.5$). Only the GG Tau parameters can be understood as resulting from resonant disk-binary interaction.

adequately represent the dynamics of disks with realistic, moderate viscosity ($\mathcal{R} \sim 10^4$).

Using the open squares with error bars, we present in Figure 4 the current estimates of orbital eccentricities and CB gap sizes for GW Ori (smaller cross) and GG Tau (larger cross). The CB disk truncation to the size $r \approx 3a$ suggested by the spectroscopic model of Mathieu et al. (1991) cannot be due to Lindblad resonances, even if the disk is nonviscous, because of the problem with exceedingly long gap opening time in that case. Proposed large gap could be easily understood if e was moderate or large which, however, is excluded by radial velocity curve. This illustrates the current difficulty with understanding the anomalously low eccentricity of this young binary; cf. Mathieu (1994). However, the GG Tau system fits comfortably within the theoretical expectations.

5. Basics of Orbital Evolution

Gravitational back-reaction of the distorted disks on stars forces the binary orbit to evolve. Gain of energy and angular momentum by a disk is accompanied by an equal loss from the binary. The orbital energy thus transferred is dissipated within the tide or the wave and may even under favorable conditions (in close binaries) be observable as thermal radiation of gas. The sign of orbital energy change of the binary determines the sense of change of its mean separation, and the ratio between the energy and angular momentum transfer rates determines the growth or decay of the orbital eccentricity. CB disks cause a to decrease, whereas the CS disks increase a. If both are present, the ratio of surface densities of CS to CB disks determines the outcome of interaction (spiraling in is normally expected because of massive CB disks). The eccentricity of the binary is excited (Artymowicz et al. 1991). To see this result consider the general formula for the eccentricity increase due to a torque T_{ml} on the disk associated with the (m,l) binary harmonic (cf. Artymowicz 1993):

$$\dot{e} = \frac{1-e^2}{e\mu(1-\mu)\Omega_B aM}\left(\frac{1}{\sqrt{1-e^2}} - \frac{l}{m}\right) T_{ml} \, . \tag{3}$$

It is straightforward to see that the strongest $m = l$ torques, which are independent of e, do not necessarily imply largest $|\dot{e}|$, if $e \ll 1$. The (2,1) harmonic of the disk density, however, does excite the binary eccentricity efficiently. The positive sign of $\dot{e}$ follows from $T_{21} > 0$, a well-known property of the OLRs (Goldreich & Tremaine 1979). Artymowicz et al. (1991) indicated that all but $\sim 10\%$ of the $\dot{e}$ measured in their SPH simulation with the generic binary parameters ($\mu = 0.3$, $e = 0.1$) was produced by this harmonic. Of course, this is consistent with (2,1) harmonic totally dominating the disk response, as seen above.

The circular or slightly eccentric binary surrounded by a CB disk is subject to a rapid growth of eccentricity. For a fixed density profile of the disk, the growth would be exponentially fast, because in equation (3) $T_{21} \sim \phi_{21} \sim e^2$, from which it would follow that $\dot{e} \sim e$. This scaling is indeed observed at small e but breaks down already at $e \approx 0.1$, since with the increasing e the inner disk edge is subject to increasingly stronger torques and recedes, thereby moderating the torques that fall rapidly with r. In addition to that simple effect, we have considered the possibility that while the disk edge location may be controlled by a single OLR of type $(m,1)$, the eccentricity evolution may in fact be due to a small number of competing resonances. At large eccentricity, the CRs and ILRs of higher harmonics are expected to become noticeable in the disk, and torques associated with them could partially cancel the positive OLR torques.

Based on many SPH calculations, we conclude that typically $\dot{e}$ decreases with e for $e \gtrsim 0.1$, if a given binary and disk temperature are adopted, but the long time-average of $\dot{e}$ remains positive. However, we have discovered that the instantaneous values and the short-term averages are in some cases very surprising – alternatively positive and negative! In the following two sections, we present a few highlights of our work in progress.

6. Dynamically Cold Disks with Eccentric Gaps

The dynamically cold disks ($c/v_k \lesssim 0.05$) evolve on long time-scales (hundreds to thousands of P) by becoming eccentric. As a rule of thumb, the disk eccentricity may reach very approximately the level of the binary eccentricity. Although much more dramatic examples may be presented involving $e \gtrsim 0.2$, here we will keep to our generic case of $e = 0.1$, as even in that case the main trends

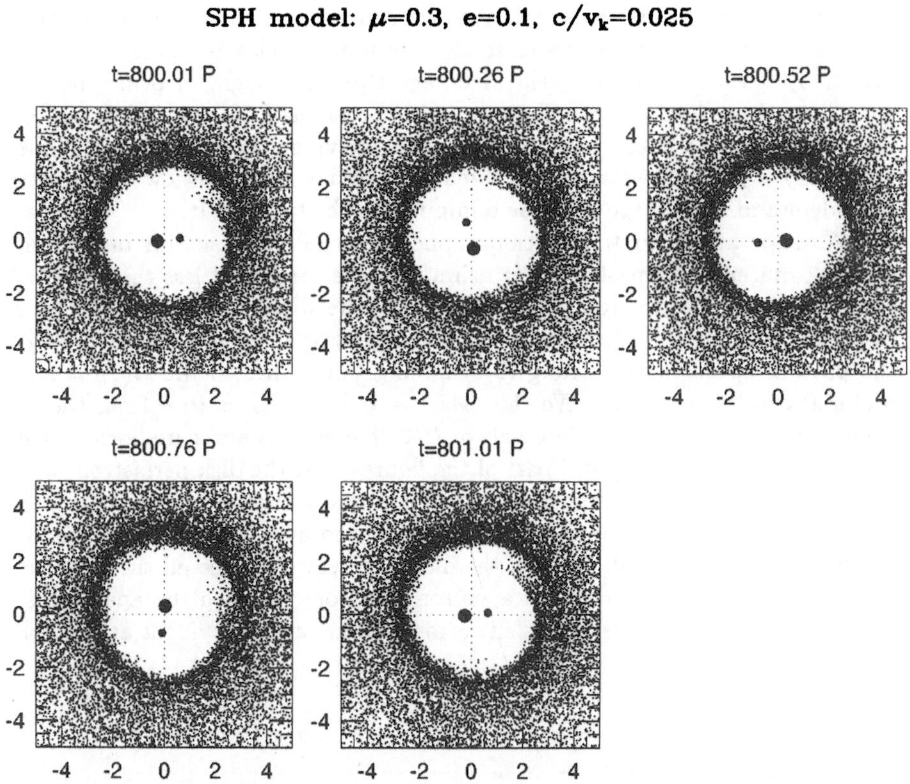

Fig. 5. Sequence of snapshots from the SPH simulation of a $\mu = 0.3$, $e = 0.1$ binary system during its 800th orbital period. The horizontal and vertical axes are the inertial (x, y) axes marked in units of a. The disk and the binary rotate in the couterclockwise direction. The tidal filaments near the elliptic disk's edge move at orbital velocity and do not represent a density wave. The $(m, l) = (2, 1)$ density wave is clearly detectable in Fourier analysis but less obvious to the eye.

and phenomena are already visible after long time. Figure 5 is a short "movie" constructed from snapshots spaced by $P/4$ of an SPH run with $\mu = 0.3$, $e = 0.1$, and $c/v_k = 0.025$ (a 3-D disk simulation this time, starting with a disk between $r = 2.6a$ and $r = 6a$). The movie presents the 800th rotation of the binary.[3] The disk's edge becomes eccentric after $\sim 250P$ and later on its long-term average eccentricity stabilizes at $e_d \sim 0.1$. The e_d exhibits moderate and regular variations in time, with a period of $230P$, equal to the precession period of a disk orbit with semi-major axis $r_d \approx 1.8a$ under the axisymmetric, non-point-like potential of a binary. Radius r_d is found to coincide with the region of maximum gas density (the black ring in Figure 5); the disk precesses as a slow $m = 1$ mode similar to the one studied by Hirose & Osaki (1994). In this mode,

[3] It is due to a *small* effective viscosity simulated by SPH that we are able to follow the system for times as long as this – and, in fact, 10 times longer! – before substantial disk spreading on viscous time scale.

pressure gradients between adjacent rings synchronize their precession. While the Hirose–Osaki mode precesses freely, our mode is clearly (at least initially) driven by the action of the central binary star. Although, in principle, more complicated mode-mode coupling may also be at play, we have evidence both from SPH and analytical calculations that the driving is mainly due to the direct forcing by a $(1,0)$ harmonic of the binary potential. It would now be intersting to understand the nature of mode damping which stabilizes it.

We draw attention to a curious appearance of the disk rim in Figure 5: the disk material at apastron is at azimuth angle somewhat less than $\pi/2$ with respect to the coordinate system drawn. At the disk rim, there are features moving with respect to the binary at the angular velocity *not* corresponding to the pattern speed of the strong (2,1) forcing but rather to the $\Omega(r) = \Omega_B/3$ material orbital frequency. We thus observe material arms torn from the disk edge by the tides at regular intervals of $P/2$. These tidal arms are born, but not yet easily visible, in the $y < 0$ part of the figures near the disk periastron, where the binary harmonics are much stronger than at apastron. Some of the material looses angular momentum and is transferred onto an elliptic orbit which after one half of its period collides with the disk. Therefore, we see all the tidal arms only between the periastron and apastron and none between the apastron and periastron (at $x < 0$). Such transient arms are particularly evident at moderate and large c.

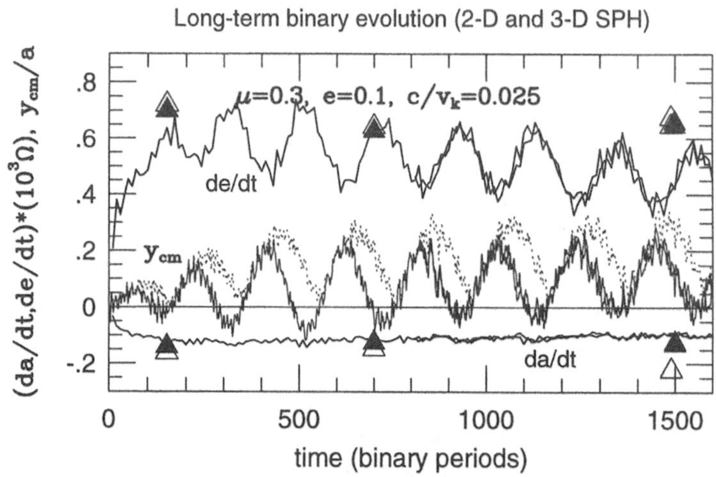

Fig. 6. The rate of orbital evolution ($\dot{e}$, $\dot{a}$) in disk-binary interaction. See text for explanation.

Figure 6 presents some of the consequences of nonaxisymmetric waves and modes in the CB disk in the SPH model discussed above (Figure 5). The vertical axis scale assumes unit disk mass. Two curves, labeled as y_{cm}, represent the y-coordinate of the center of mass of the disk – full line assuming a unit total mass for the disk – and the distance of the center of mass from the binary's center of mass (dotted line). These functions give a simple description of the growth and rotation of the slow $m = 1$ wave mode in a disk. At the same time, y_{cm} is proportional to the effect an elliptic ring surrounding a binary system has on its eccentricity evolution. Namely, positive y_{cm} corresponds to such a value of the *disk's* (1,0) potential which *damps* the binary eccentricity e. Conversely, negative y_{cm} means that the binary eccentricity *increases* under the action of the elliptic disk. Such trends in secular eccentricity evolution are also found in celestial mechanics. The upper curve on Figure 6 presents the $\dot{e}$, again, assuming unit disk mass). It shows unmistakably the precise anticorrelation of the variations in y_{cm} and $\dot{e}$. In systems with different parameters, e.g., higher e and/or smaller μ, the oscillations in $\dot{e}$ are able to reverse its sign for a part of the precessional period of the disk. This has significant observational ramifications. For example, detection of a short-term orbit circularization in a young binary does not guarantee that in the long run its eccentricity does not grow. The lowest curve in Figure 6 presents the rate of semi-major axis decrease. It is unaffected by the disk precession, in accordance with theoretical expectations. For $t > 800P$, results of a 3-D disk calculation (started from 2-D results at $t = 800P$) are overplotted and agree very well with the 2-D disk simulation.

We have found a good agreement of the long-term average of $\dot{e}$ simulated by SPH with the predictions of an analytical theory in which the steady positive contribution to $\dot{e}$ is due to (2,1) disturbance excited resonantly in the disk in the vicinity of the 1:3 orbital commensurability. We have used the SPH-calculated surface density, $\Sigma(r)$, the Goldreich & Tremaine (1979, 1980) torque formulae for OLR, and Eq. (3). The resultant values are plotted in Figure 6 as black triangles. Open triangles were obtained by adding the Goldreich-Tremaine fluxes due to all other OLRs and CRs present in disk; the modifications introduced by such harmonics are minor. The same good agreement characterizes $\dot{a}$ but becomes worse at later times, if all the OLRs are used including those which act on small gas density in the gap. Our binary system would evolve at long-term average rates given by $\dot{e} \simeq 5 \times 10^{-4} q_d \Omega_B$ and $\dot{a} \simeq -1 \times 10^{-4} a q_d \Omega_B$, where q_d stands for the mass of the disk out to $r \approx 10a$ divided by the mass of the binary. The SPH-derived ratio $\dot{e}a/\dot{a} = -5$ is in perfect agreement with the prediction of resonant theory. In a system like GW Ori ($P = 0.66$ yr), and $q_d \geq 10^{-3}$ (plausible, in view of $q_d > 0.1$ for $r \to \infty$, Mathieu et al. 1994), we obtain $\dot{e} \gtrsim 6 \times 10^{-6}$ yr^{-1} and $\dot{a}/a \lesssim -1 \times 10^{-6}$ yr^{-1}. One difficulty with the GW Ori is that at such $\dot{e}$, it could have been surrounded by a disk described by us for only 10^4 yrs, much shorter than it's age. GW Ori must look unlike our Figure 5.

7. Dynamically Warm Disks with Tidal Streams and Stellar-Mass Equalization

The question of mass transfer through the CB gap is a long-standing unresolved problem both in observations (Mathieu 1994) and theory (Artymowicz & Lubow 1994). The CB disk often coexists with the CS disks producing the Hα emission lines near the stellar surface and other indicators of active accretion onto stars. The appearance of an almost completely empty gap around the binary (Figure 6) suggests, to the contrary, that it is impossible for the CB disk to feed and sustain a CS disk from outside, and that it should disappear on its viscous spreading (accretional) time scale. We believe that the problem might be resolved with a greater theoretical recognition of the diversity among disks. For instance, as a consequence of different physical scales and fluxes of received stellar radiation, the CB disks in GG Tau and GW Ori systems are dynamically different: GW Ori's disk may have thickness $z/r \sim c/v_k \lesssim 0.05$ and is dynamically cold, whereas the GG Tau ring has an observed thickness $z/r \sim c/v_k \approx 0.157$ (Dutrey et al. 1994) and is dynamically warm. If $T(r) \approx 22(r/180\text{AU})^{-1/2}$ K is a good interpolating formula for all disks – GG Tau and the inner solar nebula providing reference points – then, in general, $z/r = 0.15(r/180\text{AU})^{1/4}$, and it follows that larger disks are warmer dynamically than the small ones, while the reverse is true of their internal thermodynamics. A thin disk like GW Ori's may lack the means of pushing the bulk of gas over the mild energetic barriers, measured by Jacobi constant, if $e = 0$, separating the CB disk from the circumstellar environments. However, a warm disk of the sort observed around GG Tau can, in principle, make use of its pressure or, more precisely, enthalpy gradients to overcome those barriers and transfer mass through the gap in a form of a stream or tidal filaments without obliterating the gap completely as in viscous gap closing.

In a simulation intended for GG Tau system, reproduced in Figure 7, we have modeled the tidal mode of mass transfer through the gap (cf. Artymowicz & Lubow 1995). Some of the parcels of gas lose angular momentum and orbital energy under the gravity of the eccentric binary and form kinematically collimated streams – inner parts of "arms" – falling inwards onto the black dots – stars, or CS disks. The cycle of arm/stream formation and destruction, and the low/high mass transfer intervals repeat with the binary period P (Figure 7). Assuming that the CB ring around GG Tau comprises $m_r = 0.1\ \text{M}_\odot$ (cf. Dutrey et al. 1994), the mass accretion rate onto the binary shown in Figure 7 equals $\dot{m} \simeq 10^{-6}$ $\text{M}_\odot$/yr. GG Tau may thus have started as much less massive system ~ 1 Myr ago (estimated age). The mass accretion ratio equals $\dot{m}_2/\dot{m}_1 = 1.055 \pm 0.02$, i.e. the less massive star, or star+CS disk subsystem, receives more mass than the more massive component. For comparison, $m_2/m_1 = \mu/(1-\mu) = 0.79$. Why is that the case? By analogy with the topology of critical Roche surfaces in a restricted 3 body problem, the communication between inside and outside of the binary is easiest through the vicinity of the outer Lagrange point on the side of the *less massive* star; i.e., through this very point a contact binary starts losing mass. In our case the mass transfer proceeds in the reverse direction[4]. We refrain from calling the less massive component "secondary", because given enough $\dot{m}_2$ it will actually dominate the system brightness with its larger accretional luminosity – and thus be called primary by the observer! Tracing back the evolution of GG Tau, based on the SPH estimate of $\dot{e} \simeq 10^{-6}(m_r/0.1\ \text{M}_\odot)\ \text{yr}^{-1}$,

[4] The accretional mass equalization has been first studied in free particle approximation by Artymowicz (1983).

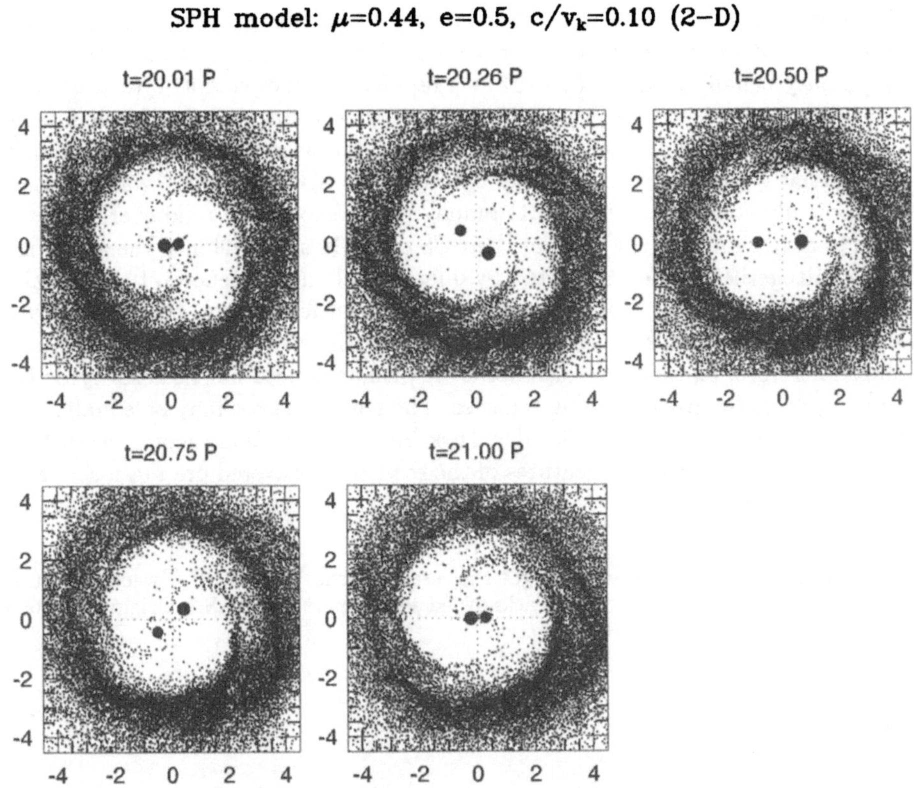

Fig. 7. SPH simulation of a binary with $\mu = 0.44$, $e = 0.5$ relevant to GG Tau, a classical T Tau-type, pre-main-sequence binary star with $a \sim 70$ AU, $P \sim 480$ yr, and $z/r \sim 0.15$. All rotation is counterclockwise; the coordinate system is at rest. The modeled size of the gap and the presence of a dense ring at the disk's edge agree well with observations by Dutrey et al. (1994). But does GG Tau have streams of gas from CB disk refueling the possible CS disks, as in our model?

we conclude that GG Tau might have been born with very unequal components ($m_2/m_1 = 0.1 - 0.5$) on a nearly circular orbit.

8. Conclusion

A number of potentially observable effects arise in a purely gravitational interaction between binary stars and circumstellar (CS) and circumbinary (CB) accretion disks. Strong tides and weaker, more distant resonances open gaps in the circumstellar medium surrounding the binary system. A theory predicts how large a gap to expect for any gas viscosity and binary parameters. Density waves produced at Lindblad resonances may be observable in close binaries transferring much orbital energy to the disk. Tidal forces, especially in eccentric systems, disturb the rims of the disks and continually material quasi-spiral produce arms.

The outer parts of CS disk and inner parts of CB disks may become markedly eccentric and precess as a slow $m = 1$ mode. Under the gravitation of the disks, the binary usually tightens ($\dot{a} < 0$) and rapidly becomes eccentric ($\dot{e} > 0$). Evidence for such a trend during the pre-main-sequence stage may be hidden in the orbital statistics of main-sequence binary stars. The accretion from cold CB disks is essentially terminated by the gap creation. CB disks which are warm enough dynamically (typically $c/v_k \gtrsim 0.1$) manage to overcome the mild energetic barriers to accretion onto the stars or their CS disks. In such systems, the gap is not closed altogether but is penetrated by quasi-spiral and, in general, time-variable streams of gas falling inside. More matter is transferred onto the less massive star which may acquire a larger accretional luminosity than the more massive star. Therefore, photometric or bolometric primaries need not necessarily be the primary stars by mass in young binaries. The stellar masses may be equalized, if the total mass increases significantly. Disk-binary interaction theory successfully explains the main observed features of the recently discovered pre-main-sequence binary GG Tauri.

Acknowledgement. P. A. acknowledges the research grant from NFR (Swedish Natural Science Research Council). S. L. acknowledges support by NASA Origins of Solar Systems Grant NAGW-4156.

References

Adams, F.C., & Lin, D.N.C. 1993, in Protostars and Planets III, Eds. E. Levy & M.S. Matthews (Tuscon: Univ. Arizona Press), p. 721
Adams, F.C., Ruden, S.P., & Shu, F.H. 1989, ApJ, 347, 959
Artymowicz, P. 1993, PASP, 105, 1032
Artymowicz, P., Clarke, C.L., Lubow, S.H., & Pringle, J.E. 1991, ApJ Lett., 370, L35
Artymowicz, P., & Lubow, S.H. 1994, ApJ, 421, 651
Artymowicz, P., & Lubow, S.H. 1995, in the proc. of 10th IAP meeting "Circumstellar Dust Disks and Planet Formation", Eds. R. Ferlet & A. Vidal-Madjar (Frontieres: Gif sur Yvette)
Bodenheimer, P., Ruzmaikina, T., & Mathieu, R.D. 1993, in Protostars and Planets III, Eds. E. Levy & M.S. Matthews (Tuscon: Univ. Arizona Press), p. 367
Bonnell, I.A. 1994, MNRAS, 269, 837
Bonnell, I.A., & Bastien, P. 1992, ApJ, 401, 654
Boss, A.P. 1991, Nature, 351, 298
Clarke, C.J., & Pringle, J.E. 1991, MNRAS, 249, 584 & 588
Duquennoy, A., & Mayor M. 1992, A&A, 248, 485
Dutrey, A., Guilloteau, S., & Simon, M. 1994, A&A, 286, 149
Dutrey, A., & Guilloteau, S., 1995, proceedings of the 10th IAP meeting "Circumstellar Dust Disks and Planet Formation", Eds. R. Ferlet & A. Vidal-Madjar (Edition Frontieres: Gif-sur-Yvette)
Ghez, A., M., Emerson, J., P., Graham, J., R., Meixner, M, & Skinner, C.J. 1994, ApJ, 434, 707
Goldreich, P., & Tremaine, S. 1979, ApJ, 233, 857
Goldreich, P., & Tremaine, S. 1980, ApJ, 241, 425
Hirose, M., & Osaki, Y. 1994, PASJ, 45, 243

Jensen E.L.N., Mathieu, R.D., & Fuller, G.A. 1994, ApJ Lett, 429, L29

Kawabe, R., Ishiguro, M., Omodaka, T., Kitamura, Y., & Miyama, S.M. 1993, ApJ Lett, 404, L63

Kitamura, Y., Omodaka, T., Kawabe, R., Yamashita, T., & Handa, T. 1993, Publ. Ast. Soc. Jap., 45, L27

Koresko, C.D., Beckwith, S.V.W., Gehz, A.M., Matthews, K., Neugebauer, G. 1991, AJ, 102, 2073

Lin, D.N.C., & Papaloizou, J. 1979a, MNRAS, 186, 799

Lin, D.N.C., & Papaloizou, J. 1979b, MNRAS, 188, 191

Lin, D.N.C., & Papaloizou, J. 1993, in Protostars and Planets III, Eds. E. Levy & M.S. Matthews (Tuscon: Univ. Arizona Press), p. 749

Lubow, S.H., & Artymowicz, P. 1993, in Binary Stars as Tracers of Stellar Evolution, ed. A. Dequennoy & M. Mayor, (Cambridge Press: Cambridge), 145

Malbet, F., Rigaut, F., Bertout, C., & Lena P. 1993, A&A, 271, L9

Mathieu, R.D. 1992, in Interacting Binary Stars, IAU Symp. 151, Eds. Y. Kondo, R. Sistero & R. Polidan, (Kluwer: Dordrecht), p. 21

Mathieu, R.D. 1993, in Binary Stars as Tracers of Stellar Evolution, ed. A. Dequennoy & M. Mayor, (Cambridge Press: Cambridge), p. 155

Mathieu, R.D. 1994, ARA&A, 32, 465

Mathieu, R.D., Adams, F. C., & Latham, D.W. 1991, AJ, 101, 2184

Menard F., Monin, J.L., Angelucci, F., & Rouan, D. 1993, ApJ Lett, 414, L117

Monin, J.-L., and Menard, F. 1995, in proceed. of the 10th IAP meeting "Circumstellar Dust Disks and Planet Formation", Eds. R. Ferlet & A. Vidal-Madjar (Frontieres: Gif sur Yvette)

Ostriker, E.C., Shu, F.H., & Adams, F.C. 1992, ApJ, 339, 192

Pringle, J.E. 1989, MNRAS, 239, 361

Pringle, J.E. 1991, MNRAS, 248, 754

Simon, M., Guilloteau, S. 1992, ApJ Lett, 397, L47

Skrutskie, M.F., Snell, R.L., Strom, K.M., Strom, S. E., Edwards, S., Fukui, Y., Mizuno, A., Hayashi, M., & Ohashi, N. 1993, ApJ, 409, 422

Weintraub D.A., Masson C.R., Zuckerman B. 1989, ApJ, 344, 915

Yuan, C., & Cassen, P. 1994, ApJ, 437, 338

Zinnecker, H. 1995, in proceed. of the 10th IAP meeting "Circumstellar Dust Disks and Planet Formation", Eds. R. Ferlet & A. Vidal-Madjar (Frontieres: Gif sur Yvette)

Numerical Simulations of the Formation of Binary and Multiple Protostellar Disks, Starting from Dynamic Initial Conditions

A.P. Whitworth, A.S. Bhattal, S.J. Chapman, M.J. Disney, H. Pongracic, and J.A. Turner

Department of Physics and Astronomy, University of Wales, Cardiff CF2 3YB, Wales, UK.

Summary. We present numerical simulations of the formation of protostellar disks. The disks form in shocked layers resulting from mildly supersonic collisions between molecular clumps. We describe the physical mechanisms which lead to the majority of these disks being in binary and higher multiple systems. And we suggest that the initial conditions required for these mechanisms to operate are very general, and therefore likely to occur commonly in nature. The simulations were performed using an SPH code.

1. Introduction

Whilst many of the most detailed and unambiguous observations of star formation pertain to the nearby, but relatively quiescent Taurus-Auriga region, this appears to be a rather sparse and infertile breeding ground for stars. There is considerable evidence to suggest that the majority of stars are formed in more dynamically agitated regions like Orion and 30 Doradus, where the density of young stars is much higher, and star formation appears to be self-propagating. In Cardiff, we are conducting a program of numerical simulations to explore the consequences of dynamical initial conditions for star formation, and to try to understand how binary stars form.

2. The Code

Our code uses Smoothed Particle Hydrodynamics (Lucy 1977; Gingold & Monaghan 1977; Monaghan 1992) to calculate hydrodynamic accelerations. The density-range of the code is increased by using an adaptive scale-length for the particles. Gravitational forces are calculated using a spatial tesselation tree (Barnes & Hut 1986; Hernquist 1987); the tree is also used to find neighbours for the hydrodynamic force calculation. Details of the code, and of the tests to which it has been subjected, are given in the appendices to Turner et al. (1994). There it is shown that the code is well able to handle the physical processes which occur in these simulations; in particular it can handle stable isothermal gas clouds, shocks and rarefaction waves, freefall collapse, and collapses of unstable isothermal gas clouds, both with and without rotation – additionally the results are little affected by the inward propagation of the boundary conditions.

3. Equation of State

The isothermal sound speed of the gas, $c \equiv (P/\rho)^{1/2}$, is given by

$$c^2 = \begin{cases} c_0^2, & \rho \leq \rho_0 \; ; \\ (c_0^2 - c_1^2)\left(\frac{\rho}{\rho_0}\right)^{-2/3} + c_1^2, & \rho > \rho_0 \; ; \end{cases} \tag{1}$$

where $c_0 = 0.6$ km s^{-1}, $c_1 = 0.2$ km s^{-1}, and $\rho_0 = 7 \times 10^{-22}$ g cm^{-3}. At low densities, $n \lesssim 300$ cm^{-3}, Eq.(1) corresponds to molecular hydrogen at ~ 100 K (a somewhat high temperature being adopted in order to mimic the extra support the gas is likely to receive from magnetic stresses and turbulence). At high densities, $n \gtrsim 10^4$ cm^{-3}, Eq.(1) corresponds to molecular hydrogen at ~ 10 K. At intermediate densities the temperature decreases with increasing density approximately as $T \propto \rho^{-2/3}$. We believe that Eq.(1) captures the important thermodynamic trends likely to occur in optically thin protostellar gas. We estimate that trapping of cooling radiation is just starting to become important right at the end of our simulations, but that the resulting temperature differentials are still small.

4. Initial Conditions

At the start of a simulation, two identical spherical clumps each having mass, M, and radius, R, are touching and about to collide at relative speed, v, and impact parameter, b. (For computational convenience, we neglect any distortion of the clumps due to mutual tidal forces acting as they approach one another; in any case such distortions will be small.)

The clumps are truncated isothermal spheres, embedded in a hot rarefied intercloud medium. In the notation of Chandrasekhar (1939), the outer boundaries of these isothermal spheres are at $\xi = 3.00$, and so they are stable. The intercloud medium surrounding the spheres is represented by non-gravitating SPH particles with high sound speed. We have checked the intrinsic stability of a single clump by following its evolution with our SPH code: it remains unperturbed for a thousand freefall times.

5. Formation of a Dense, Shocked Layer

Where the clumps collide, a layer of shocked, dense gas forms. The layer is compressed by the ram-pressure of the clumps. The surface-density of the layer increases more or less linearly with time. The amount of compression is greatly increased by the radiative cooling implicit in Eq.(1). The cooling also reduces the Jeans mass in the shocked layer well below what it was in the initial clumps, thereby enabling the layer to fragment gravitationally.

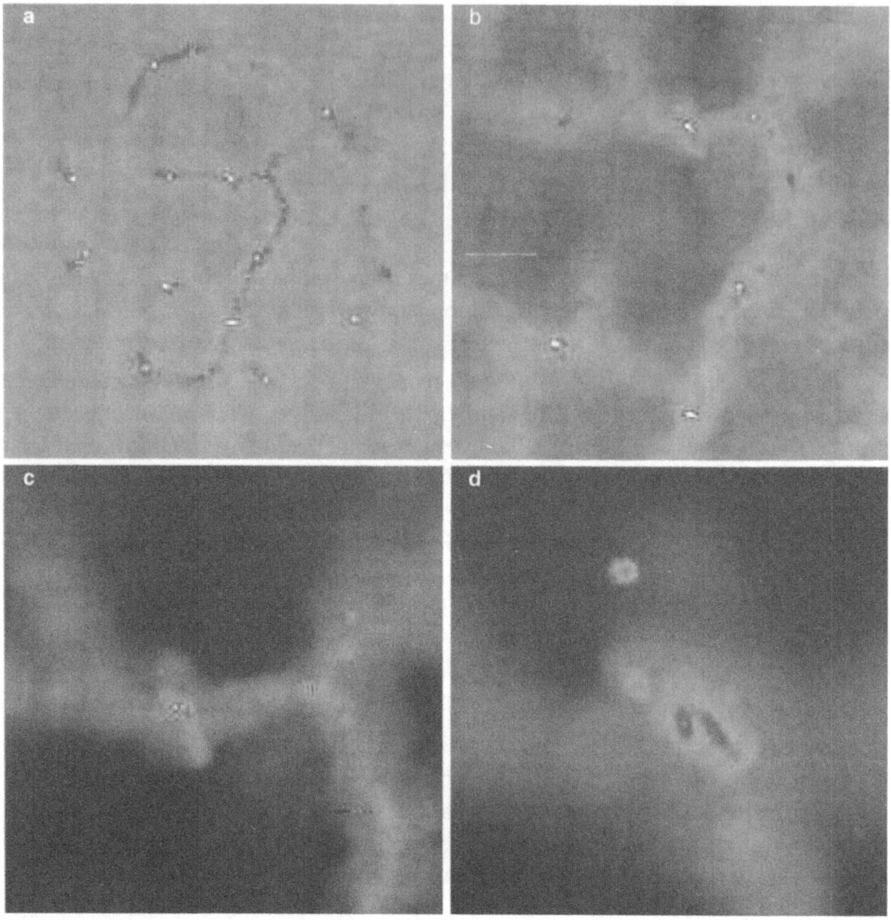

Fig. 1. Face-on view of the shocked layer produced when two clumps, having mass $M = 750\ \mathrm{M}_\odot$ and radius $R = 6.45$ pc, collide head-on at relative speed $v = 1.7\ \mathrm{km\ s}^{-1}$ at a time 2.10 Myr after the initial clumps first come into contact. The zoom-sequence is intended to show how the filaments have broken up into beads, and binary protostellar disks have condensed out of the beads. The linear sizes of the regions shown are: (a) 1.0 pc; (b) 0.4 pc; (c) 0.2 pc; (d) 0.05 pc. [All figures are presented in color on CD-ROM.]

6. Fragmentation of the Layer

Eventually the surface-density of the layer becomes so high that it starts to fragment gravitationally. For sufficiently massive layers – i.e. those resulting from collisions between massive clumps at high speed and low impact parameter – fragmentation breaks the layer up into filaments first and then into beads along the filaments. Figure 1a shows such a layer face-on. In this simulation, two clouds of mass $M = 750\ \mathrm{M}_\odot$ and radius $R = 6.45$ pc have collided head-on at relative speed $v = 1.7\ \mathrm{km\ s}^{-1}$. We are looking at the layer 2.10 Myr after the clumps first came into contact. (Similar structures are formed in collisions at finite impact parameter, $b \lesssim R$.) In this figure, and all subsequent figures, the grey-scale represents column-density from $N \sim 10^{21}$ to $N \sim 10^{27}$ H cm^{-2}.

7. An Embryonic OB Subgroup?

If we zoom in on one of the beads in Figure 1a – see Figs. 1b to 1d – we see that it is a binary system, consisting of two protostellar disks in orbit around one another. There is a total of 50 protostellar disks in the layer as a whole, and most are in binary systems or higher multiples. Many are massive (5-20 $M_\odot$), and typical separations are of order 10^4 A.U., so the overall configuration of protostars is reminiscent of a young OB subgroup.

8. Initial Conditions for Binary Formation

Before describing the mechanisms by which binary stars form in our simulations, we should first emphasize the essential geometrical features of the gas flow in which they form. First, there is a thin layer of dense cold gas which – unless $b = 0$ – is tumbling, i.e. rotating about an axis in its own plane. Second, the layer fragments initially into filaments, and because of centrifugal acceleration the filaments tend to be aligned perpendicular to the rotation axis, i.e. there are tumbling filaments. This has two fundamental consequences. (i) There is close alignment between the orbital and spin angular momenta of nearby protostellar disks that condense out of the same filament; and therefore their tidal interaction is very strong. (ii) The material accreting onto a protostellar disk tends to flow in along the filament out of which it has just condensed, and – since the filament is tumbling – brings with it ever increasing specific angular momentum, thereby causing the protostar to spin up.

In the head-on collision illustrated in Figure 1, the angular momentum of the individual filaments and the protostellar disks condensing out of them is generated by gravitational and/or hydrodynamic torques: the net angular momentum remains very close to zero.

9. Binary Formation by Capture

In this mechanism, two protostellar disks form more or less independently, but in the same filament and therefore quite close to one another. Because the gas has now lost much of its pressure support, and because the strength of gravity has been somewhat increased in the plane of the layer by compression, the two protostellar disks fall towards one another on highly eccentric orbits. Around periastron, the two disks interact viscously and tidally (throwing out spiral arms which carry away angular momentum and binding energy), thereby greatly hardening their orbit and increasing the chances of their remaining as a binary system.

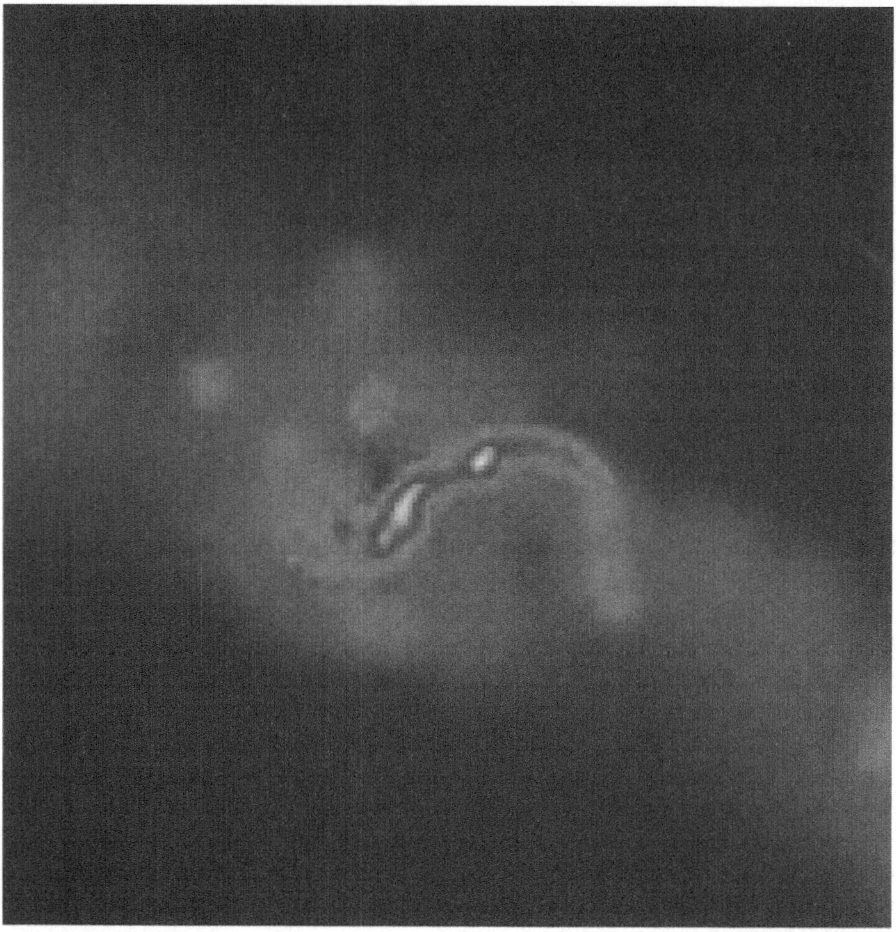

Fig. 2. Interaction of two protostellar disks, leading to capture. The two protostellar disks have formed as a result of two clumps, having mass $M = 225\ \mathrm{M}_{\odot}$ and radius $R = 1.94$ pc, colliding at relative speed $v = 1.7\ \mathrm{km\ s}^{-1}$ and impact parameter $b = 1.55$ pc. The disks have then fallen together and are pictured here close to periastron, at a time 2.42 Myr after the initial clumps first came into contact. The linear size of the region shown is $L = 0.16$ pc; the whole computational domain is 50 times bigger.

Figure 2 shows this process happening. In this simulation, the two protostellar disks have been formed as a result of two clumps, having mass, $M = 225\ \mathrm{M}_{\odot}$, and radius, $R = 1.94$ pc, colliding at relative speed $v = 1.7\ \mathrm{km\ s}^{-1}$ and impact parameter, $b = 1.55$ pc. The strength of the tidal interaction is due to the fact that the orbital and spin angular momenta are closely aligned, because all three angular momenta derive from the global angular momentum of a tumbling filament, which in turn derives from the global angular momentum of the initial clump/clump collision. The union of the two disks is therefore an arranged mar-

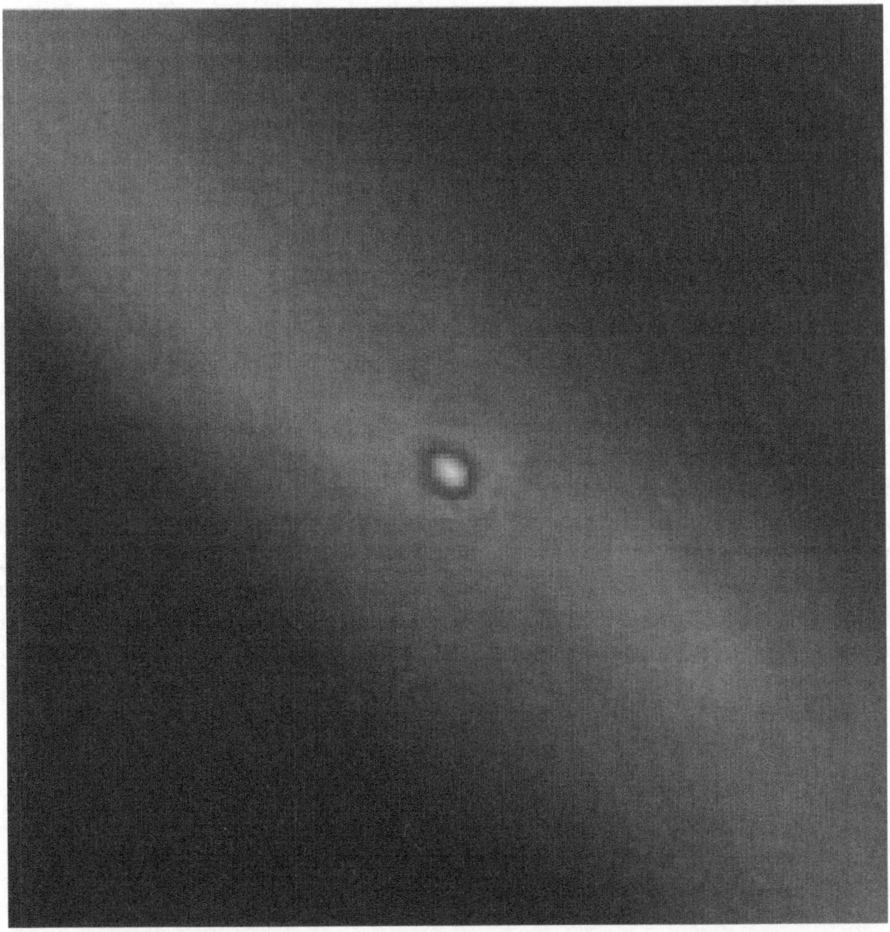

Fig. 3. The accretion flow from a tumbling filament onto a single protostellar disk. Note that there is a significant offset between the opposing accretion streams; this offset grows steadily with time. The protostellar disk has formed as a result of two clumps, having mass, $M = 75\ M_{\odot}$, and radius, $R = 0.446$ pc, colliding at relative speed, $v = 1.7$ km s^{-1}, and impact parameter, $b = 0.268$ pc. The disk is pictured here at a time, 0.477 Myr, after the initial clumps first came into contact. The linear size of the region shown is $L = 0.10$ pc; the whole computational domain is 20 times bigger.

riage. We call this mechanism Shock-induced Gravitational Fragmentation plus Capture (SGF+C).

10. Merger

If too much energy and angular momentum are removed from a pair of interacting protostellar disks, the result is a merger. However, even when this happens, the resulting single protostellar disk may eventually spawn a companion by rotational fragmentation, as described below.

11. Accretion-Induced Spin-Up

In our simulations, the majority of binary systems form from a single protostellar disk, which is spun up by its accretion flow until it becomes unstable against rotational fragmentation.

Figure 3 shows this process happening. The protostellar disk shown here has been formed as the result of two clumps, having mass, $M = 75\ \mathrm{M}_\odot$ and radius, $R = 0.446$ pc, colliding at relative speed, $v = 1.7$ km s^{-1}, and impact parameter, $b = 0.268$ pc. Because the protostellar disk has condensed out of a filament, its accretion flow is mainly along the filament. And because the filament is tumbling, the specific angular momentum of the material accreting onto the protostellar disk increases with time; the material arriving later has come from further up the filament. The upshot is that the protostellar disk is spun up, and may eventually become unstable against rotational fragmentation.

There are two distinct mechanisms of rotational fragmentation. Which one occurs depends on the rate of spin-up.

12. Rapid Spin-Up: Binary Formation by Bar-Mode Rotational Fragmentation

In this context, rapid spin-up means that $t_{spin-up} \lesssim t_{spin}$, where $t_{spin-up} \equiv (d\ln[\beta]/dt)^{-1}$, β is the ratio of rotational to gravitational energy, and t_{spin} is the spin period. Under these circumstances, the protostellar disk undergoes a bar-mode instability, *i.e.* it extends to form a bar before cleaving into two pieces of comparable mass. We call this mechanism Accretion-induced Bar-mode Rotational Fragmentation (ABRF).

Figure 4 shows a protostellar disk in the process of cleaving in two. This protostellar disk has been formed as the result of two clumps, having mass, $M = 750\ \mathrm{M}_\odot$, and radius, $R = 6.45$ pc, colliding at relative speed, $v = 1.7$ km s^{-1}, and impact parameter, $b = 7.73$ pc. The material in the region shown is almost entirely from one clump.

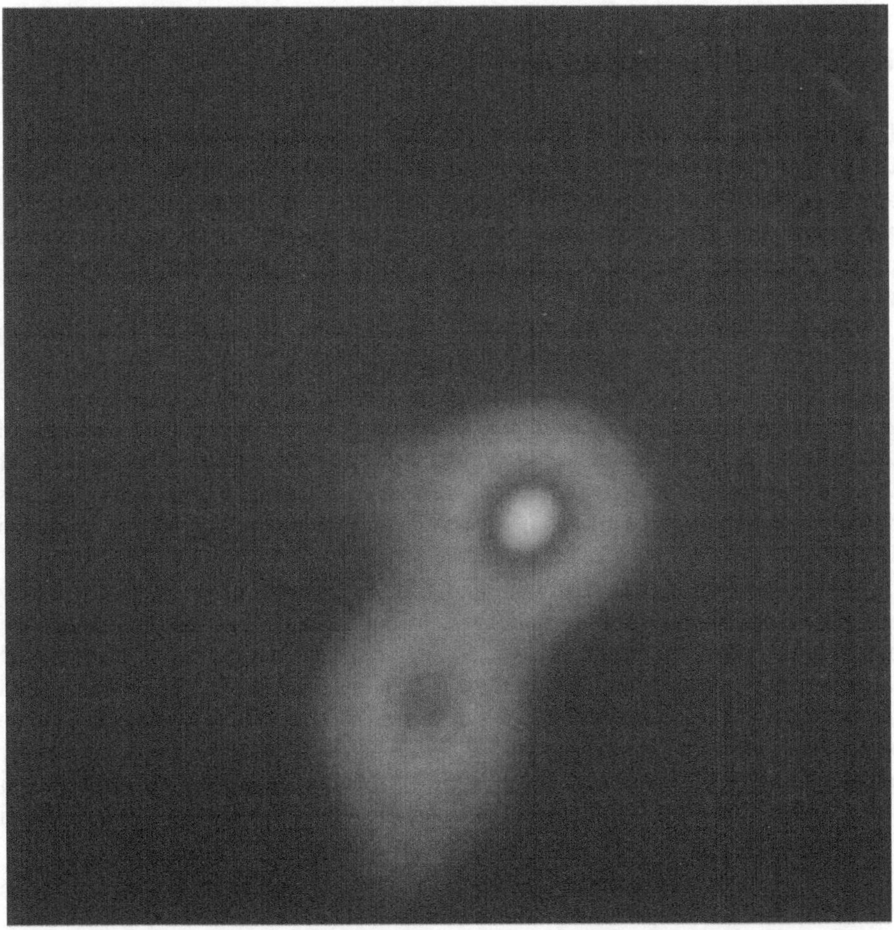

Fig. 4. A single protostellar disk in the process of cleaving into a binary system, following rapid spin-up induced by accretion. The initial single protostellar disk has formed as the result of two clumps, having mass, $M = 750\ M_{\odot}$, and radius, $R = 6.45$ pc, colliding at relative speed, $v = 1.7$ km s^{-1}, and impact parameter, $b = 7.73$ pc. The disk is pictured here at a time, 9.07 Myr, after the initial clumps first came into contact. The linear size of the region shown is $L = 0.05$ pc; the whole computational domain is 480 times bigger.

When rapid spin-up occurs, it is frequently the result of an accretion flow which is composed of a stream of self-gravitating lumps. Spin up is then rather impulsive, and the process of spin-up is closely related to merger.

13. Slow Spin-Up: Binary Formation by Spiral-Mode Rotational Fragmentation

In this context, slow spin-up means $t_{spin-up} \gtrsim 2t_{spin}$. Under these circumstances, the primary protostellar disk becomes unstable to the formation of spiral arms, and eventually one (or more) of these spiral arms may become so massive that self-gravitating lumps condense out of it and then detach to form secondary protostellar disks. We call this mechanism Accretion-induced Spiral-mode Rotational Fragmentation (ASRF).

Figure 5 shows three secondaries in the process of forming. The primary protostellar disk shown here has been formed as the result of two clumps, having mass, $M = 75\mathrm{M}_{\odot}$, and radius $R = 0.446$ pc, colliding at relative speed, $v = 1.7$ km s^{-1}, and impact parameter, $b = 0.268$ pc. At the top lefthand extreme of the primary protostellar disk, a secondary (C) is starting to condense out. At the bottom righthand extreme of the primary protostar, a secondary (B) has already condensed out and separated, and another secondary (B′) is about to separate.

Secondaries formed in this way tend to grow relative to the primary, firstly because, if more than one secondary is formed, they often merge. (B and B′ merge soon after the frame shown in Figure 5). Secondly, as long as (i) the accretion flow persists and (ii) the secondary remains much less massive than the primary, the secondary orbit is much larger than the primary orbit, and so the secondary repeatedly intercepts the accretion streams (which are symmetrically offset relative to the primary – see Figure 3); this means that the material in the accretion flow has less angular momentum relative to the secondary, and is therefore more readily accommodated by the secondary.

14. Long-Term Evolution

The different binary-formation mechanisms described here do not always occur distinctly in our simulations; sometimes they occur together, either simultaneously or consecutively. Because the processes involved are highly non-linear, and the initial conditions are dynamic, the precise sequence of events occurring in any given simulation depends very sensitively on the initial conditions. In other words, it is chaotic, in a mathematical sense.

Moreover, even with limitless computational resources, our simulations cannot determine the final state of a binary or multiple system, because our code has limited spatial resolution. Consequently, it is unable adequately to model the internal evolution of a protostellar disk, i.e. the process which converts a

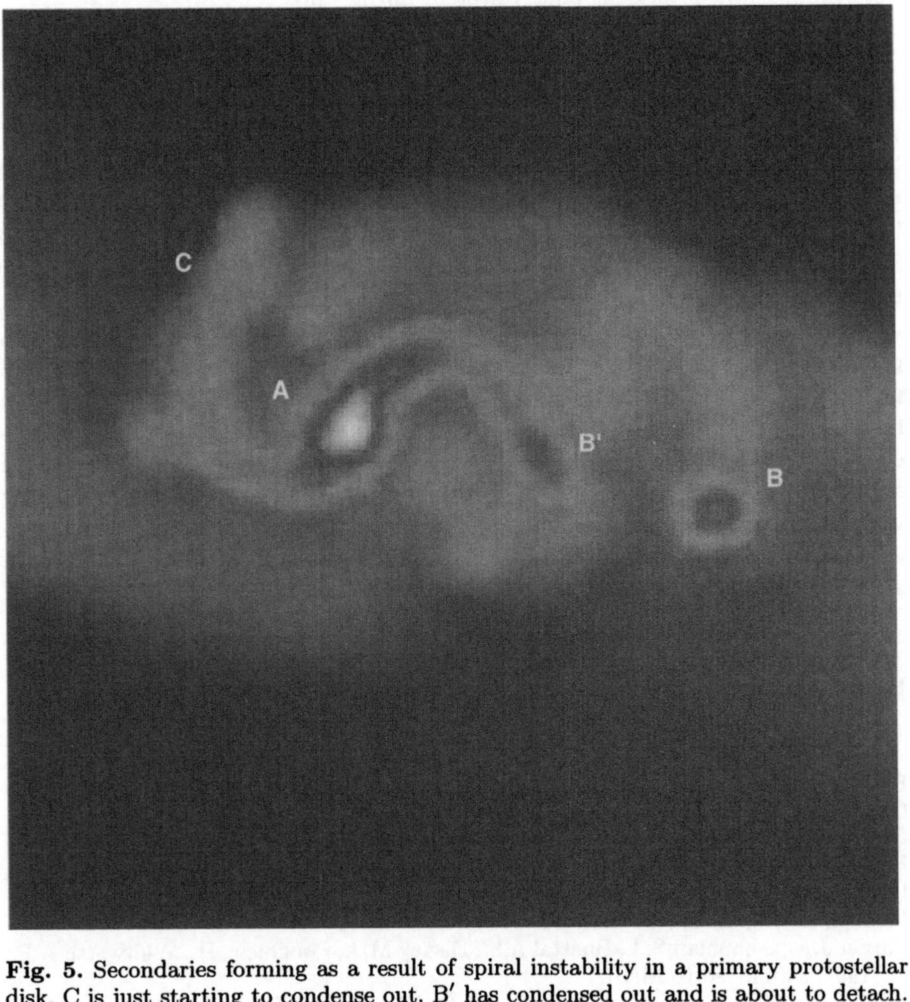

Fig. 5. Secondaries forming as a result of spiral instability in a primary protostellar disk. C is just starting to condense out. B′ has condensed out and is about to detach. B has already detached, and will soon merge with B′. The primary has formed as the result of two clumps, having mass, M = 75 $M_{\odot}$, and radius, R = 0.446 pc, colliding at relative speed, v = 1.7 km s^{-1}, and impact parameter, b = 0.268 pc. The disk is pictured here at a time, 0.579 Myr, after the initial clumps first came into contact. The linear size of the region shown is L = 0.10 pc; the whole computational domain is 20 times bigger.

plump protostellar disk (of the type produced in our simulations) into a star containing most of the mass plus an emaciated circumstellar disk. Since this emaciation of the disk is ultimately the process which terminates instabilities in self-gravitating disks and interactions between disks, it should play a key rôle in determining how a nascent binary system finishes up.

15. Conclusions

We have proved analytically (Turner et al. 1994) that there will be time for the processes occurring in our simulations to occur also in nature, and therefore we believe we have demonstrated two viable mechanisms for binary formation. In our simulations, rotational fragmentation induced by accretion seems to occur more commonly than capture, but this may not be universally true.

However, it does seem likely that under a wide variety of circumstances, protostellar disks will be fed by accretion flows in which the specific angular momentum increases with time; the material arriving later will normally come from further away and – for any credible rotation law – will tend to have higher specific angular momentum. Therefore, accretion-induced spin-up will be a common experience for protostellar disks, and this may go some way towards explaining why binary stars are so common.

More detailed descriptions of the work presented here are given in Turner et al. (1995) and Whitworth et al. (1995).

References

Barnes J., Hut P., 1986, Nature, **324**, 446.
Chandrasekhar S., 1939, An Introduction To The Study Of Stellar Structure, Dover.
Gingold R.A., Monaghan, J.J., 1977, MNRAS, **181**, 375.
Hernquist L., 1987, ApJSS, **64**, 715.
Lucy L.B., 1977, AJ, **82**, 1013.
Monaghan J.J., 1992, ARA&A, **30**, 543.
Turner J.A., Chapman S.J., Bhattal A.S., Disney M.J., Pongracic, H., Whitworth A.P., 1995, submitted to MNRAS.
Whitworth A.P., Chapman S.J., Bhattal A.S., Disney M.J., Pongracic H., Turner J.A., 1995, submitted to MNRAS.

The Quest for Evolutionary Diagrams of Young Stellar Objects

Francesco Palla

Osservatorio Astrofisico di Arcetri, Largo E. Fermi 5, 50125 Firenze, Italy
and
Groupe de Radioastronomie, Ecole Normale Superieure, 24 rue Lhomond, 75005 Paris, France

Summary. Young stellar objects (YSOs) are reviewed with an emphasis on their global evolutionary properties. The main classification schemes of YSOs based on the observed continuum spectral energy distributions and the recent suggestion of using a bolometric temperature to characterize the evolutionary status of all YSOs will be first discussed. The impact of protostar theory on pre-main-sequence evolution will be then addressed. The revised evolution differs in significant aspects from the traditional one, in particular for stars more massive than the sun. A new H-R diagram is presented in which the observed distribution of young stars of low and intermediate mass is naturally explained. The major sources of uncertainties for the location of evolutionary tracks, and hence for mass and age estimates of YSOs, are also discussed together with the effects of mass accretion and the presence of companions.

1. Introduction

The main theme of this Conference is the description of the phenomena that take place *around* young stars. Less attention is being paid to what happens *in* the star itself. However, it is clear that only a combination of these two aspects can provide a full and coherent picture of the early evolution of young stellar objects. In this contribution, the emphasis will be placed on the second aspect, the structural properties of YSOs from an evolutionary viewpoint. The importance of evolutionary diagrams in astrophysics cannot be overemphasized. For decades, the Hertzsprung-Russell (H-R) diagram has been a unique tool for the understanding of stars on and beyond the main sequence. It is less clear how such a diagram can be fully exploited when dealing with younger stars whose basic physical parameters cannot be so easily specified and determined by the observations as in the case of fully mature stars. However, it would be extremely fruitful if the field of star formation and early stellar evolution could provide a synthetic, yet accurate, description of young stars even before they settle down on the main sequence. Thanks to the wealth of data now available on large samples of YSOs, such a synthesis can now be attempted.

In the first part of this review, I will describe some of the recent efforts to classify in physically meaningful schemes YSOs in a wide variety of evolutionary stages, from the highly embedded sources to the optically visible T Tauri and Herbig Ae/Be stars close to the main sequence. A critical discussion will then follow of the modifications to the classical view of pre-main-sequence (PMS) evolution introduced by the use of more realistic protostellar initial conditions.

The revised H-R diagram and the various sources of uncertainty in the location of the evolutionary tracks will be analysed in the last section.

2. General Properties of YSOs

Studies of the *shape* of the spectral energy distribution (SED) of YSOs have proved very useful to determine their nature and evolutionary state. Since the emitted spectrum of a YSO depends on the distribution and physical properties (density, temperature and composition) of the surrounding dust and gas, it is natural to expect a dependence of its shape on the evolutionary state of the exciting source. A protostellar object deeply embedded in the parent cloud should have an infrared signature markedly different from that of a mature PMS star, for which most of the circumstellar material has been accumulated onto the central object. On these premises, various ways of characterizing YSOs have been devised in recent years. It is the purpose of this section to review them in some detail.

2.1 Extended Classification Schemes: from IR to mm

The most successful classification scheme has been developed by Lada and collaborators based on the slope of the SED longward of 2.2 μm (Lada 1987; Wilking et al. 1989). For each SED, a spectral index $\alpha \equiv -\mathrm{dlog}(\lambda\ F_\lambda)/\mathrm{dlog}\ \lambda$, with F_λ the flux density at wavelength λ, is computed between $\lambda = 2.2$ and 10–25 μm, and the resulting morphological classes can be broadly classified in three distinct groups. Each class in turn corresponds to a well defined evolutionary stage, according to current models of protostellar and early stellar evolution (e.g. Shu et al. 1993), as shown in the three lower panels of Figure 1. Class I sources have $\alpha > 0$, indicating a rise in the SED all the way up to $\lambda \sim$100 μm. The IR excess is very conspicuous and the SED is much broader than that of a single temperature blackbody function. From the evolutionary viewpoint, Class I sources are thought to represent accreting protostars, surrounded by luminous disks, with radii of $\sim 100 - 1000$AU, and by infalling, extended envelopes with sizes of $\sim 10^4$AU.

Class II sources have $-2 < \alpha < 0$. Their SEDs fall towards longer wavelenghts but are still broad due to a significant amount of circumstellar dust. A class II source results from the clearing of the circumstellar envelope, due to the action of a powerful stellar wind, and corresponds to a classical T Tauri or Herbig Ae/Be star surrounded by a spatially thin, optically thick circumstellar disk of radius $\sim$ 100AU. Finally, class III sources have $\alpha < -2$, and the SED resembles that of a normal, reddened stellar photosphere. The absence of an IR excess indicates the disappearance of the circumstellar structures, disks and envelopes, and the approach to the conditions of a normal main sequence star.

In recent years, most of the observational and theoretical effort has been directed at testing the goodness of this scheme, and in particular class I sources

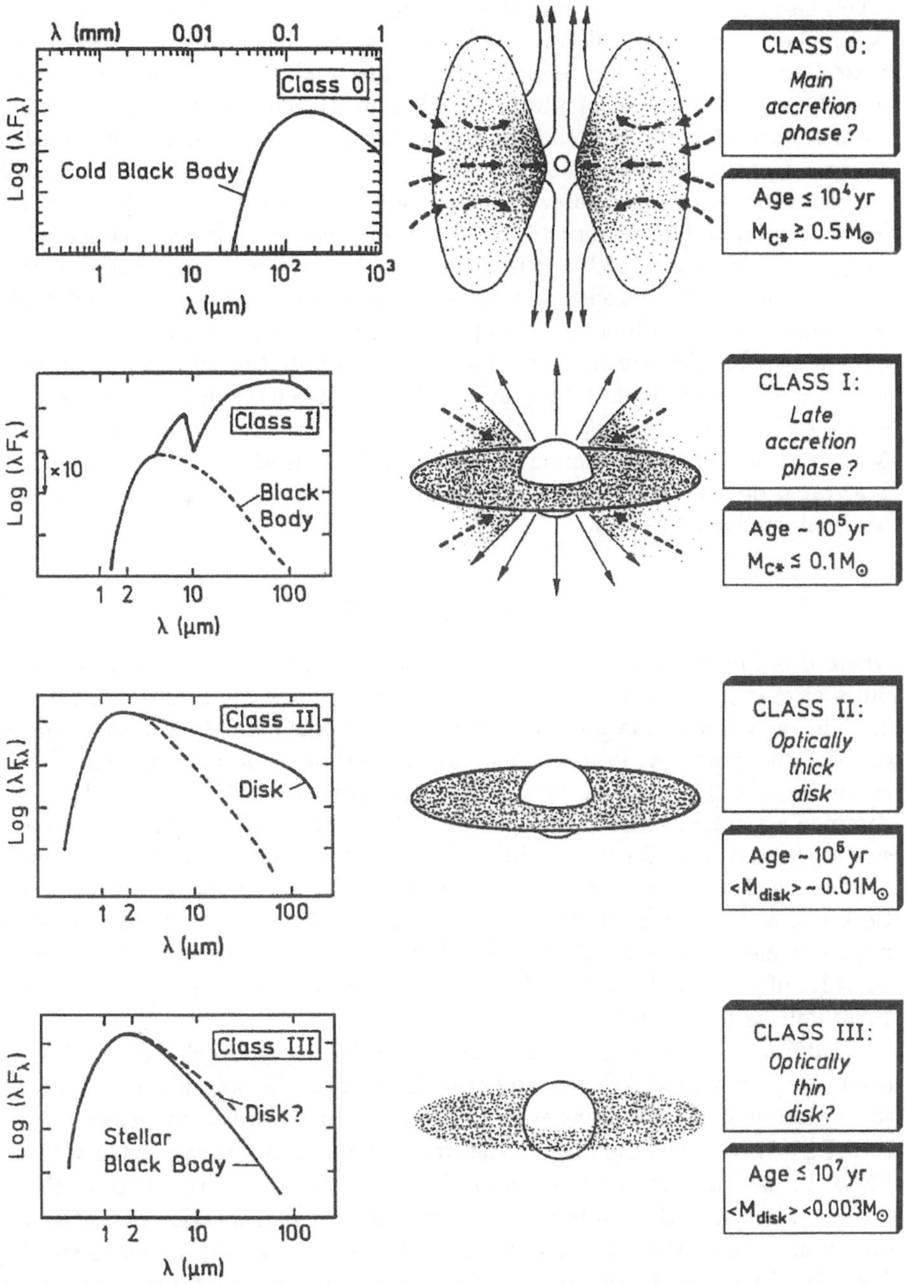

Fig. 1. The extended evolutionary classification of YSOs according to the shape of the spectral energy distribution (left) and to the mass of circumstellar matter, indicated by M_{c*} (right) (from André 1994).

have been studied in the hope of finding among them the elusive protostars. The reason for the lack of success in the identification of a *bona fide* protostar is now becoming more clear. It has been suggested by André et al. (1993) that class I sources are in fact more evolved objects than true protostars, having already assembled most of the final stellar mass. Support to this view comes from studies at millimeter and sub-millimeter wavelengths of deeply embedded sources. Continuum observations at these wavelengths offer several advantages: the emission is in most cases optically thin (a column density of order 10^{26} cm^{-2}, corresponding to more than 10^4 magnitudes of extinction, is needed for $\tau_{1mm} \sim 1$; Hildebrand 1983) and is sensitive to dust at lower temperatures located farther away from the exciting source(s). Also, the angular resolution achievable with current (sub)millimeter telescopes is much higher than in the far-IR. Thus, one can probe the density and temperature distribution of the circumstellar material and estimate its total (gas and dust) mass either in the envelope (with single dish) or in the disks (with interferometers) (e.g. Beckwith et al. 1990; Mezger et al. 1990). Assuming an optically thin, isothermal dust source, the integrated millimeter flux density, S_ν, can in fact be converted to the total mass of circumstellar matter by

$$M_{\rm env} = \frac{S_\nu d^2}{\kappa_\nu B_\nu(T_d)}, \qquad (1)$$

where d is the distance to the source, κ_ν is the dust absorption coefficient (in units of cm^2 g^{-1}), and $B_\nu(T_{\rm d})$ is the Planck function for a dust temperature $T_{\rm d}$. The mass determinations depend critically on the knowledge of the behavior of the dust opacity at long wavelengths, a poorly known quantity. Thus, these estimates are still affected by large uncertainties (a factor of 2 to 10). It is commonly assumed that $\kappa_\nu \propto \nu^\beta$, with $\beta \sim$1.5-2 for dust located in the extended envelopes, and $\beta \sim 1$ for the conditions appropriate to circumstellar disks, where a substantial increase of dust size, and hence of the opacity, is expected (e.g. Beckwith & Sargent 1991; Mannings & Emerson 1994). The recent laboratory measurements of Agladze et al. (1994) indicate values of κ_ν at $\nu \sim 1$ mm about an order of magnitude greater than those normally assumed (e.g. Draine & Lee 1984; Pollack et al. 1994).

André & Montmerle (1994) have measured the continuum flux at $\lambda = 1.3$ mm from a large sample of YSOs in and near the ρ Ophiuchi molecular cloud, i.e. the same sources used by Wilking et al. (1989) for their IR classification system, and used Eq. (1) to derive the mass of the circumstellar material associated with the YSOs. In their approach, the *circumstellar mass* rather than the shape of the IR spectrum becomes the fundamental variable for the classification of YSOs. The survey has shown that the amount of circumstellar material decreases steadily from class I to class III sources (by a factor of about 100) and that class I sources have resolved, extended structures, while class II sources appear unresolved on scales of ~2000 AU (the angular resolution of the IRAM 30 m telescope at an assumed distance of d = 160 pc). This is consistent with the predictions of the evolutionary sequence based only on the infrared emission properties.

The novel aspect is that the typical circumstellar masses of class I sources (M_{env} ~0.06 $M_\odot$) remain much smaller than the inferred stellar masses, $M_* \geq 0.1$ $M_\odot$. The latter can be evaluated as follows. If class I sources were true protostars, then the observed bolometric luminosity, L_{bol}, would be entirely due to the conversion of the kinetic energy of collapse at the protostellar accretion shock. In such a case, the stellar mass M_* can be obtained from the relation $M_* = L_{bol}(R_*/G\dot{M}_{acc})$, provided the protostellar radius, R_*, and mass accretion rate, $\dot{M}_{acc}$, are known. Typical values of $\dot{M}_{acc}$ are in the range 2×10^{-6}–10^{-5} $M_\odot$ yr^{-1}, while $R_* \sim 5$ $R_\odot$ for a 1 $M_\odot$ protostar. During the main accretion phase, most of the matter is still in the envelope and therefore one would expect for class I objects envelope masses *bigger* than that of the central star. On the contrary, the observations have indicated that in the majority of the sources $M_{env} < M_*$. André & Montmerle conclude that class I objects are in a more advanced evolutionary stage, and that genuine protostellar candidates are to be found in a different category of YSOs which they identify in the so-called *class 0* objects. This conclusion represents the major revision to the IR classification scheme of Lada and collaborators, as indicated in the upper panel of figure 1. On the other hand, the revised age ordering does not affect the later stages.

Class 0 objects are readily distinguishable from class I sources, since they are undetected at wavelengths shortward of 25 μm, and since their SEDs resemble those of a single temperature blackbody at T~15-30 K, (see André 1994 and Bontemps et al. 1995 for a general discussion). Note also the very short dynamical timescale ($t < 10^4$ yr) estimated for this phase. Thus, protostars in the main accretion phase should be very rare, in agreement with the paucity of candidates found in several star forming regions (see André 1994). The idea that class I sources, on the contrary, represent a later stage of the evolution is consistent with their relatively high frequency of occurrence.

Very recently, Saraceno et al. (1995) have used the millimeter flux observed towards a sample of class 0 and class I sources with and without outflows as an additional means to discriminate the evolutionary status. The results of their survey is shown in the bolometric luminosity vs. millimeter flux diagram of Figure 2. In this diagram, the youngest objects (class 0) have large mm-fluxes but low L_{bol}, and compact molecular outflows. They show no correlation between F_{mm} and L_{bol}. However, for class I sources with outflows the correlation is clearly present and extends over several decades along the outflow strip (solid line). This relationship does not hold for class I sources without outflows, which have less millimeter flux at the same L_{bol}. An evolutionary sequence can thus be traced in this way, but the physical origin of the correlation is still not clear.

2.2 A Generalized H-R Diagram for YSOs: L_{bol} vs. T_{bol}

There have been several suggestions as to how to characterize YSOs in addition to the shape of the SEDs. These include the amount of visual extinction, A_V, derived from models of the emitted spectrum (Adams 1990); the mean frequency ν of the spectrum (Ladd et al. 1991); and the ratio of the (sub)millimetric to the bolometric luminosity (André et al. 1993). We have seen already how the

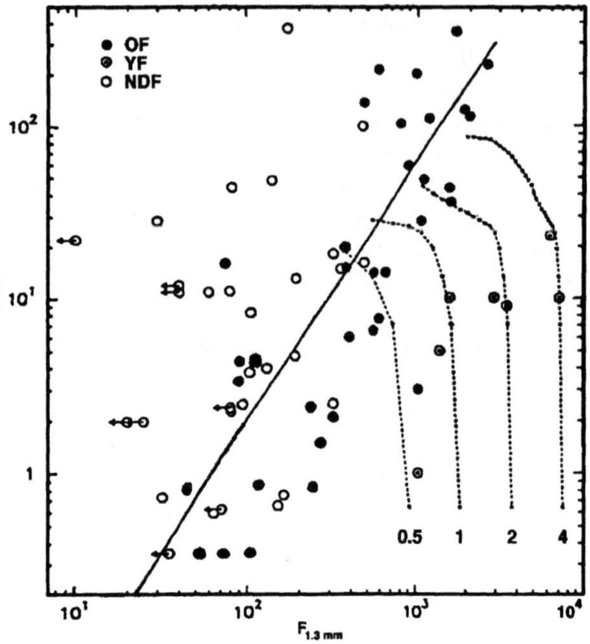

Fig. 2. Bolometric luminosity vs. 1.3 mm flux density for class 0 (labelled YF) and class I sources with (filled circles) and without (empty circles) outflows. The evolution of a YSO, as measured by the strength of the millimeter flux, is from right to left. The solid line shows the correlation line for the outflow sources. The dashed lines indicate the path followed by protostellar cores of different mass accreting at a rate of 10^{-5} $M_\odot$ yr $^{-1}$ (from Saraceno et al. 1995).

latter quantity allows to recognize the youngest class 0 sources. As for the other diagnostics, all of them suffer from some major disadvantages. For example, the use of A_V as an age indicator although correct in principle (since the extinction towards the central object is expected to decrease with time as more material is depleted from the parent cloud), is, however, of limited application. The determination of A_V in real sources is, in fact, very uncertain due to the unknown geometric distribution of the dust with respect to the line of sight and is thus highly model dependent.

A more global approach has been taken by Myers & Ladd (1993) who have introduced the concept of a *bolometric temperature*, T_{bol}, as a means to characterize YSOs. Accordingly, T_{bol} is defined as the temperature of a blackbody having the same mean frequency ν as the observed continuum spectrum; ν represents the ratio of the first and zeroth frequency moments of the SED and measures the departure of the spectrum from a blackbody, or its redness. The advantage of this definition is that it uses all of the available data of the spectrum of a given source instead of a limited wavelength region, as in the Lada or André & Montmerle schemes. Of course, in order to be of any utility, the range of observed frequencies should be large enough to include most of the emitted source luminosity, L_{bol}. Since such data are now available for a very large number of YSOs, Myers & Ladd (1993) have proposed to use the L_{bol} vs. T_{bol} diagram as the analog of the classical H-R diagram to describe the evolutionary properties of YSOs (see also Chen et al. 1995 for an extension of the data base). In such a diagram, T_{bol} plays the same role as T_{eff} for sources with no or little infrared excess, but it becomes vastly different for more embedded sources. For example, class I sources have $T_{bol} \sim 300$ K; class II have $T_{bol} \sim 1000 - 2000$ K, and only the weak T Tauri stars, which show very little IR and UV excesses, have $T_{bol} \approx T_{eff} \sim 3600$ K. In this scheme, class 0 objects are the coldest ones with $T_{bol} \leq 100$ K. The main point, however, is that T_{bol} shows a smooth increase with the inferred age of the YSOs and can then be taken as a useful age indicator.

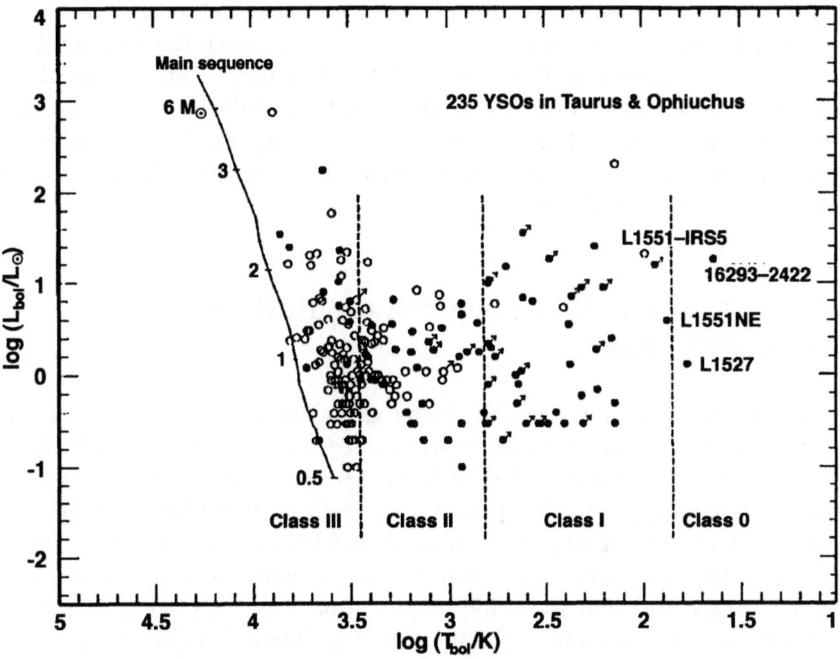

Fig. 3. The L_{bol} vs. T_{bol} diagram of YSOs in the Taurus-Auriga region. Sources with (empty circles) and without (filled circles) known spectral types are indicated. The zero-age-main-sequence for the indicated mass interval is shown by the solid line (from Chen et al. 1995).

An example of the distribution of the YSOs in the Taurus-Auriga region is given in Figure 3. Note that use of T_{bol} allows the placement in the diagram even of sources without a known spectral type, as in the case of the highly embedded class 0 and I sources! Of course, the generality of such a diagram is of the highest interest, but some words of caution about its validity are in order at this point. First of all, although the *empirical* definition of T_{bol} is perfectly correct, the *physical* basis on which it rests appears much less clear. According to Myers & Ladd (1993), T_{bol} represents a weighted average of the effective stellar and bolometric disk temperatures, with the weight being provided by the fraction of the stellar luminosity intercepted and reradiated by the disk. To derive all but T_{eff} requires some modeling. Although a certain consensus has been reached towards the models *à la* Shakura-Sunyaev for circumstellar disks, the understanding of the actual physical processes governing their structure and evolution (the mechanisms of angular momentum transport and mass accretion) is far from complete (see Adams, this volume). Also, the use of these simplified disk models with a large number of free parameters to reproduce the observed SEDs of YSOs is plausible but not unique, as demonstrated by several authors in the case of both class I and class II sources (e.g. Hartmann et al. 1993; Kenyon et al. 1993; Mitskevich 1994). Moreover, in a real young stellar system, the anisotropic distribution of the circumstellar matter causes the SED, and hence the bolometric temperature, to depend strongly on the viewing angle (Yorke et al. 1995). Finally, the scatter in the data points on which the relation T_{bol}-*age* is derived (cf. Figure 3 of Chen et al. 1995) is so large that it is questionable whether T_{bol} can indeed be taken as a valid age indicator. All these factors should be considered with great care before adopting the L_{bol}-T_{bol} diagram as a new standard tool, but it is nevertheless very encouraging that such a general classification can now be attempted.

3. The Standard H-R Diagram for PMS Stars: New Developments

The preceding discussion naturally leads us to the next subject of this review, the evolutionary properties of the more evolved, but optically visible PMS stars. The location of these stars in the H-R diagram combined with the use of theoretical evolutionary tracks has played a central role in stellar evolution studies since the early work of Hayashi, Cameron and Iben in the early '60s. However, the realization that many stars show excess emission at optical and ultraviolet wavelengths and that radiation longward of 1 μm represents an important contribution to the bolometric luminosity (e.g. Rucinski 1985; Rydgren & Zak 1987; Basri & Bertout 1989) has casted some doubts about the possibility of determining the true positions of PMS stars in the H-R diagram. It is now recognized that T Tauri and Herbig Ae/Be stars are surrounded by, and interact with, circumstellar disks of appreciable mass. The presence of disks can affect the interpretation of the observations since the system luminosity is much more complex, resulting from the contribution of the direct stellar radiation *plus* that due to the reprocessing of absorbed starlight and to accretion of matter through the disk, *plus* the radiation at the boundary layer (Cohen et al. 1989; Strom et al. 1989a; Hartmann & Kenyon 1990). Also, the accretion of matter can modify the mass of the star, and hence its position in the H-R diagram (Kenyon & Hartmann 1990). Whether these effects are substantial depends on the mass accretion rate, which must be inferred from the observations. We will return below to a quantitative assessement of these effects. For the moment, let us discuss first another point which, in my view, represents an even more important development of the classical theory of PMS evolution: the inclusion of realistic initial conditions as derived from protostellar theory and the impact on the classical H-R diagram.

3.1 The Influence of Protostellar Evolution

Hayashi and his contemporaries lacked a theory of the protostar phase, and, therefore, could not specify with any degree of confidence the initial conditions for PMS contraction. A simple argument indicated that if the interstellar cloud radiated away little energy during its collapse, the starting values of the stellar radius would be some two orders of magnitude greater than the corresponding main-sequence values (Hayashi 1966). The contraction time, given by the Kelvin-Helmholtz time, $\tau_{\rm KH} = \frac{GM_*^2}{R_* L_{\rm surf}}$, is so short for these large radii and high surface luminosities that the poorly understood early collapse would be much too brief to affect the subsequent evolution. Adopting the large initial radii, Iben (1965) and Ezer & Cameron (1967) followed numerically the contration of stars with masses from less than 1 $M_\odot$ to over 100 $M_\odot$. The contraction of all stars was thought to be qualitatively the same, with quantitative differences scaling in a simple manner with the stellar mass M_* (notice that this approach still persists nowadays in most PMS calculations!).

The development of protostar theory has altered the conception of PMS evolution, in particular for stars more massive than solar. Many such stars undergo a transient period of thermal relaxation prior to settling down to the type of evolution envisioned earlier. Furthermore, the union of protostar and PMS theory implies that stars of 100 $M_\odot$ or even 20 $M_\odot$ should have no PMS phase at all. To follow these changes in detail one must know how stars of various masses are built up from collapsing cloud. An exhaustive account of this phase can be found in the review by Palla (1994). Here, it suffices to point out that the main result of the calculations is the establishment of a *mass-radius relation* for the accreting protostar, due to the nuclear burning of deuterium in the stellar interior (Stahler 1988; Palla & Stahler 1990). The large radii predicted by Hayashi are never achieved; instead the protostellar radius remains limited to a few solar radii by the efficient radiation losses at the accretion shock (e.g. Stahler et al. 1980). As a consequence, the internal structure departs significantly from the standard assumption of thermal convection: in low mass stars, convection is due to nuclear burning in the center; in stars more massive than ~2 $M_\odot$ convection is confined to subsurface regions, while the deep interior is depleted of deuterium and radiatively stable.

The calculation of the buildup of a protostar to a mass, M_*, automatically yields the initial PMS structures for all stars of lower mass, provided only that the main accretion process ends relatively quickly (compared to the Kelvin-Helmoltz time scale). Since the contraction time is of order 10^7 yr for a solar-type star, while the typical clearing time due to a wind/outflow is 10^5 yr (cf. Figure 1), this assumption is well satisfied. The salient features of the initial phases of PMS evolution can be discerned by comparing several characteristic luminosities as functions of stellar mass (see Figure 4). The figure depicts the surface luminosity, $L_{\rm surf}$, at the start of the PMS phase, obtained from

$$L_{\rm surf} = 4\pi R_*^2 \sigma T_{\rm surf}^4. \qquad (2)$$

Here, R_* is the protostellar radius, and T_{surf} is the surface temperature at the end of accretion, which is constrained in the narrow interval 4000-5000 K. Also, Figure 4 depicts L_{rad}, the luminosity transported by radiation within the star, which can be shown to be

$$L_{rad} = L_0 \left(\frac{M_*}{M_\odot}\right)^{11/2} \left(\frac{R_*}{R_\odot}\right)^{-1/2}, \tag{3}$$

where L_0 is a luminosity of order 1 $L_\odot$. Finally, Figure 4 gives the protostellar accretion luminosity

$$L_{acc} = \frac{G M_* \dot{M}_{acc}}{R_*}. \tag{4}$$

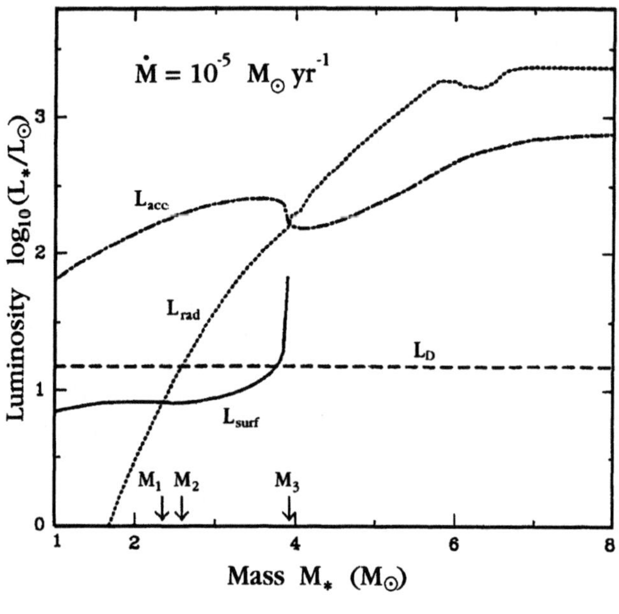

Fig. 4. The mass dependence of luminosities in protostars accreting at 10^{-5} $M_\odot$ yr^{-1}. The arrows on the horizontal axis indicate the critical masses where L_{rad} intersects L_{surf}, L_D, and L_{acc}, respectively. At M_3, L_{surf} climbs steeply to join L_{rad} (from Palla & Stahler 1993).

Let us first apply Figure 4 to the low mass T Tauri stars ($M_* < 2$ $M_\odot$). According to the figure, such stars begin contracting with L_{surf} greater than L_{rad}. In other words, they radiate more energy per unit time from their surfaces than can be supplied by photons from the interior. The excess luminosity must be carried by thermal convection. Thus, T Tauri stars begin their evolution as fully convective objects with nearly fixed surface temperatures, in the same manner described by Hayashi.

More massive stars, the Herbig Ae/Be stars ($M_* > 2$ $M_\odot$), are not initially convective from surface cooling; a large fraction start the PMS contraction with L_{surf} less than L_{rad}. In this case, more energy is transported per unit time through the deep interiors than can be radiated into space, and the star is in a thermally unrelaxed state. In the H-R diagram, this star appears first with a surface luminosity and temperature that place it *below* the classical evolutionary tracks for that mass. During subsequent relaxation, the stellar radius, stellar luminosity, and surface temperature all increase, while the outer convections disappears. Then, the star joins onto the radiative portion of its classical track, which it then follows to the main sequence.

Not all the intermediate mass stars undergo thermal relaxation. If a star is sufficiently massive, it will have already relaxed prior to the end of protostellar accretion, i.e. before the star is optically visible. This condition is met when $L_{\rm rad}$ from Eq. (3) becomes larger than $L_{\rm acc}$ from Eq. (4). In Figure 4, the critical mass is about 4 $M_\odot$. More massive Herbig Ae/Be stars appear immediately on their classical radiative tracks.

3.2 A New H-R Diagram

The results of the calculations of PMS evolution incorporating the protostellar initial conditions are shown in the H-R diagram of Figure 5 for stars from 0.6 to 6.0 $M_\odot$. While low mass stars descend from the birthline (the starting locus from which stars begin the PMS phase) on convective paths, stars more massive than $\sim$4 $M_\odot$ move to the left on radiative ones. Stars between 2 and 4 $M_\odot$ undergo thermal relaxation, looping back slightly behind the birthline and recrossing it before pursuing horizontal tracks. Compared to the classical set of PMS tracks, those shown in Figure 5 occupy a much more limited portion of the diagram. The reason is that the older tracks correspond to stellar radii too large to have been attained during protostellar accretion. Thus, the starting luminosities are much lower than previously envisioned. In addition, the surface temperature of these stars begin *higher*. For example, a 5 $M_\odot$ star starts with $T_{\rm eff}$ = 12 000 K in contrast to the value of 4400 K obtained by Iben (1965). Notice also that the birthline intersects the main sequence at a mass $M_* \sim$8 $M_\odot$: this point represents the critical stellar configuration in which hydrogen burning has stopped gravitational contraction while the star is still growing in mass. More massive stars, therefore, have no PMS phase at all, but appear directly on the main sequence once they are optically revealed.

The most important result of the new calculations is that observed PMS stars should not be located to the right of the birthline. For both T Tauri and Herbig Ae/Be stars the agreement is excellent, despite the rather strong assumption of a constant mass accretion rate of 10^{-5} $M_\odot$ yr^{-1} used in the models (see Palla & Stahler 1994 for a display of the actual location of PMS stars in the H-R diagram). This fact indicates that a limited dispersion in the properties of molecular cloud cores (which determine the magnitude of $\dot{M}_{\rm acc}$) somehow produces an impressive range of stellar masses. The case for even more massive stars is complicated by the difficulty in determining the appropriate stellar parameters to place them in the H-R diagram (see the controversial case for IRAS08159-3543 discussed by Staude and Neckel in this volume).

Finally, the other important result is the drastic reduction in age obtained for Herbig Ae/Be stars. The same is true, of course, for T Tauri stars, but the quantitative impact is minor. Even a large drop in the initial radius of a typical T Tauri star hardly changes its total contraction age, since τ_{KH} scales as R_*^{-3}. For the Herbig stars, however, Eq. (3) implies that τ_{KH} only increases as $R_*^{-1/2}$. The age reduction is far more significant for masses well below the upper cutoff at 8 $M_\odot$. Such a reduction is consistent with the observations of surface activity in these stars (e.g. Catala 1994).

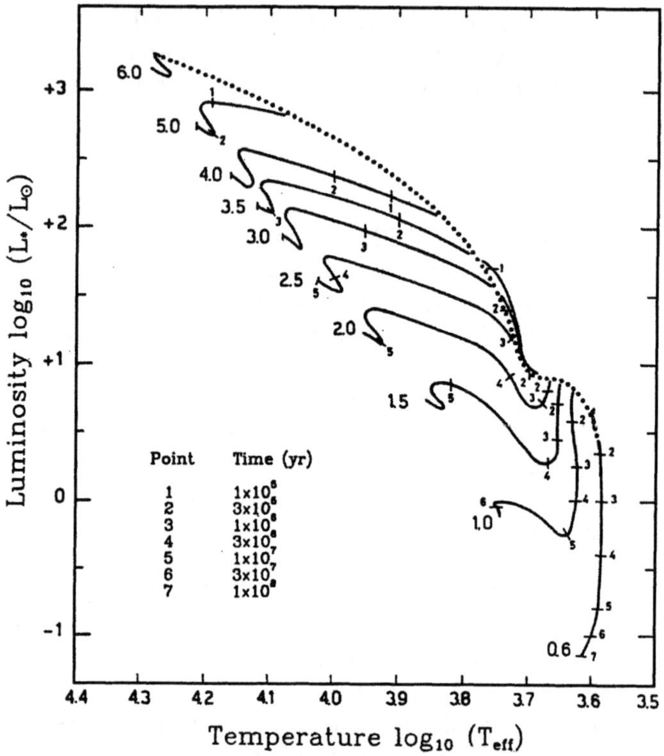

Fig. 5. Evolutionary tracks in the H-R diagram for low- and intermediate-mass stars. Each track is labeled by the corresponding mass, in solar units. Tick marks indicate points in the evolution whose time is given in the table. For each track, the evolution starts at the birthline (dotted curve) and ends at the ZAMS (not shown) (from Palla & Stahler 1993).

3.3 Nuclear Burning

The largest departure from the standard theory of PMS stars concerns the history of deuterium burning. Salpeter (1954) was the first to recognize that deuterium ignition via the reaction $^2H(p,\gamma)^3He$ at a temperature of about 1×10^6 K could slow down contraction or, for the least massive stars, stops it entirely for a relatively brief time. This effect was confirmed by the detailed calculations of

Grossman & Graboske (1971) and Mazzitelli & Moretti (1980) who introduced the concept of a "deuterium main sequence" to denote the curve in the H-R diagram crossed by the PMS tracks at the point of ignition. However, Stahler (1988) has shown that this concept is obsolete, since deuterium burning occurs already during the accretion phase for stars with masses above a few tenths of a solar mass (as a consequence of the small protostellar radii and of the correspondingly high central temperatures). Thus, the locus of the deuterium main-sequence is nothing else than the birthline shown in Figure 5. T Tauri stars, having consummed most of the available deuterium during the accretion phase, begin the contraction phase by simply descending along the vertical portion of the Hayashi track. The same is true for a Herbig Ae/Be star, which begins with a severely depleted deuterium supply. While this supply lasts, the outermost layers remain convectively unstable. But the luminosity provided by deuterium burning, L_D, is now so minor compared to L_{rad} (cf. Figure 3) that even this limited convection vanishes by the time the star has joined the classical radiative portion of its evolutionary track. Finally, for stars of mass smaller than $\sim$0.3 $M_\odot$, protostar theory still implies deuterium burning during the PMS phase, while for masses as low as $M_* = 0.018\ M_\odot$ degeneracy effects set in and prevent nuclear burning (D'Antona & Mazzitelli 1994).

In addition to deuterium, other light elements ^{7}Li ^{9}Be and ^{10}Be, are depleted in the course of PMS contraction, when the central temperature becomes sufficiently high (in excess of about 3×10^6 K). However, their relatively tiny abundances render the effect energetically and structurally unimportant: in order to halt PMS contraction the abundances should be enhanced by a factor 10^5 or more over the observed ones. On the other hand, these trace elements are of cosmological interest, and their depletion histories (especially for ^{7}Li) provide insight into problems as diverse as the age estimates of young clusters (Balachandran 1995) and galactic chemical evolution (D'Antona & Matteucci 1991; Galli et al. 1995a).

We will return later on to the important problem of lithium depletion in young stars as a test of PMS theoretical models. For now, let us consider the interesting case of ^{3}He. This isotope results from the nuclear conversion of deuterium and is destroyed by the reaction ^{3}He(^{3}He,2p)^{4}He at temperatures greater than 7×10^6 K. These conditions are encountered just prior to the arrival on the main sequence. The reaction rate is very sensitive to temperature ($\sim T^{18}$), so that the burning is concentrated into a small convective core ($\approx$10% in mass, according to Forestini 1994). However, if the abundance of this isotope is sufficiently high, the energy liberated can cause a partial halt of the contraction: the existence of a ^{3}He main-sequence was first postulated by Ulrich (1971) to explain some features of the anomalous color-magnitude diagrams of young clusters (Orion, NGC 2264). Depending on the adopted abundances, the ^{3}He main-sequence would lie 1 or 2 magnitudes above the normal main-sequence for stars more massive than the Sun, while stars less massive are not affected by the ^{3}He. Such an effect would cause a cluster to appear younger than it really is.

This suggestion has lost ground in the past since the required abundances of ^{3}He are a factor of 10-100 above what is considered the typical interstellar value [^{3}He/H] = 2.4×10^{-5}, by number. Note that because any initial deuterium is converted to ^{3}He during PMS evolution, the ^{3}He abundances are actually the sum of the initial ^{2}H and ^{3}He abundances. The reason for reconsidering the role of ^{3}He during PMS evolution is that the actual ^{3}He abundance in the interstellar medium is still very uncertain and subject to large fluctuations. Also, models of the galactic evolution of deuterium and ^{3}He, that are commonly used to obtain bounds on cosmological parameters, do indeed predict large abundances, such as those needed to create a ^{3}He main sequence (Galli et al. 1996a). Since the color-magnitude diagrams of young clusters are presently scrutinezed with higher accuracy than in the past (Strom, private communication), it is important to keep this possibility in mind for their interpretation. At the same time, new PMS models with high ^{3}He abundances are currently being computed (Galli et al. 1996b).

4. Tests to PMS Evolutionary Diagrams

The location of young stars in the H-R diagram is commonly used to estimate their masses and approximate ages. When young clusters or associations are studied, the information on the mass of individual members can yield direct information on the rate and efficiency of star formation and, eventually, on the initial mass function to compare to that observed locally. Also, the amount of depletion of the light elements due to nuclear activity in these early stages sheds light on the internal structure and on the mechanisms of angular momentum transport. It is important at this point to discuss the uncertainties that affect the computation of modern evolutionary tracks, such as those described earlier. The sources of uncertainty can be due to the uncertain knowledge of certain aspects of the input physics used in the theoretical models, or to the complex phenomena that accompany the evolution of young stars, such as the interaction with circumstellar disks and the accretion of residual matter, or the presence of companions and so on. In the following, I will first overview the uncertainties *intrinsic* to the models and then discuss the effects of the environment.

4.1 Comparison of Tracks

In three recent studies, D'Antona & Mazzitelli (1994), Forestini (1994), and Swenson et al. (1996) have reconsidered the PMS evolution of stars with population I compositions in the mass range 0.015 to 3.0 $M_\odot$. The main motivation for these studies is the progress in the fields of the equation of state, radiative opacities and the treatment of convection, each of which greatly influences the surface properties, i.e. the observables, of PMS stars. Before summarizing the main properties of the new models, let me first stress the following point: the average error in the determination of the observable quantities, L_{bol} and T_{eff}, is of the same order of magnitude than the theoretical ones. Typically, $\Delta T_{eff} = \pm 300$ K and $\Delta \log (L_*/L_\odot) = \pm 0.25$ dex, for stars observed in the nearby star forming regions (e.g. Strom & Strom 1994). On the other hand, the mean spacing between Hayashi tracks is of the order of ~130 K for $M_* = 0.5\ M_\odot$, and of only ~60 K for a 1 $M_\odot$ star (Forestini 1994). Thus, for very young stars still on the vertical portion of the tracks, the mass assignment based on their position can lead to errors of a factor of 2.

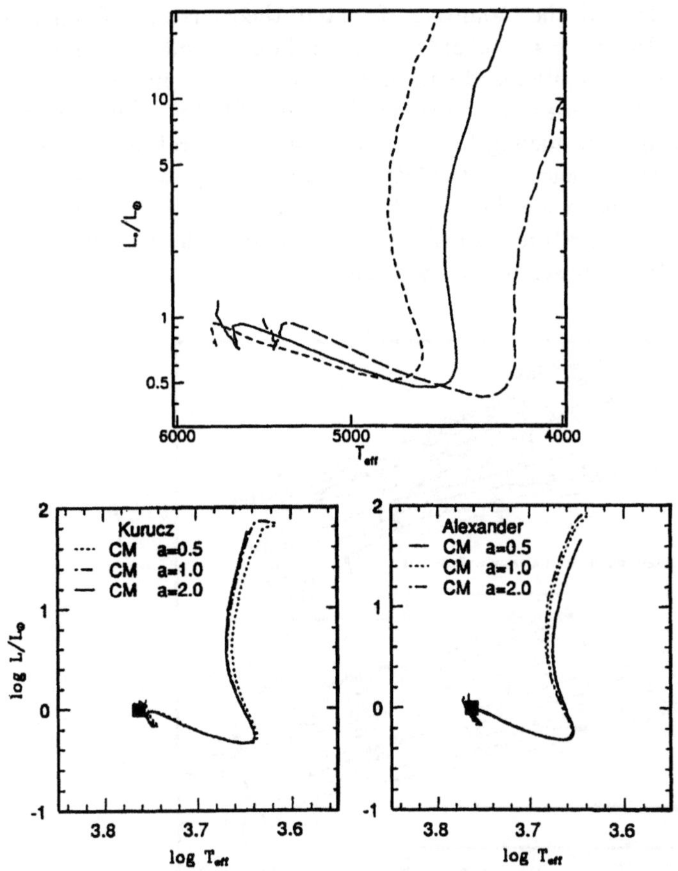

Fig. 6. (Top panel) Effects of the variation of the free parameter α in the mixing-length theory on the evolutionary tracks of a $M_* = 1\ M_\odot$ star. The solid line is for a model with $\alpha = 1.5$ which fits the sun, while the other curves are for $\alpha = 1$ (redshifted) and 2 (blueshifted), respectively (from Forestini 1994). (Bottom panel) Comparison of tracks computed with the Canuto & Mazzitelli (1990) prescription for turbulent convection and different external opacities. The tracks refer to a 1 $M_\odot$ star (from D'Antona & Mazzitelli 1994).

The largest source of uncertainty in the position of the tracks in the H-R diagram is still due to the poor description of turbulent convection. If the mixing-length theory is adopted, then the difference in effective temperature can amount to ~700 K for a variation of the free parameter α from 1 to 2 (the standard value for a $M_* = 1\ M_\odot$ star that fits the sun is $\alpha = 1.5$). Such a variation is shown in the top panel of Figure 6. The difference is thus comparable to the observational error. However, the discrepancy is greatly reduced if one follows the nonlocal treatment of convection developed by Canuto & Mazzitelli (1990) and used by D'Antona & Mazzitelli (1994) in their new models. As shown in the bottom panel of Figure 6, the variation in T_{eff} is lower than 3% at the top of the PMS tracks and becomes insignificant along the radiative portion. Note that the parameter a in the Canuto & Mazzitelli theory is not a free parameter as in the mixing-lenght sense, but it is introduced only to allow for fine tunings in the calculations of the standard sun and for internal comparisons. The important aspect to underline is that the value of α which fits the solar T_{eff} does not result in a Hayashi track having the same effective temperatures as for D'Antona & Mazzitelli (in the sense that the latter obtain *blueshifted* tracks). The differences, however, are within the theoretical uncertainties associated with the inaccuracy of the low temperature opacities, and, therefore, the two convection theories cannot be tested observationally in this way.

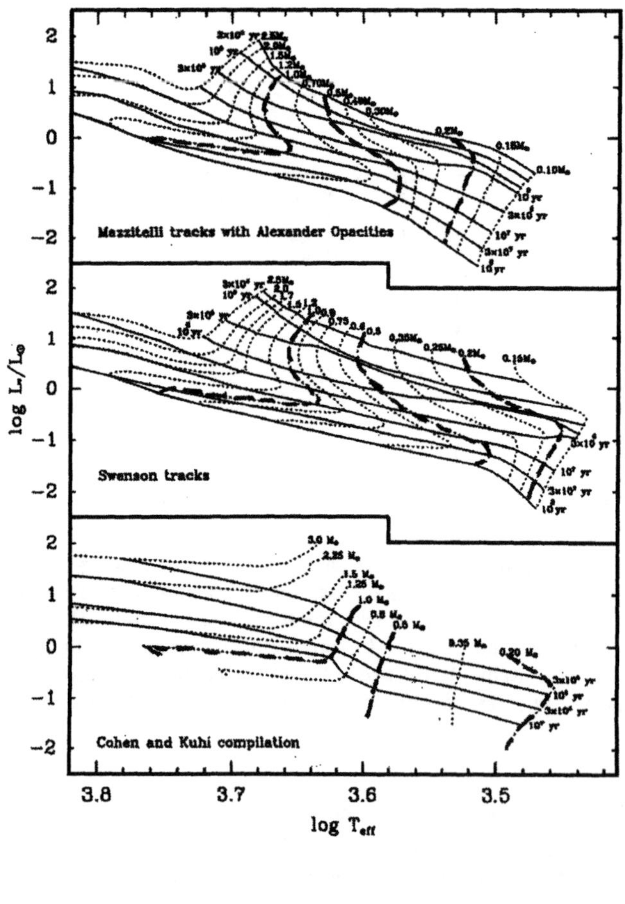

Fig. 7. Comparison of PMS evolutionary tracks of low mass stars computed by different authors. The solid lines reproduce the isochrones computed assuming the star to have zero age on the birthline (from Hartigan et al. 1994).

The overall level of uncertainty expected from the existing theoretical models can be appreciated by inspection of the different evolutionary tracks shown Figure 7, taken from the work of Hartigan et al. (1995). It is evident that at fixed T_{eff} the tracks yield substantially different values of M_*, and that the discrepancy is largest for the old tracks of Cohen & Kuhi (1979), based on the classical results of Iben (1965). In particular, the most recent calculations (upper panels) have smaller values of the stellar mass. This shift in mass has an impact on the correct location of the birthline in the H-R diagram. Figure 8 shows the variation produced by using the tracks of D'Antona & Mazzitelli in place of Cohen & Kuhi. The shift occurs at $\log_{10} T_{eff} > 3.58$, corresponding in the old Cohen & Kuhi tracks to the mass where deuterium ignites in the center ($M_* \sim 0.3\ M_\odot$), setting the conditons for an increse of stellar radius with mass, responsible for the upward rise of the luminosity. The new tracks imply that this condition is met only at $T_{eff} \sim 3.62$. At even higher temperatures, the two birthlines join again.

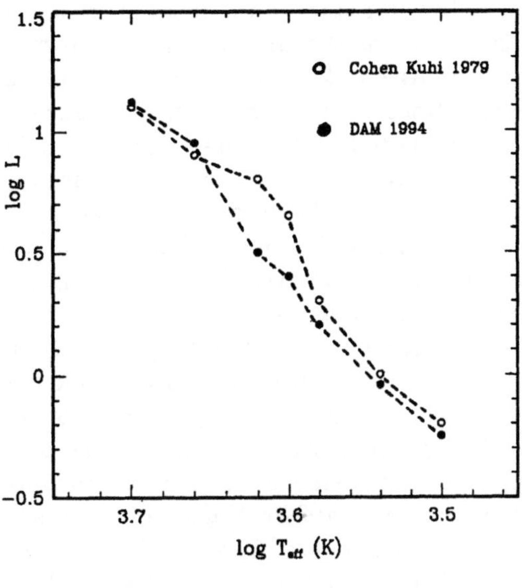

Fig. 8. Comparison of the birthlines obtained using Cohen & Kuhi (1979) or D'Antona & Mazzitelli (1994) PMS evolutionary tracks.

4.2 Lithium Depletion

The depletion of lithium in both young and evolved stars is considered one of the strongest constraints imposed on stellar evolution theory. The surface abundances, in fact, provide a critical test for the physical processes (convection, opacity, rotation, mass loss) which ought to be included in the theoretical models in order to reproduce the observed level of depletion. Due to its small binding energy, ^{7}Li (the most abundant of the two stable isotopes of lithium) is easily destroyed in stellar interiors via the reaction $^7\mathrm{Li}(p,\alpha)^4\mathrm{He}$ at temperatures around 2.5×10^6 K. According to standard models, depletion occurs only when the outer convection extends down to that temperature. Thus, it appears quite unavoidable that some ^{7}Li depletion due to nuclear burning takes place already during PMS contraction. Other processes, associated with the mixing due to transport of angular momentum through the star, are effective on much longer time scales and dominate the depletion history during the main sequence phase (see Zahn 1994).

The theoretical predictions are rather well established, independently of the adopted models. Stars more massive than 1.7 $M_\odot$ should *not* show any ^{7}Li depletion at all; solar-type stars are significantly depleted, while for sub-solar stars there should be be no ^{7}Li left over (e.g. Forestini 1994). The limiting mass for ^{7}Li-burning is $M_* \geq 0.06\ M_\odot$, since degeneracy effects in lower mass stars prevent the onset of nuclear reactions (Pozio 1991; D'Antona & Mazzitelli 1994). This value is below the substellar mass limit, usually taken between 0.08 and 0.07 $M_\odot$. Therefore, it has been suggested that the presence of lithium lines in the observed spectra can be used as a powerful probe to search for brown dwarf candidates (Magazzú et al. 1993; Rebolo et al. 1995). The evolution of the ^{7}Li surface abundances as a function of time for stars in the mass range 0.6 to 1.3 $M_\odot$, computed under different assumptions on the input physics, is shown in Figure 9. The level of depletion is most sensitive to the value of the parameter α of the mixing-length theory, yielding variations of about one order of magnitude for a solar-type star; the sensitivity on the radiative opacities or on the model atmosphere is much less (up to 40% differences). Clearly, the predictions vary so widely that one is cautioned against the use of the observed abundances of ^{7}Li to derive age estimates! Again, the results of D'Antona & Mazzitelli show a much reduced scatter in the level of depletion, but the warning still remains.

From the observational viewpoint, we know that the cosmic abundance of lithium is [Li/H] = 10^{-9}, or log N(Li) = 3 (in a scale with log N(H) = 12), as derived from observations of T Tauri stars, F stars in young clusters, and the interstellar medium. The situation concerning PMS stars has greatly improved in the last few years thanks to both the high quality of the spectroscopic data of the λ6707 Å LiI resonance doublet and to refinements in the procedures for the determination of the abundances (Strom et al.1989b; Magazzú et al.1992). Recently, Martin et al. (1994) have observed a sample of 53 weak T Tauri stars ranging from K0 to M3 ($\sim$1.3 to 0.3 $M_\odot$) and found that the initial Li abundance of is remarkably uniform with log N(Li) = 3.11±0.06, very close to the cosmic value. There is also clear evidence in their results for an increase of the amount of depletion towards stars of lower luminosity, as predicted by the models. The results for the abundances are shown in the plot of Figure 10. The overall agreement with the theoretical evolution is rather good, but there are differences in detail. In particular, Li starts being severely depleted at higher luminosity than predicted for low mass stars (0.2-0.4 $M_\odot$). Similar results have been found by García-Lopéz et al. (1994) on a smaller sample of PMS stars in the Pleiades. Furthermore, it has been found that there are no fast rotators with *low* Li abundance, in analogy to the abundance-rotation correlation known in young open clusters (e.g. Soderblom et al. 1993). This point is particularly

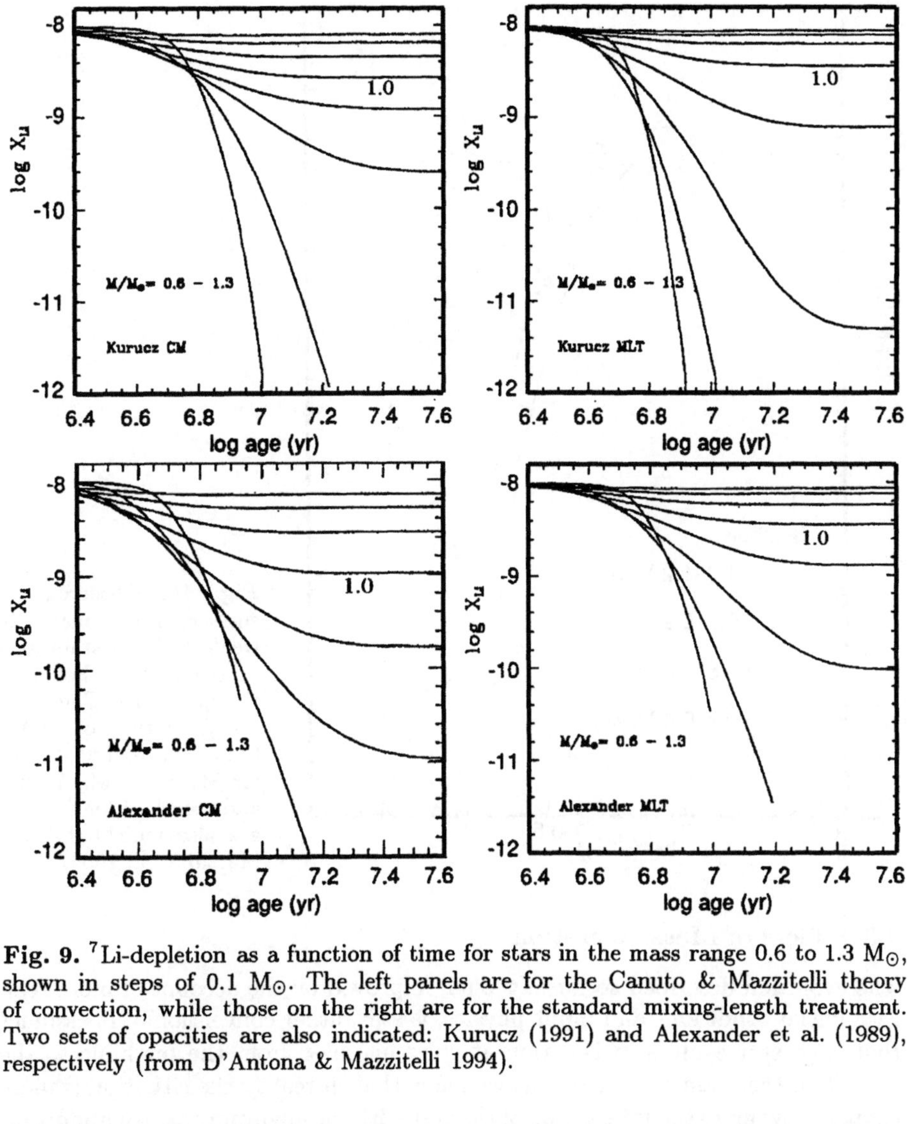

Fig. 9. ^{7}Li-depletion as a function of time for stars in the mass range 0.6 to 1.3 $M_\odot$, shown in steps of 0.1 $M_\odot$. The left panels are for the Canuto & Mazzitelli theory of convection, while those on the right are for the standard mixing-length treatment. Two sets of opacities are also indicated: Kurucz (1991) and Alexander et al. (1989), respectively (from D'Antona & Mazzitelli 1994).

important, since the role played by rotation and angular momentum transport is still controversial. Unfortunately, the available data do not appear sufficient to draw firm conclusions.

It would be very interesting to check whether the results described above also hold in the case of the classical T Tauri stars. In the few cases known, no appreciable Li-depletion has been observed, but the observations refer only to rather high luminosity objects. Priority should be given to the study of low luminosity classical T Tauri stars, even though these stars suffer from several problems (line saturation, optical veiling, non-LTE effects) that make the interpretation and the determination of abundances not straightforward (see Magazzù et al. 1994).

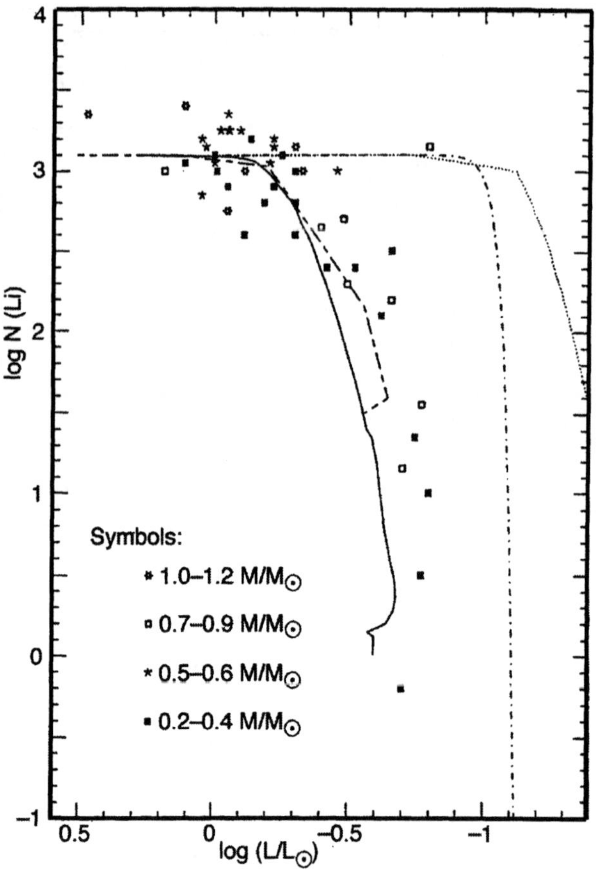

Fig. 10. Observed ^{7}Li surface abundances in weak T Tauri stars as a function of the bolometric luminosity. The theoretical curves of D'Antona & Mazzitelli (1994) for $M_* = 0.8\ M_\odot$ (solid) and $0.4\ M_\odot$ (dot-dashed) are also indicated (from Martin et al. 1994).

4.3 Effects of Mass Accretion

The evolution described so far, although improved by the account of more realistic initial conditions and input physics, is still rather conventional. It assumes that each star evolves in isolation at constant mass from the birthline to the arrival on the main sequence. But we know that in reality the situation is more complex, owing to the interaction of the star with the environment. An important effect to consider is the occurrence of residual mass accretion during the PMS phase. As we have already anticipated, accretion alters the total (star+disk) luminosity, the evolutionary path and, hence, the age estimate of a star.

Current estimates of mass accretion rates derive from the analysis of the excess optical continuum radiation in T Tauri and Herbig stars that also show an infrared excess. The latter is interpreted as being due to the presence of an active circumstellar disk. In this case, the disk luminosity is the sum of two contributions,

$$L_{\rm disk} = \frac{GM_*\dot{M}_{\rm acc}}{2R_*} + L_{\rm rep}, \tag{5}$$

where the first term represents the accretion luminosity, while the second term is due to the absorption and reradiation of light from the central star. Typically, $L_{\rm rep}$ ~0.25-0.5 L_*. Thus, unless $L_{\rm disk} \geq L_*$, the identification of the excess luminosity as *accretion* is not secure. On the other hand, observations indicate that at most $L_{\rm disk}$ ~0.5 L_*.

If the optical excess is assumed to arise in a boundary layer (BL) between the star and a disk, then the standard theory of viscous accretion disks predicts that up to half of the accretion luminosity is emitted as radiation,

$$L_{\rm BL} \leq \frac{GM_* \dot{M}_{\rm acc}}{2R_*}. \tag{6}$$

Thus, measurements of $L_{\rm BL}$ can provide a direct estimate of the amount of mass actually falling onto a star. Measurements of the veiling in a large sample of T Tauri stars yield values in the range $\dot{M}_{\rm acc} = 10^{-8}$–$10^{-7}$ $M_\odot$ yr^{-1} (Hartmann & Kenyon 1990; Bertout & Regev 1992). Higher values of $\dot{M}_{\rm acc}$, up to 100 times, have been derived for the more massive Herbig Ae/Be stars (e.g. Hillenbrand et al. 1992). Note, however, that in the absence of angular momentum loss, accretion at these rates would spin up the central star to half the break-up velocity in a few million years. Since young stars are known to rotate at typically 1/10-1/5 of the break-up speed (e.g. Bouvier 1994), it is likely that strong winds carry away the angular momentum excess. Therefore, it is still not clear what fraction of the accreted mass is actually incorporated into the star.

The question, then, is whether the inferred rates are high enough to modify the evolutionary tracks. Hartmann & Kenyon (1990) argue that, indeed, the path is greatly affected, and estimate from qualititative arguments that about 10% of the final mass of a T Tauri is accreted during the PMS phase. To test this effect, the evolution of stars with initial mass of 0.6 and 1.0 $M_\odot$ has been followed allowing for steady mass accretion at different rates for the first 3 million years. The resulting evolution is displayed in Figure 11. As we can see, as long as the value of $\dot{M}_{\rm acc}$ remains below ~10^{-7} $M_\odot$ yr^{-1}, the new tracks do not differ substantially from those computed at constant mass. Moreover, if the *global* effect of mass accretion is limited to about 10% of the final mass, it would take only 1 million years for a 1 $M_\odot$ star to accrete such an amount of mass at $\dot{M}_{\rm acc} = 10^{-7}$ $M_\odot$ yr^{-1}. From Figure 11, we see that after this time the evolutionary tracks are undistinguishable from the case with no mass accretion. Small departures can be appreciated only after at least 2 million years for $\dot{M}_{\rm acc} = 10^{-7}$ $M_\odot$ yr^{-1}, implying that, unless accretion continues for a substantial part of the PMS evolution at high rates (or that most the final mass is gathered during PMS evolution), the uncertainty in the location of a star in the H-R diagram due this effect is small.

As a final comment against the notion of high accretion rates (especially in the case of more massive stars), consider that the disk accretion time cannot be much shorter than $\tau_{\rm KH}$. If it were, the accreting star would remain so distended that it would always be on the birthline once optically revealed. Moreover, even a rough equality of the two timescales is implausible. We know that any disk mass $M_{\rm d}$ must be less than M_* to avoid gravitational instability (Adams, Ruden

& Shu 1989). The disk depletion time $M_d/\dot{M}_d$ would therefore be less than the stellar lifetime in such a marginal case. For the disk to survive, some external reservoir - the surrounding cloud - must supply the mass lost to the star. The presence of such a cloud would lead to obscuration of the star, the very problem that motivated the disk hypothesis in the first place.

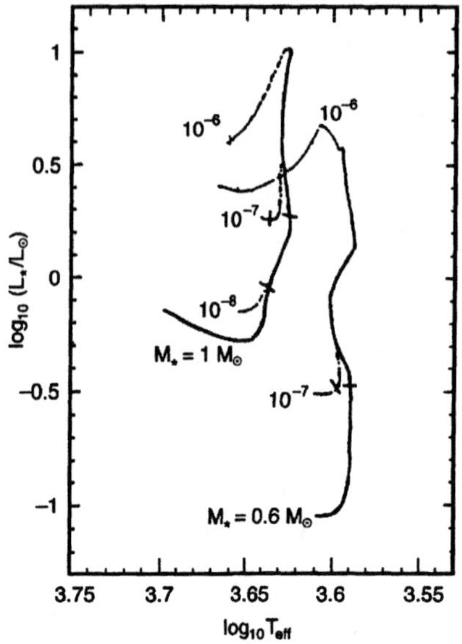

Fig. 11. Effects of mass accretion on the PMS evolutionary tracks of stars of 0.6 and 1 $M_\odot$, respectively. Three values of $\dot{M}_{acc}$ are considered, 10^{-8}, 10^{-7} and 10^{-6} $M_\odot$ yr^{-1}. The modified tracks are shown by the dotted lines, while the solid line is for the evolution at constant mass. Evolutionary times are indicated by tick marks (from Parigi 1992).

4.4 Effects of Binaries

The study of PMS binaries offers the best way to test the theoretical evolutionary tracks. The measurement of the dynamical mass of the components of any binary system would yield in fact an absolute mass calibration of the tracks and thus provide a direct answer for the exact location of the tracks in the H-R diagram. The progress in the field of PMS binaries has been tremendous (see the excellent review by Mathieu 1994). The population of binary systems in the nearest star forming regions has now been studied and a binary frequency of at least 50% derived for systems with angular separations in the range 3 to 1400 AU (see Simon, this volume). It appears that there is even an excess of such systems in comparison to what found among nearby solar-type stars, possibly implying that binary formation is the main mode of star formation (Simon et al. 1995). Unfortunately, only in one case it has been possible to derive the dynamical mass of the components of an eclipsing binary, EK Cep (Martin & Rebolo 1993). However, the stars are located in the H-R diagram very close to the zero-age main sequence so that they cannot be used to constrain the theoretical tracks. Only the discovery of an extremely young, less than $\sim 10^6$ yr, PMS eclipsing system would provide the critical test.

Similarly, double-lined systems enable the determination of mass ratios, but absolute mass estimates are prevented by the unknown inclination angle. From an analysis of the presently known systems, Lee (1992) and Padgett & Stapelfeldt (1994) conclude that their positions compare well with the reference tracks of D'Antona & Mazzitelli (1994). Despite the present limitations, it is clear that this field of research will prove very useful in the near future.

The other important effect to consider is the incorrect assignment of the star luminosity that affects the position in the H-R diagram and the age estimate. Since in the past PMS stars were not known to have companions, their luminosity has been overstimated. For the youngest stars on the vertical portion of the Hayashi track, the higher luminosity corresponds to a younger age. However, to correct for this effect is not an easy task since binary surveys are usually made photometrically and, therefore, provide only monochromatic magnitudes (in particular, K-magnitudes). Thus, in order to derive an estimate of the luminosity of the secondary, one has to extrapolate to longer wavelengths which may result in large errors, since young stars have large infrared excesses.

By considering only the correction for the K-flux ratios, Simon et al. (1993) have derived the contribution of the secondary to the luminosity in a large sample of binary stars in Taurus. They find that the average age of these stars is 2-3 times older than previously thought, and that the apparently different distribution in the H-R diagram of singles and binaries is only an artifact of the luminosity overestimate. Figure 12 clearly shows that binaries are uniformly distributed throughout the digram in the same way as the singles. The other interesting property is that stars of all periods have been found near the birthline, strongly suggesting that binaries are formed before they become optically visible PMS objects. These findings pose severe constraints on the formation mechanisms and tend to favor the fragmentation hypothesis over capture or fission. Additional support to this view comes from the recent study of Hartigan et al. (1995) who derived *relative* ages of 26 wide (separations greater than 400 AU) systems of known bolometric luminosity and effective temperatures. From the comparison with theoretical tracks, they concluded that in two-thirds of the cases the two components lie on the same isochrone and that therefore there is no measurable age difference. The next obvious step is to extend the study to binaries with smaller separations.

5. Conclusions

The main points discussed in this overview can be summarized as follows.

- YSOs can be conveniently classified in an evolutionary sequence by their spectral energy distributions. The strength and distribution of the emission, however, suggest a revision of the original classification scheme based on the infrared properties only. In the new scheme, candidate protostellar objects (*class 0*) are barely detectable at mid-infrared wavelenghts and become prominent only in the (sub)millimeter.

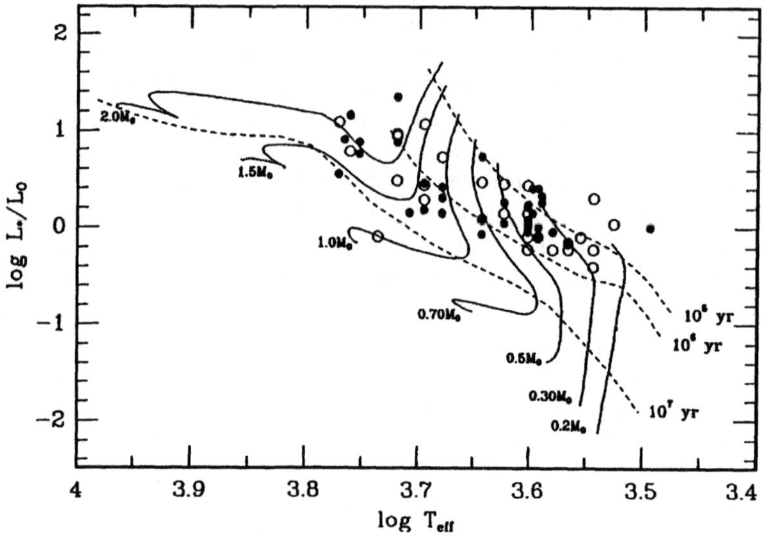

Fig. 12. The distribution of single (filled circles) and binary (open) PMS stars in Taurus-Auriga and Scorpius-Centaurus (from Mathieu 1994).

- The evolutionary sequence of YSOs can also be parametrized in terms of a mean frequency of the SED, or *bolometric temperature*. The youngest embedded sources are characterized by the lowest values of T_{bol}, while optically revealed PMS stars have $T_{bol} \sim T_{eff}$.
- Protostellar evolution has a big impact on the classical H-R diagram. The initial phases of PMS evolution greatly differ from what was envisioned earlier and depend sensitively on the stellar mass, with stars more massive than solar showing the largest deviations. The use of more realistic initial conditions yields a much better agreement between the predictions of protostellar evolution and the location of optically visible stars in the range $0.3 \leq M_*/M_\odot \leq 10$. More massive stars are expected to show no PMS phase at all, and a theory for their formation is clearly needed.
- Comparison of evolutionary tracks computed under different assumptions on the input physics shows that the uncertainty in the mass and age estimate of young stars is still large (about a factor of 2). The largest differences are found for subsolar stars. The accurate determination of ^{7}Li surface abundances of T Tauri stars can be used as a powerful test to PMS models.
- Other important effects that alter the evolution of PMS stars, such as residual mass accretion and the presence of companions, can be now incorporated in theoretical models in a quantitative fashion. Mass accretion at the currently derived rates does not affect the evolutionary tracks substantially, but it does change the amount of ^{7}Li surface depletion. The best test to calibrate the tracks comes from the mass estimates of PMS binary systems.

Acknowledgement. It is a pleasure to thank Prof. S. Beckwith and the SOC for their invitation to present these ideas during the meeting in honor of Prof. H. Elsässer. The preparation of this review has benefitted by the input of many colleagues, among which I would like to thank in particular P. André, D. Galli, A. Magazzù, T. Montmerle, A. Natta, R. Pallavicini, P. Saraceno, S. Stahler and H. Zinnecker.

References

Adams, F.C. 1990: ApJ 363, 578
Adams, F.C., Ruden, S.P., Shu, F.H. 1989: ApJ 347, 959
Aglazde, N.I., Sievers, A.J., Jones, S.A., Beckwith, S.V.W. 1994: Nature 372, 243
Alexander, D.R., Augason, G.C., Johnson, H.R. 1989: ApJ 345, 1014
André, P. 1994: in *The Cold Universe*, eds. T. Montmerle, C.J. Lada, I.F. Mirabel & J. Tran Thanh Van, Editions Frontières, p.179
André, P. & Montmerle, T. 1994, ApJ 420, 837
André, P., Ward-Thompson, D. & Barsony, M. 1993): ApJ 406, 122
Balachandran, S. 1995, in *The Abundances of the Light Elements*, ESO/IAPC Conf., ed. P. Crane (Heidelberg: Springer), p. 252
Basri, G., Bertout, C. 1989, ApJ 341, 340
Beckwith, S.V.W., Sargent, A.I. 1991, ApJ 381, 250
Beckwith, S.V.W., Sargent, A.I., Chini, R.S., Gusten, R. 1990, AJ 99, 294
Bertout, C., Regev, O. 1992, ApJ 399, L163
Bontemps, S., André, P., Ward-Thompson, D. 1995, A&A 297, 98
Bouvier, J. 1994, in The Eight Cambridge Workshop on *Cool Stars, Stellar Systems, and The Sun*, ed. J.-P. Caillault, A.S.P. Conf. ser., p. 151
Canuto, V.M., Mazzitelli, I. 1990, ApJ 370, 295
Catala, C. 1994, in *The Nature and Evolutionary Status of Herbig Ae/Be Stars*, eds. P.S. Thé, M. Perez & J.P.E. van den Heuvel P.A.S.P. Conf. Ser., p.188
Chen, H., Myers, P.C., Ladd, E.F., Wood, D.O.S. 1995, ApJ 445, 377
Cohen, M., Kuhi, L.V. 1979, ApJS 41, 743
Cohen, M., Emerson, J.P., Beichman, C.A. 1989, ApJ 339, 455
D'Antona, F., Matteucci, F. 1991, A&A 248, 62
D'Antona, F., Mazzitelli, I. 1994, ApJS 90, 467
Draine, B.T. & Lee, H.M. 1984, ApJ 285, 89
Ezer, D., Cameron, A.G.W. 1967, Can. Jour. of Physics 45, 3429
Forestini, M. 1994, A&A 285, 473
Galli, D., Palla, F., Straniero, O., Ferrini, F. 1996a, ApJ , in press
Galli, D., Palla, F., Straniero, O. 1996b, in preparation
García Lopéz, R.J., Rebolo, R., Martin, E.L. 1994, A&A 282, 518
Grossman, A.N., Graboske, H.C. 1971, ApJ 164, 475
Hartigan, P., Strom, K.M., Strom, S.E. 1995, ApJ 427, 961
Hartmann, L., Kenyon, S.J. 1990, ApJ 349, 190
Hartmann, L., Kenyon, S.J., Calvet, N. 1993, ApJ 407, 219
Hayashi, C. 1966, ARA&A 4, 171
Hildebrand R.H. 1983, QJRAS 24, 267
Hillenbrand, L.A., Strom, S.E., Vrba, F.J., Keene, J. 1992, ApJ 397, 613
Iben, I. 1965, ApJ 141, 993
Kenyon, S.J., Hartmann, L. 1990, ApJ 349, 197
Kenyon, S.J., Hartmann, L., Calvet, N. 1993, ApJ 414, 676
Kurucz, R.L. 1991, in *Stellar Atmospheres: Beyond the Classical Models*, eds. L. Crivellari, I. Hubeny & D.G. Hummer (Dordrecht: Kluwer), p.441

Lada, C.J. 1987, in *Star Forming Regions*, IAU Symp. 115, eds. M. Peimbert & J. Jaguku (Dordrecht: Reidel Publ. Co.), p.1
Ladd, E.F., Adams, F.C., Casey, S., Davidson, J.A., Fuller, G.A., Harper, D.A., Myers, P.C., Padman, R. 1991, ApJ 366, 203
Lee, C.-W. 1992, Ph.D. Thesis, University of Wisconsin-Madison
Magazzù, A., Rebolo, R., Pavlenko, Ya.V. 1992, ApJ 392, 159
Magazzù, A., Martin, E.L., Rebolo, R. 1993, ApJ 404, L17
Magazzù, A., Martin, E.L., Rebolo, R., García Lopéz, R.J., Pavlenko, Ya.V. 1994, in *The Abundances of the Light Elements*, ESO/IAPC Conf., ed. P. Crane (Heidelberg: Springer), p. 260
Mannings, V. & Emerson, J.P. 1994, MNRAS 267, 361
Martin, E.L., Rebolo, R. 1993, A&A 274, 274
Martin, E.L., Rebolo, R., Magazzù, A., Pavlenko, Ya.V. 1994, A&A 282, 503
Mathieu, R.B. 1994, ARA&A , p. 465
Mazzitelli, I., Moretti, M. 1980, ApJ 235, 955
Mezger, P.G., Wink, J., Zylka, R. 1990, A&A 228, 95
Mitskevich, A.S. 1994, A&A 281, 471
Myers, P.C., Ladd, E.F. 1993, ApJ 413, L47
Padgett, D.L., Stapelfeldt, K.R. 1994, AJ 107, 720
Palla, F. 1994, in *Infrared Astronomy*, eds. A. Mampaso & F. Sanchez (Cambridge: Cambridge University Press), p.1
Palla, F., Stahler, S.W. 1990, ApJ 360, L47
Palla, F., Stahler, S.W. 1993, ApJ 418, 414
Palla, F., Stahler, S.W. 1994, in *The Nature and Evolutionary Status of Herbig Ae/Be Stars*, eds. P.S. Thé, M. Perez & J.P.E. van den Heuvel P.A.S.P. Conf. Ser., p.391
Parigi, G. 1992, Diss., Univ. of Florence
Pollack, J.B., Hollenbach, D., Beckwith, S.V.W., Simonelli, D.P., Roush, T., Pong, W. 1994, ApJ 421, 615
Pozio, F. 1991, Mem. Soc. Astron. Ital. 62, 171
Rebolo, R., Magazzù, A., Martin, E.L. 1995, in *The Bottom of the Main Sequence and Beyond*, ESO Workshop, ed. Ch.G. Tinney (Heidelberg: Springer), p. 159
Rucinski, S.M. 1985, AJ 90, 2321
Rydgren, A.E., Zak, D. 1987, PASP 99, 141
Salpeter, E.E. 1954, Mem. Roy. Soc. Liège, 14116
Saraceno, P., André, P., Ceccarelli, C., Griffin, M., Molinari, S. 1995, A&A , in press
Shu, F.H., Najita, J., Galli, D., Ostriker, E., Lizano, S. 1993, in *Protostars and Planets III*, eds. E.H. Levy & J.I. Lunine (Tucson: Univ. of Arizona Press), p.3
Simon, M., Ghez, A.M., Leinert, Ch. 1993, ApJ 408, L33
Simon, M., Ghez, A.M., Leinert, Ch. et al. 1995, ApJ 443, 625
Soderblom, D.R., Jones, B.F., Balachandran, S., Stauffer, J.R., Duncan, D.K., Fedele, S.B., Hudon, J.D. 1993, AJ 106, 1059
Stahler, S.W. 1988, ApJ 332, 804
Stahler, S.W., Shu, F.H., Taam, R.E. 1980, ApJ 241, 637
Strom, K.M., Strom, S.E. 1994, ApJ 424, 237
Strom, K.M., Strom, S.E., Edwards, S., Cabrit, S., Skrutskie, M.F. 1989a, AJ 97, 1451
Strom, K.M., Wilkin, F.P., Strom, S.E., Seaman, R.L. 1989b, AJ 98, 1444
Swenson, J.H., Faulkner J., Rogers, F.J., Iglesias, C.A. 1996, ApJ , in press
Ulrich, K.H. 1971, ApJ 165, L95
Wilking, B.A., Lada, C.J., Young, E.T. 1989, ApJ 340, 823
Yorke, H.W., Bodenheimer, P., Laughlin, G. 1995, ApJ 443, 199
Zahn, J.-P. 1994, A&A 288, 829

Neutral Gas in the Vicinity of Herbig AeBe Stars: What Can We Learn from Red-Shifted Sodium Absorption Lines?

V. Grinin[1], C. Sorelli[2], and A. Natta[3]

[1] Crimean Astrophysical Observatory, Crimea, 334413, Nauchny, Ukraine
[2] Dipartimento di Astronomia, Università di Firenze, Largo Fermi 5, 50125 Firenze, Italy
[3] Osservatorio Astrofisico di Arcetri, Largo Fermi 5, 50125 Firenze, Italy

Summary. Variable, high-velocity red-shifted absorption components have been observed recently in the Na D lines of some Herbig Ae/Be stars with non-periodic Algol-type minima. The existence of neutral sodium in the environment of relatively hot stars, such as UX Ori (spectral type A3) cannot be easily explained. We present here the preliminary results of calculations aimed at exploring under which physical conditions a sizable amount of neutral sodium can survive in the immediate vicinity of such a star. Spectral lines of circumstellar origin of other elements are also discussed.

1. Introduction

The sporadic infall of gas onto young stars is an interesting phenomenon, well known since many years but not yet understood. Infall is usually discussed in the framework of the classical scenario, proposed by Lynden–Bell and Pringle (1974), of viscous disk accretion (see, for example, Bertout 1987). In this scenario the transfer of angular momentum is due to turbulent viscosity. The chemical composition of the infalling gas is the same (or about the same) as the chemical composition of the young star itself.

The recent spectral observations of β Pictoris and their interpretation (see Lagrange-Henri et al. 1988; Ferlet et al. 1993, and references therein) show that another scenario of sporadic accretion of circumstellar (CS) matter is possible in stars with proto–planetary disks: the infall onto the star of evaporating star–grazing bodies. In this case the transport of angular momentum occurs at the time of the star–grazing orbit formation due to the gravitational perturbations produced by planets (Beust et al. 1991), or, in the case of a double system, by the secondary component; in this case, the gas accreting onto the star very likely has an excess of heavy elements with respect to the atmospheric composition.

In the case of β Pictoris two observational facts favour the star–grazing bodies hypothesis: a) the existence in the immediate vicinity of the hot (Sp = A5) star of relatively cool gas, shown by redshifted absorption components in the CaII resonance lines; and b) the variability of these components on time scales from few days to few hours.

Similar spectral features have been observed recently in the spectra of the hot young variables UX Ori (Grinin et al. 1994a; de Winter 1995), BF Ori (de Winter 1994; Kozlova 1994) and RR Tau (Grinin et al. 1995). In particular, all these stars show high-velocity red-shifted absorption components in the resonance lines of neutral sodium at 5900 Å. In two cases (UX Ori and BF Ori) strong variations of their shape from night-to-night were observed. In all three cases the stars were in their "normal" (bright) state. Red-displaced absorption features were also found in the UV lines of other metals from archival IUE observations of some Herbig Ae/Be stars (Grady et al. 1995).

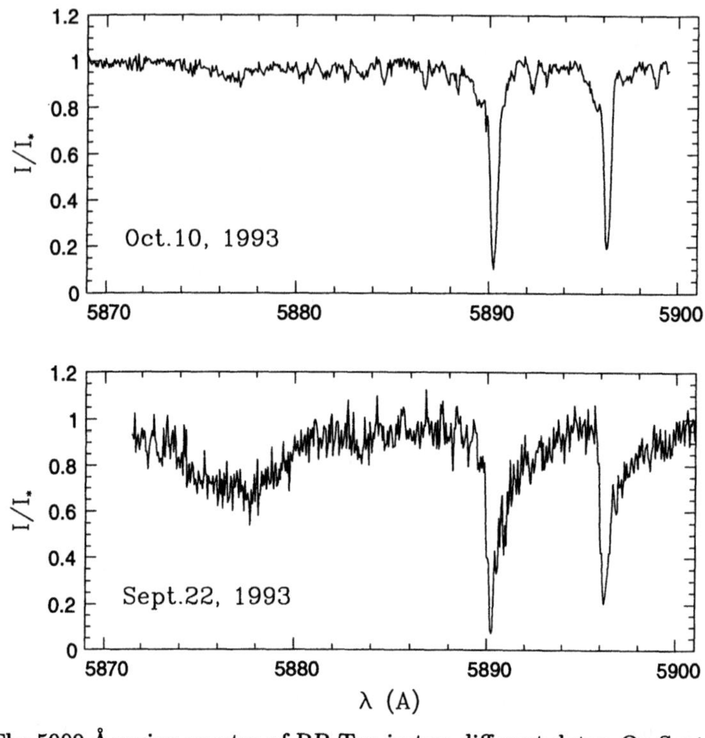

Fig. 1. The 5900 Å region spectra of RR Tau in two different dates. On Sept.22, 1983, one can clearly see the extended red wings of the Sodium doublet, as well as a broad absorption at the positon of the HeI 5876 Å line; from Grinin, Kozlova and Rostopchina (1994b).

According to the General Catalogue of Variable Stars (Kholopov et al. 1987) the above mentioned Herbig Ae/Be stars are members of a small group of stars (UX Ori is the prototype) with *non-periodic Algol-type brightness minima.* There are reasons to believe (see Grinin et al. 1991 and references therein) that their unusual photometric activity and observed anti-correlation between the brightness and the intrinsic linear polarization are the result of the orientation of their proto-planetary disk-like envelope, which is seen edge-on, and of the presence of a variable amount of CS matter along the line of sight. This makes attractive the suggestion (Grinin 1992) that the UX Ori type stars are the young progenitors of β Pic systems.

In this paper we present preliminary results of a non-LTE analysis of the expected physical conditions of a gas cloud in the radiation field typical of UX Ori type stars. The aim is to predict the expected optical depth in a number of lines of CS origin as a function of the distance from the star. These calculations, as we will show in the following, provide evidence that the appearance of red-shifted sodium absorption components in the spectra of Herbig AeBe stars very likely occurs in the context of the solid bodies evaporating scenario.

2. The Model

We consider a cloud of gas of constant density at distance r_c from the star. The cloud physical conditions have been computed using the code CLOUDY (Ferland 1993). This code calculates the temperature balance inside the cloud (taking into account gas and dust heating and cooling processes), the ionization state of the most important atoms, and the optical depth in selected spectral lines.

For typical values of r_c of the order of 10 stellar radii, the assumption of quasi-stationary conditions in the cloud is verified as long as $N_e \geq 10^7 cm^{-3}$, so that the hydrogen recombination time is shorter than the free fall time.

The code requires the specification of a number of parameters, many of which can be constrained by the existing observations, as described below.

2.1 Model Parameters

2.1.1 The Central Source of Radiation. We assume that the energy distribution of UX Ori can be described by the Kurucz (1991) model with effective temperature $T_{\mathrm{eff}} = 9000$ K and lg $g = 4$. This agrees with the spectral type of the star and the values of mass and radius ($M_* = 3.3 M_\odot$ and $R_* = 3.2 R_\odot$) estimated by Hillenbrand et al. (1992).

Some Herbig AeBe stars are known as weak X-ray sources with luminosities $L_x = 10^{30} - 10^{32}$ ergs/s and $T \simeq 10^6 K$ (Zinnecker and Preibisch 1994). To our knowledge, there are no specific observations for the stars of interest here. However, in the case of RR Tau the observations have shown that hot gas which produces the He I 5876 A line is present at the same time than the redshifted absorption components in the D NaI lines (Fig. 1) and is therefore likely to be the product of gas accretion onto the stellar surface (Grinin et al. 1995). Therefore, we decided to add a "hot" component ($L_x = 10^{31}$ erg/sec; $T = 10^6 K$) to the photospheric radiation field, which, however, turned out to affect only marginally the results of interest here.

2.1.2 The Cloud Parameters. The observed properties of the redshifted D NaI components set several constraints on the cloud parameters.

a) The distance r_c of the cloud to the star must be of the order of 10 R_*, to account for the observed maximum radial velocities (up to 300 km/s in case of RR Tau) and the timescale of their variability. Note that the region inside this sphere is the region of intensive sublimation of solids in the radiation field of UX Ori type stars (Grinin and Tambovtseva 1994).

b) The characteristic size of the cloud D can be estimated from the observations of saturated absorption red wings in the D NaI resonance lines: $D \leq 0.3R_* \approx 10^{11}$ cm (Grinin et al. 1995).

c) If the cloud is formed by gas evaporating from solid matter, then its chemical composition is not normal. We explore this possibility by varying the metallicity index (i.e., the ratio of elements heavier than helium to hydrogen) from the value currently observed in HII regions (m=1) to m=1000. In all cases we assume a dust-to-gas mass ratio of 0.1 and grain properties as observed in the Orion star-forming region.

d) An upper limit to the cloud column density N is set by the fact that the redshifted absorption D NaI components have been observed when the stars were in their *bright state.* This means that the CS cloud, responsible for the line absorption, was optically thin in the continuum, or, in other words, that, for normal ratio of the dust to gas, the column density of the hydrogen atoms was $<10^{21} cm^{-2}$. For the characteristic cloud size $D = 10^{11}$ cm, this gives an approximate limit to the gas density $n \leq 10^{10}$ cm^{-3}.

e) The observed width of the D NaI red-shifted absorption components can be as large as 100 km/s. It is likely that this broadening is caused by velocity gradients inside the cloud. We simulate the effects of this (unknown) velocity field using a turbulent velocity v_t.

In summary, we have computed models for the following values of the parameters:

Radiation field:
Star: $T_{\rm eff} = 9000$ K; lg $g = 4$; $L = 60$ L⊙;
Hot plasma radiation: $L_x = 10^{31}$ erg/sec; $T_x = 10^6$ K;

Cloud: Distance to the star $r_c = 10\ R_*$;
Size $D = 10^{11}$ cm;
Density of hydrogen atoms: n = 10^8, 10^9, 10^{10} cm^{-3};
Turbulent velocity $v_t = 50$ km/sec;
Excess of metals relative to the HII regions chemical composition:
$m = 1$, 10, 100 and 1000;
Orion dust mixture with normal gas to dust ratio.

3. Results

Figure 2 shows the optical depth of a selected number of resonance lines, namely the D_1 NaI at 5895Å, the K CaII at 3934Å, the OI line at 1304Å and the MgI line at 2852Å, as a function of the metallicity index m. Different symbols refer to different values of the density n.

The main result displayed in Fig. 2 is that a cloud at distance of about 10 stellar radii from UX Ori is optically thick in the D NaI lines only if *it has strong excess of heavy elements*. This conclusion (which holds even in the densest cloud models we have computed, e.g., with $n = 10^{10}$ cm^{-3}) provides the main support to the hypothesis that the neutral gas in the immediate vicinity of UX Ori type stars is produced locally by sublimation of solid bodies.

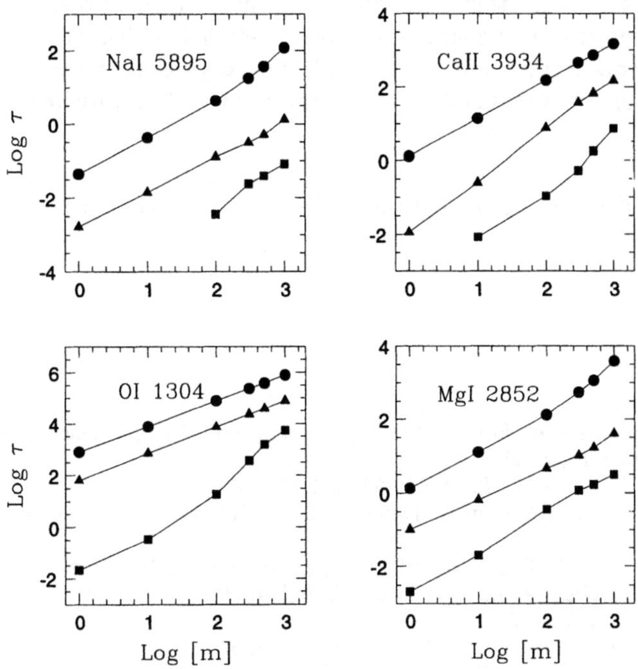

Fig. 2. Optical depth of selected resonance lines as a function of the metallicity index m for different values of the density of hydrogen atoms n. The dots are for $n = 10^{10}$ cm^{-3}, the triangles for $n = 10^9$ cm^{-3}, the squares for $n = 10^8$ cm^{-3}

The optical depth of the cloud in the K CaII and the MgI 2852Å lines is sistematically larger than that in the D NaI line, so that values of $\tau \simeq 1$ can be obtained also in clouds of normal chemical composition ($m = 1$), provided that n is large enough. Other UV resonance lines, such as the OI 1304Å and the MgII 2798Å (not shown in Fig. 2), have even larger optical depths. Therefore, while these lines are very sensitive to the presence of CS gas, they do not provide any strong evidence either in favour or against the comet-like bodies infall scenario.

The models discussed in this paper are very simple, and may not cover all the possible range of the parameters. We think in particular of heating mechanisms, not included in CLOUDY, and of possible variations of the dust-to-gas ratio. However, preliminary calculations seem to indicate that the results shown in Figure 2 are quite insensitive to reasonable variations of these parameters.

4. Conclusion

The calculations presented in this paper predict that UX Ori type stars can easily show red-shifted absorption components in the resonance lines of various heavy elements, as soon as infalling gas exists in their CS environment. The strongest lines (resonance lines of elements such as MgII, CI, CII, OI, etc.) are expected in the UV region of the spectrum. These lines are very sensitive to even small amounts of CS matter on the line of sight. However, only the NaI resonance lines can provide a meaningful test of the solid-body infall scenario, since they can only be observed when the gas metallicity is strongly enhanced with respect to normal values. This circumstance, if confirmed by more detailed models, stresses the importance of systematically studying the NaI lines as possible tracers of evaporated solid bodys and planetesimals in the vicinity of hot stars with young protoplanetary disks.

Acknowledgement. One of the authors (V.G.) thanks Professor F. Pacini for his extended hospitality in Arcetri.

References

Bertout, C., 1987, in *Circumstellar Matter*, IAU Symp. 122, eds. I.Appenzeller and C.Jordan, Reidel, Dordrecht, p. 23
Beust, H., Vidal-Madjar, A., & Ferlet, R., 1991, A&A, **247**, 505
Ferland, G.J., 1993, University of Kentucky, Department of Physics and Astronomy, Internal Report.
Ferlet, R., Lagrange–Henri, A.-M., Beust, H., Vitry, R., Zimmerman, J-P., Martin, M., Char, S., Belmahdi, M., Clavier, J.P., Coupiac, P., Foing, B., Sevre, F. & Vidal–Madjar, A., 1993, A&A, **267**, 137
General Catalogue of Variable Stars, IV, 1985, ed. P.N.Kholopov.
Grady, C.A. et al., 1995, in *Circumstellar Dust and Planet Formation*, eds. R.Ferlet and A.Vidal-Madjar, Editions Frontieres, in press
Grinin V.P., Kiselev N.N., Minikulov N.Kh., Chernova G.P. & Voshchinnikov N.V., 1991, Astrophys. Space Sci. **186**, 283
Grinin V.P., 1992, Astr.Astrophys.Transact., **3**, 17
Grinin V.P., Thé P.S., de Winter D., Giampapa M.S., Rostopchina A.N., Tambovtseva L.V., van den Ancer M., 1994, A&A, **292**, 165
Grinin V.P., Kozlova O.& Rostopchina A., 1995, in *Circumstellar Dust and Planet Formation*, eds. R.Ferlet and A.Vidal-Madjar, Editions Frontieres (in press).
Grinin V.P.& Tambovtseva L.V., 1994, ibid
Kozlova, O.V., 1994, private communication
Kurucz, R.L., 1991, in Proceedings of the Workshop on *Precision Photometry: Astrophysics of the Galaxy*, eds. A.C.Davis Philip, A.R.Upgren, and K.A.James, p 27.
Lagrange–Henri, A.M., Vidal–Madjar, A. & Ferlet, R., 1988, A&A, **190**, 275
Lynden–Bell, D. & Pringle, J.E., 1974, MNRAS, **168**, 603
de Winter, D., 1994, private communication
de Winter, D. et al., 1995, in *Circumstellar Dust and Planet Formation*, eds. R.Ferlet and A.Vidal-Madjar, Editions Frontieres (in press)
Zinnecker, H. & Preibisch, T., 1994, in Proceedings of First Conference on *The Nature and Evolutionary Status of Herbig Ae/Be Stars*, eds. P.S.Thé, M.Pérez, E.van den Heuvel, p. 281

The Stellar Populations of Deeply Embedded Young Clusters: Near-Infrared Spectral Classification *

Michael R. Meyer[1], Suzan Edwards[2], Kenneth Hinkle[3], Michael F. Skrutskie[1], and Stephen E. Strom[1]

[1] Department of Physics and Astronomy, University of Massachusetts, Amherst, MA 01003
[2] Astronomy Department, Smith College, Northampton, MA 01063
[3] National Optical Astronomy Observatories, Tucson, Arizona, 85726

Summary. Deeply embedded clusters of young stars offer a unique opportunity to study stellar evolution at the very earliest phases. We describe a program aimed at understanding the formation of young stellar objects (YSOs) through the use of near-infrared spectroscopy. Over the past 18 months, we have carried out an extensive survey of fundamental MK spectral standard stars in the near-infrared spanning luminosity classes I-II, III, and IV-V from spectral types O5 through M5 at a resolving power of R=3500. We summarize the results of this survey and present a two-dimensional classification scheme derived from H-band spectra. We have also observed a set of well-studied optically-visible young stars in the Taurus molecular cloud at high spectral resolution in the H-band in order to quantify the effects of accretion activity on near-infrared classification spectroscopy of young stellar objects. Our sample covers a range of spectral types and disk accretion rates including classical and weak-lined T Tauri stars. Results suggest that spectral types derived from infrared spectra agree with their optically derived counterparts to within a few subclasses for late-type stars. Further, weak-lined T Tauri star spectra closely resemble dwarf star spectra although they appear to have somewhat lower surface gravities. We also use these data to calculate the level of infrared excess emission in some classical T Tauri stars. Such measurements of the infrared "veiling" will provide crucial constraints on models of the distribution of warm circumstellar material surrounding some YSOs.

1. Introduction

Near-infrared array observations of nearby molecular clouds have revealed rich stellar clusters in regions obscured by several to tens of magnitudes of visual extinction. These deeply embedded young clusters have unique attributes which make them very useful for studies of the initial mass function (IMF).

– No evolutionary corrections needed from present day distribution of high mass stars to IMF.

* Based in part on observations obtained at Kitt Peak National Observatory, which is operated by the Association of Universities for Research in Astronomy, Inc. (AURA) under cooperative agreement with the National Science Foundation. Data also obtained at 1.3m McGraw Hill Telescope of the Michigan-Dartmouth-MIT Observatory.

- Youth precludes wandering of cluster members far from their birthsites due to peculiar motions.
- Small area encompassed by dense rich clusters minimizes contamination by foreground stars.
- Cloud core extinction minimizes contamination by background stars.
- Greater sensitivity to substellar mass objects than in older open clusters.

We wish to study mass and age distributions of single star-forming events, associated with individual molecular cloud cores at the earliest stages of stellar evolution. Our first approach is to utilize near-IR photometry alone in order to estimate rough mass and age distributions of embedded clusters. More reliable values of stellar masses are obtained by combining photometry with spectral types derived from near-IR spectroscopy and placing stars directly in the HR diagram. We have targeted two distinct categories of embedded clusters; those with evidence for high mass star formation and those thought to be forming primarily intermediate and low mass stars. These clusters are compared below in Table 1, which gives the mass of the most massive star, the core mass as derived from observations of the molecular gas, sampled within the confines of the dense cores ($\sim$ 0.25 pc diameter). Stellar densities are estimated from K-band source counts in near-IR surveys with comparable sensitivity limits ($M_K \sim 7.5^m$). Fractions of stars with obvious NIR excess are taken from the most recent literature. Although many of these quantities are extremely uncertain, relative comparisons should be meaningful.

Table 1. Properties of Target Clusters.

Cluster	**Most Massive Star**	**Core Mass**		**Stellar Density** [$N_* \ pc^{-3}$]	**NIR Excess** [%]
1. R CrA	2 $M_\odot$	20 $M_\odot$	(H_2CO)	$\sim 500 - 1000$	> 40
2. ρ Oph A	3 $M_\odot$	50 $M_\odot$	(H_2CO)	$\sim 500 - 1000$	> 70
3. Mon R2	10 $M_\odot$?	210 $M_\odot$	(CS)	$\sim 2000 - 4000$	> 90
4. NGC 2024	15 $M_\odot$?	430 $M_\odot$	(CS)	$\sim 2000 - 4000$	> 70

References: *NIR Survey Data:* 1. Taylor and Story (1984); Wilking *et al.* (1995). 2. Wilking and Lada (1983); Strom *et al.* (1995). 3. Dougados *et al.* (1995). 4. Lada *et al.* (1991a); this work. *Molecular Survey Data:* 1&2. Loren *et al.* (1983). 3. Wolf *et al.* (1990). 4. Lada *et al.* (1991b).

2. Photometric Techniques

Armed with J-, H-, and K-band photometry, what can we say about the embedded stellar populations found in molecular cloud cores? We have developed a technique to analyze our near-IR photometry which provides a reconnaissance

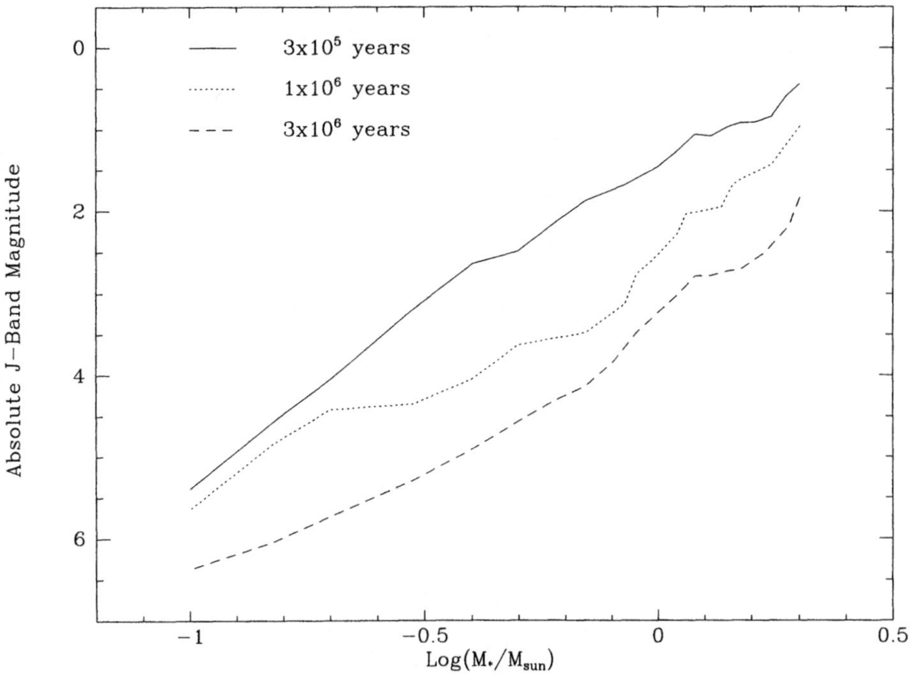

Fig. 1. Theoretical PMS tracks of D'Antona and Mazzitelli (1994) transformed into absolute J-band magnitudes from $0.1 - 2.0M\odot$ for three different ages. Note the luminosity evolution from bright to faint with age.

of stellar masses in the absence of spectral types (Meyer *et al.*, 1995). We adopt a set of theoretical PMS evolutionary tracks (eg. D'Antona and Mazzitelli, 1994) which provide the luminosity and temperature for stars of different mass as a function of their age. Using tabulated intrinsic colors and bolometric corrections, we translate luminosity and temperature into absolute J-band magnitude. This results in a unique relationship between stellar mass and absolute J-band magnitude for each isochrone (Figure 1). Because extinction can range from A_v=1-20^m or more, we must deredden each source individually. We assume the embedded population has circumstellar properties similar to optically-visible T Tauri stars, which allows us to deredden each star and obtain estimates of absolute magnitudes. Classical T Tauri stars occupy a well-defined region of the $(H - K)$ versus $(J - H)$ infrared color-color diagram (Figure 2). For stars within the reddened locus of classical T Tauri stars, we deredden the stars in this color-color diagram which takes into account their *intrinsic* IR excess (Figure 3A). For stars within the reddened main sequence in the color-color diagram, we deredden in the color-magnitude diagram (Figure 3B). Assuming a common age for a cluster, we can then infer the mass of each star. Errors in the technique are estimated by applying it to optically-visible T Tauri stars in the Taurus and Lupus dark clouds. In these regions, optical spectroscopy is available to check the accuracy of the solutions from the algorithm. At present, age estimates are made from spectro-

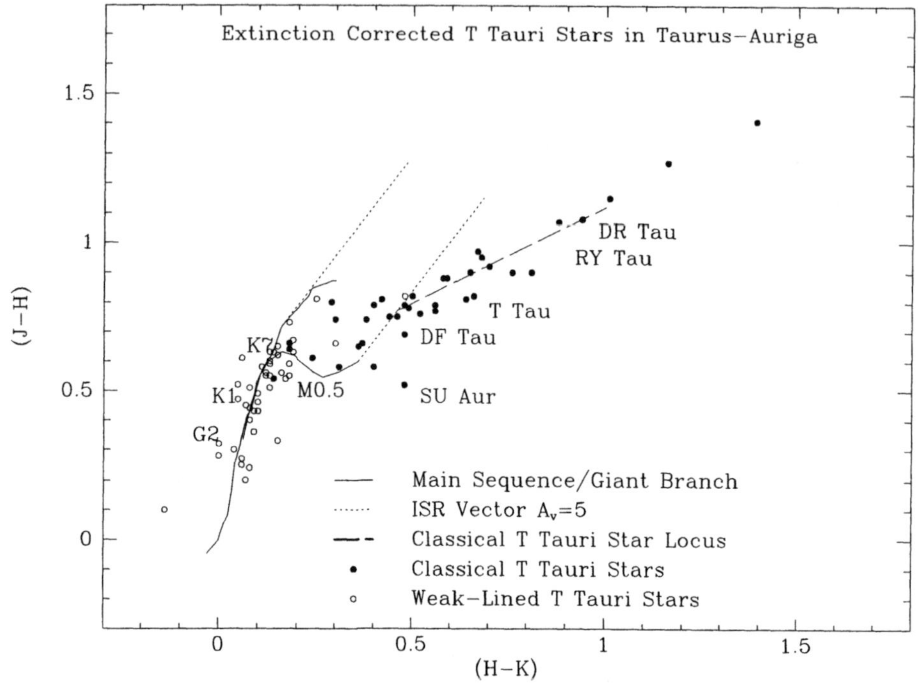

Fig. 2. Near-IR color-color diagram for *dereddened* T Tauri stars in Taurus-Auriga. Note the locus of classical T Tauri stars. The five objects indicated are those for which IR veiling measurements have been obtained (see section IV).

scopic studies of the optically-visible population associated with the embedded cluster.

3. Infrared Classification Spectroscopy

However photometric techniques have limitations: i) they provide no information about cluster age; ii) errors in derived masses are difficult to estimate; and iii) they are not applicable to YSOs with extreme IR excesses, whose near-IR spectral energy distributions cannot be modeled in a standard way as circumstellar disks. Such infrared excesses may be indicative of rapid disk accretion and/or infalling envelopes (ie flat spectrum or Class I sources; Calvet *et al.*, 1994; Kenyon *et al.*, 1993). To remove ambiguities associated with photometric results, we need to place optically-invisible objects on the HR diagram in order to compare them directly with PMS evolutionary tracks. We can combine extant infrared photometry and infrared spectra to determine luminosities and effective temperatures (and thus stellar masses and ages) for optically-obscured YSOs. The IR spectra enable us to determine the spectral type of each YSO providing stellar effective temperatures, intrinsic stellar colors for dereddening, and bolometric corrections appropriate for determining stellar luminosities. Such

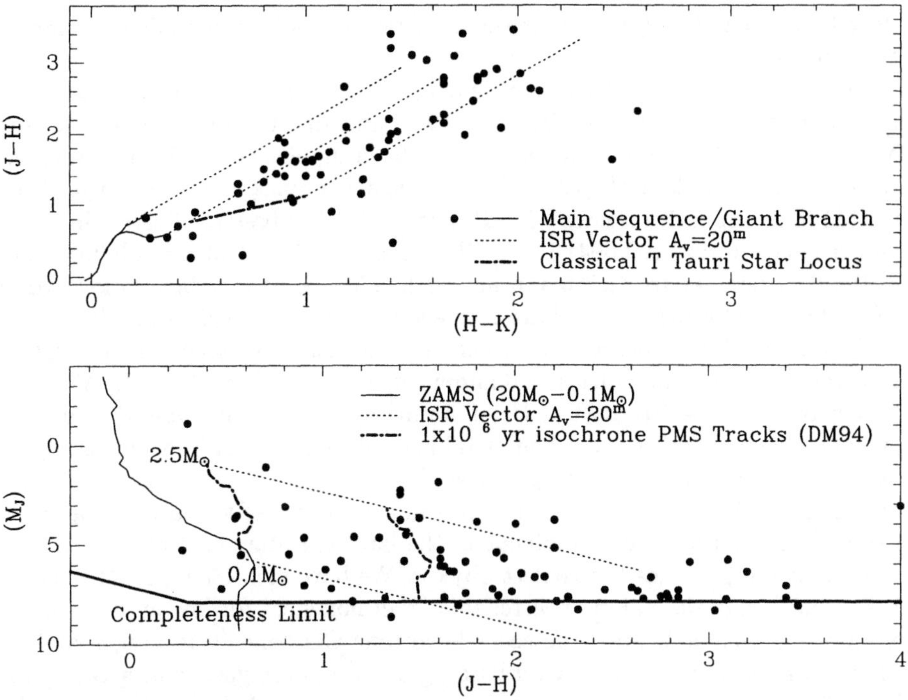

Fig. 3. Near-IR photometry for the cluster associated with the NGC2024 nebula taken with the NICMASS camera at the MDM Observatories 1.3m McGraw Hill Telescope: A.) Near-IR color-color diagram. Intrinsic colors of main sequence, giant, and classical T Tauri stars are shown with the direction and magnitude of the reddening vector for $A_v=20^m$. Many of the embedded cluster members appear to be heavily reddened classical T Tauri stars. B.) Near-IR color-magnitude diagram. Shown is the 1x10^6 yr isochrone from the PMS tracks of D'Antona and Mazzitelli (1994) and the main sequence for comparison. The interstellar reddening vector is projected from the 2.5M$_\odot$ and 0.1M$_\odot$ mass limits up to the points of $A_v=20^m$. The *completeness* limit of the survey ($J=16.5^m$; $H=15.9^m$) is indicated.

data allow i) direct estimates of the embedded cluster age essential for deriving stellar masses based on photometric techniques; ii) error estimates for photometrically derived stellar masses necessary for the proper statistical comparison of mass spectra to the solar neighborhood IMF; and iii) insight into the nature of the Class I/flat-spectrum sources.

K-band spectroscopy has been successfully used in pioneering studies of L1641 North (Hodapp and Deane, 1993) and ρ Oph (Greene and Meyer, 1995). We have explored the H-band in order to test its suitability as a spectral region for stellar classification by observing a large number of spectral standards using the Fourier Transform Spectrometer (FTS) on the Kitt Peak National Observatory (KPNO) 4m telescope. The choice of the H-band reflects a compromise between minimizing extinction (smaller at K; larger at J) and complications due to IR excess emission (smaller at J; larger at K). Previous work in the H-band done by Origlia *et al.* (1993) at lower resolution was a helpful guide in designing this survey.

The source list was drawn from fundamental lists of MK standards; for early-type stars Morgan, Abt, and Tabscott (1978) and for late-type stars Keenan and McNeil (1989). For stars later than M3 we have supplemented this list from the work by Kirkpatrick *et al.* (1991). This list spans a range of spectral types from O5-M5 and luminosity classes I-II, III, and IV-V. We observed 88 stars in the J- and H-bands during 1993-94. The S/N ranges from 75-100 at a resolving power of R=3500 (similar to Kleinmann and Hall, 1986, K-band atlas). One-third of the data are fully reduced and are represented in the following analysis.

The H-band contains several atomic and molecular photospheric absorption features which are temperature and luminosity sensitive (Figs. 4A, 4B). All early-type stars show the Brackett series of atomic hydrogen, while only O and early B stars have helium features. The metal lines of Mg, Al, and Si begin to appear at G0 and tend to grow in strength toward late-type stars. The molecular bands of OH and CO are the strongest features in the spectra of the latest-type stars. However, not all features prominent in the late-type stars continue to grow in EW as one approaches the coolest objects. We have identified an OH band at 1.688 μm that does, making it a reliable "thermometer" for low mass stars. As a surface gravity indicator, the second-overtone CO band $[\nu, \nu' = 6, 3]$ at 1.62 μm is most sensitive. This is a very strong feature in the giant star spectra, yet is barely detectable in the late-type dwarf stars. Similar behavior is seen in the first overtone CO features in the K-band (Kleinmann and Hall, 1986).

We summarize these trends in Figure 5 plotting $\frac{EW[OH]}{EW[HI]}$ vs. $\frac{EW[CO]}{EW[Mg]}$. Here the ordinate measures the surface gravity and the abscissa tracks photospheric temperature. We offer this as a preliminary two-dimensional classification plane.

Table 2. Photospheric Absorption Features.

Central Wavelength [μm]	Prominent Features	Comments
1.556-1.736	HI	Brackett Series; seen in early-type stars, strongest at A0.
1.575	Mg	temp. sensitive; appears at G0, increasing in strength toward later-type stars.
1.589	Si	temp. sensitive; appears at G0, increasing in strength toward later-type stars.
1.618	CO [ν,ν'=6,3]	lum. sensitive; appears at K0, increasing in strength toward later-type giants.
1.673	Al	temp. sensitive; appears at G0, strongest at K7.
1.688	OH	temp. sensitive; appears at K0, increase in strength toward the coolest stars.
1.700	He	seen only in O and early B stars.
1.709	Mg	temperature sensitive; appears at G0, strongest at K7.

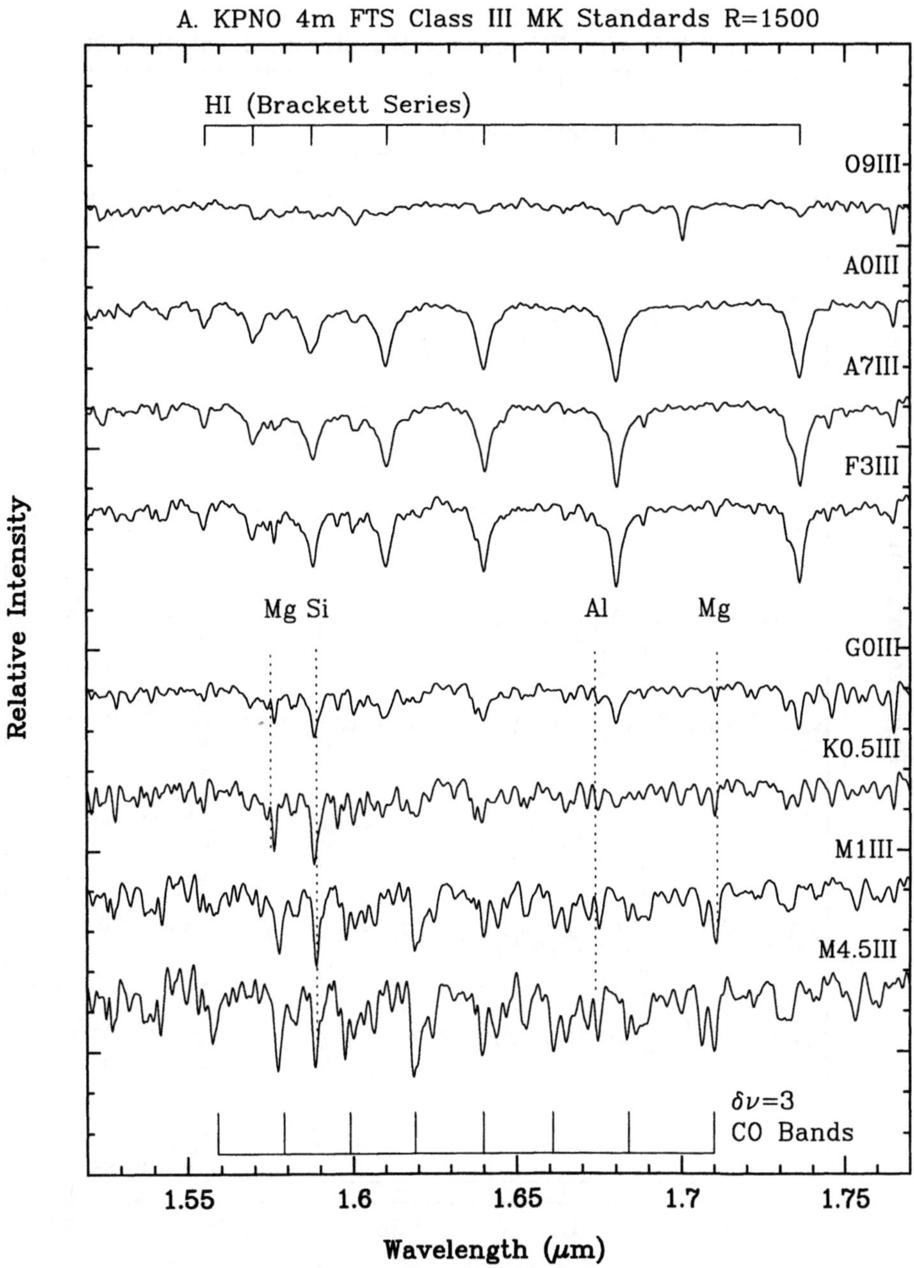

Fig. 4. a) KPNO 4m FTS MK Spectral Standards Convolved to $R = 1500$: These data have been corrected for telluric absorption and continua have been normalized. Luminosity Class III Sequence.

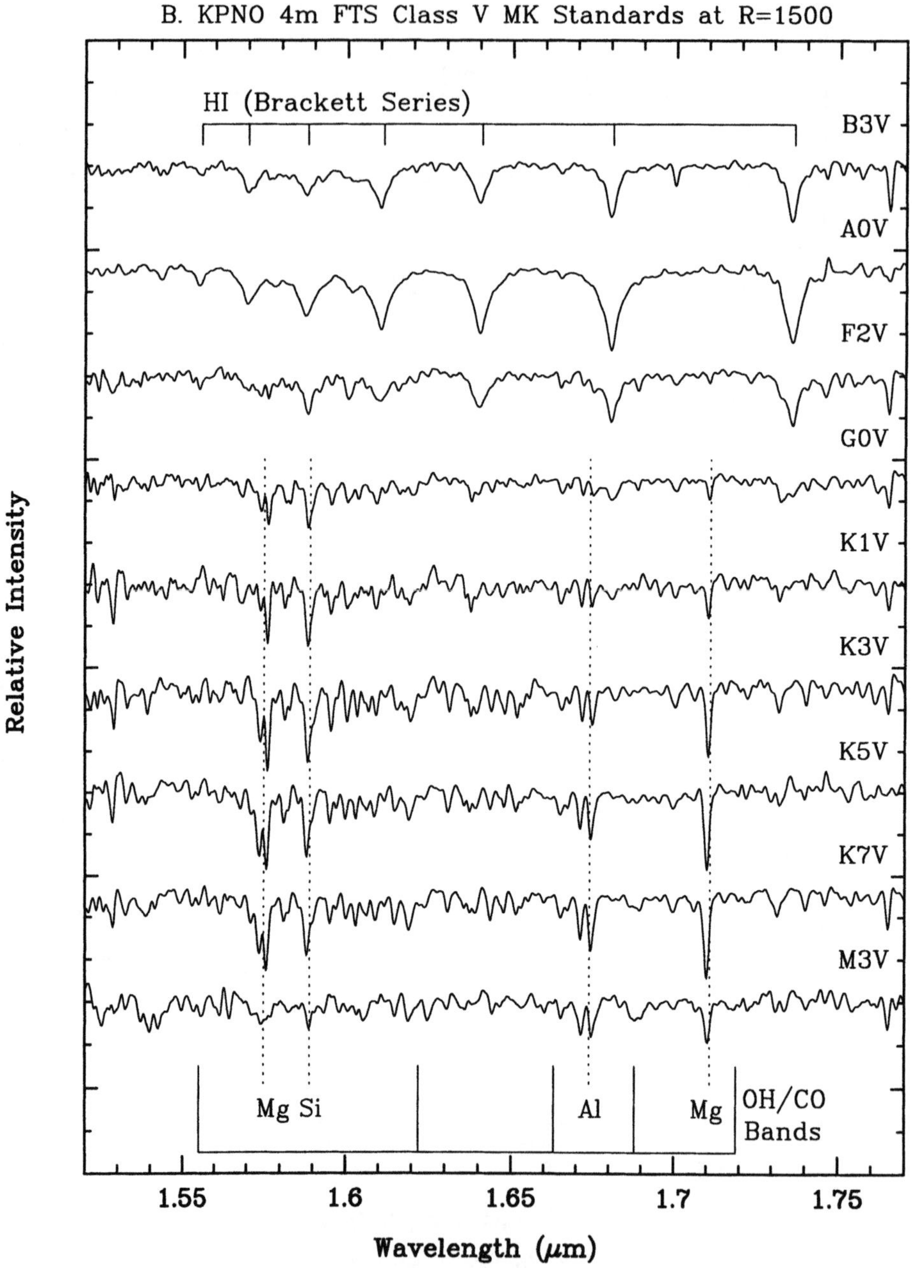

Fig. 4. b) Same as Figure 4a, Luminosity Class V Sequence.

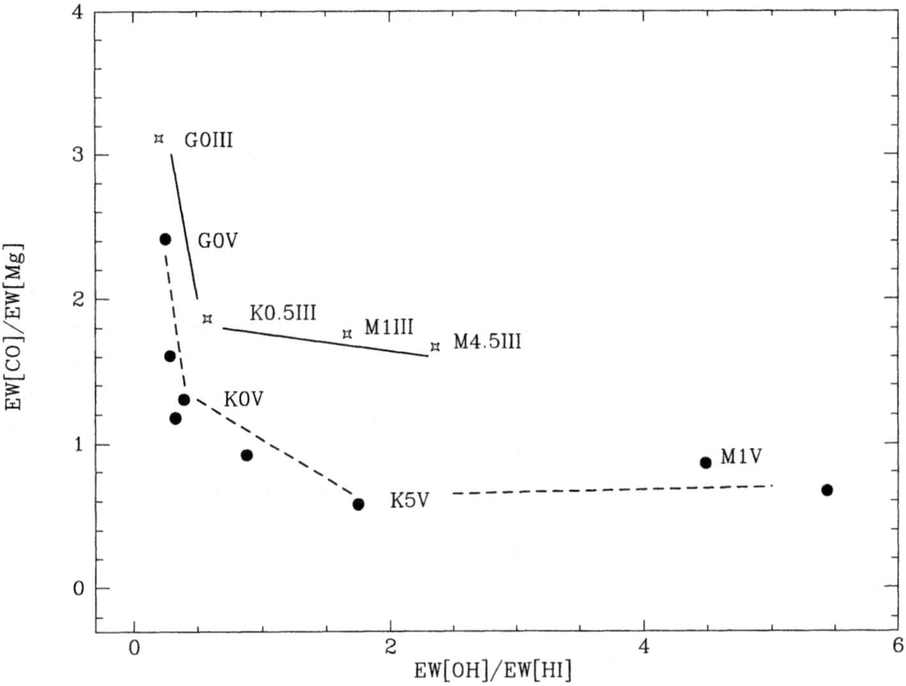

Fig. 5. Two-Dimensional Classification of Late-Type Stars in the H-Band. Here the x-axis traces photospheric temperature while the y-axis indicates luminosity class.

4. High-Resolution Infrared Spectroscopy

In the optical spectra of T Tauri stars, continuum excess emission associated with accretion related phenomena complicates the derivation of spectral types from moderate resolution data. Excess emission is also observed in the near-IR spectral energy distributions of young stellar objects. If there are strong gradients in this near-IR continuum emission, or if it is accompanied by strong emission lines, the line ratios identified above could give erroneous results when used to infer spectral types. In order to investigate complications that arise due to the PMS nature of sources found in embedded clusters we obtained $R = 7100$ spectra in the 1.61–1.64 μm region of several well-studied optically-visible young stars along with a grid of spectral standards. We used our IR camera "NICMASS" (Skrutskie *et al.*, 1995), as the focal plane detector with the 0.9m Coudé Feed telescope/spectrograph at Kitt Peak National Observatory to acquire spectra of both weak-lined T Tauri stars (WTTS) which lack evidence of accretion and classical T Tauri stars (CTTS) with a variety of accretion rates.

Comparison of the PMS stellar spectra with our MK standards confirms that spectral types derived from IR spectra agree with those derived in the optical for young stars. In Figure 6A we show a late-type WTTS spectrum along with dwarf and giant star spectra of similar spectral type. Note the features in the WTTS are very similar to those in the dwarf star, however, there are some recognizable "giant-like" characteristics indicating lower surface gravity. We observe this in stars as early as K0 while in earlier-type stars, the giant and dwarf star spectra are not distinguishable.

We can also use these data to study the infrared excess emission by looking at the continuum "veiling" of photospheric lines. The veiling parameter, r_λ is the ratio of excess emission to pure stellar emission such that:

$$r_\lambda = (EW_{\lambda,unveiled}/EW_{\lambda,veiled}) - 1$$

In Table II we present the veiling measurements from 1.61 – 1.64μm for five CTTS. The veiling estimates from optical spectra are given for wavelengths near 0.6 μm (Hartigan *et al.*, 1995; "*" indicates Basri and Batalha, 1990) and the presence or absence of IR companions is also noted (Simon *et al.*, 1995). WTTS of similar spectral type were used as comparison stars for the equivalent width analysis; for DF Tau and DR Tau we used Hubble 4 (K7V), for T Tau and RY Tau we used V773 Tau (K2V), and for SU Aur we used HDE283572 (G5IV). Although V773 Tau does have evidence of accretion activity (Hartigan *et al.*, 1995), it does *not* have a measurable near–infrared excess. The features used in this analysis are visible in Figure 6B.

Table 3. IR Veiling Measurements.

CTTS	Spectral Type	Double	$r_{0.6\mu m}$	$r_{1.62\mu m}$
SU Aur	G2	No	0.0*	0.87 ± 0.18
T Tau	K1	Yes	0.0*	0.61 ± 0.16
RY Tau	K1	No	0.1	3.2 ± 0.8
DR Tau	K7	No	9.2	4.0 ± 2.0
DF Tau	M0.5	Yes	0.66	0.59 ± 0.19

How do these spectroscopic measurements of continuum excess compare with estimates made from near-IR photometry? In Figure 2 we showed the locations of *dereddened* T Tauri stars in Taurus-Auriga in the near-infrared (H-K) vs. (J-H) color-color diagram. Extinction values were calculated from color excesses derived from observed (R-I) indices. The extinction corrected classical T Tauri stars lie along a well-defined locus in this diagram. This correlation extends beyond the maximum IR excess that can be understood through standard models of accretion disks (e.g. Lada and Adams, 1992). The veiling measurements tabulated above are well-matched to the intrinsic IR excesses of the sources labeled in Figure 2. These data provide crucial constraints on future models for the SEDs of these sources. For example, the 1.62 μm excess emission probes the inner regions of the circumstellar environment where one might expect to find evidence for holes in the disks of T Tauri stars.

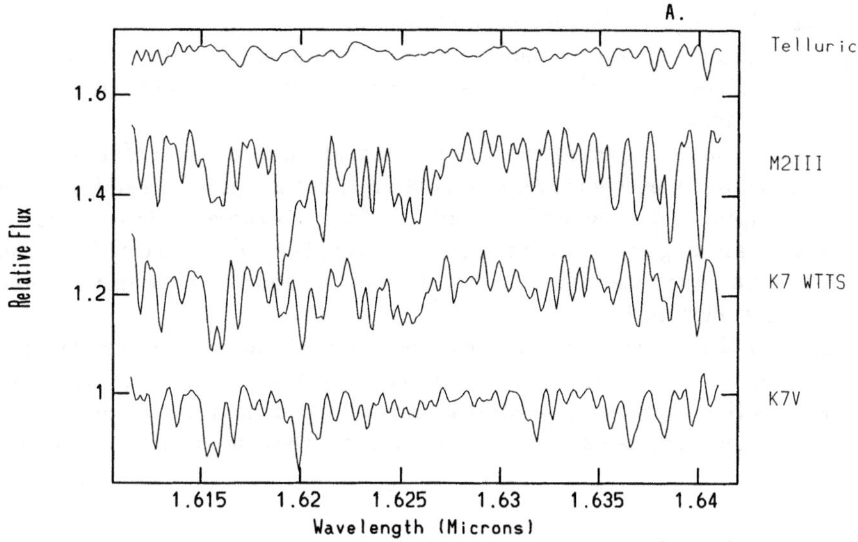

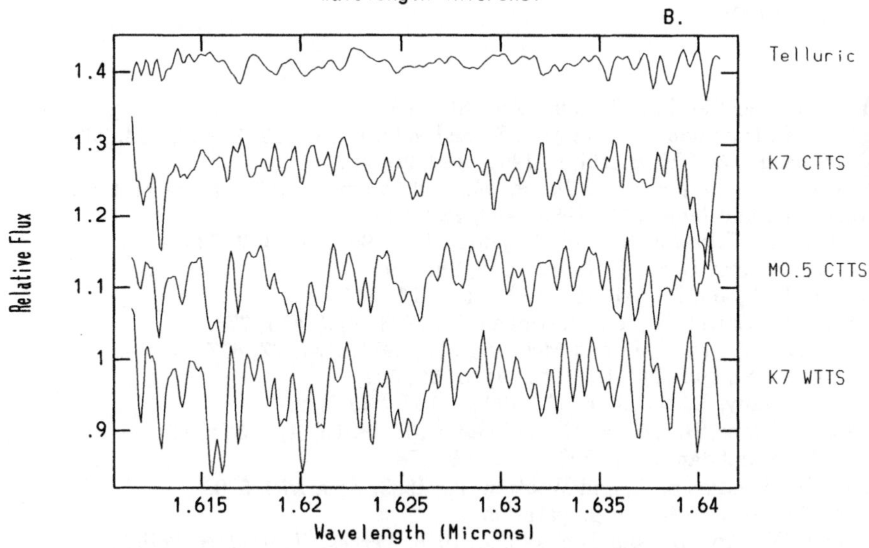

Fig. 6. High resolution $R = 7100$ near-infrared spectra obtained with "NICMASS" at the KPNO Coudé Feed telescope/spectrograph: A.) The spectrum of Hubble 4 (K7 WTTS) plotted with a dwarf and giant star of similar spectral type. The CO band at 1.62 μm is prominent in the giant star and the OH features are visible in all spectra. B.) Along with Hubble 4 are plotted two classical T Tauri stars for which veiling estimates were obtained; DF Tau (M0.5) and DR Tau (K7). The Ca doublet at 1.617 μm is visible along with several OH features from 1.63–1.64 μm.

5. Summary

1. Near-IR photometric survey data can provide an initial reconnaissance of mass distributions in embedded clusters.
2. The H-band photometric spectral region contains a variety of temperature and luminosity sensitive features that are suitable for two-dimensional spectral classification.
3. High resolution spectroscopy indicates that spectral types derived in the infrared agree with spectral types obtained in the optical. In addition, PMS stars appear to have surface gravities closer to those of dwarf stars than giants, although they are intermediate. From these data we can derive near-IR veiling estimates and place constraints on models for the distribution of circumstellar dust.
4. H- and K-band spectroscopy now underway will allow us to construct HR diagrams for deeply embedded young clusters. When compared with theoretical PMS isochrones, these HR diagrams will yield mass and age distributions for clusters of stars likely to be found in the earliest stages of PMS evolution.

References

Basri, G., and Batalha, C., 1990, *ApJ*, **363**, 654.
Calvet, N., Hartmann, L., Kenyon, S., and Whitney, B., 1994, *ApJ*, **434**, 330.
D'Antona, F., and Mazzitelli, I., 1994, *ApJ*, **90**, 467.
Dougados, C., Carpenter, J., Meyer, M., and Strom, S., 1995, in preparation.
Greene, T., and Meyer, M., 1995, *ApJ*, **450**, 233.
Hartigan, P., Edwards, S., and Ghandour, L., 1995, *ApJ*, **452**, 736.
Hodapp, K., and Dean, J., 1993, *ApJS*, **88**, 119.
Keenan, P.C., and McNeil, R., 1989, *ApJS*, **71**, 245.
Kenyon, S., Calvet, N., and Hartmann, L., 1993, *ApJ*, **414**, 773.
Kirkpatrick, J., Todd, J., and McCarthy, D., 1991, *ApJ*, **77**, 417.
Kleinmann, S., and Hall, D., 1986, *ApJS*, **62**, 501.
Lada, E., Bally, J., and Stark, A., 1991a, *ApJ*, **368**, 432.
Lada, E., Evans, N., DePoy, D., and Gatley, I., 1991b, *ApJ*, **371**, 171.
Lada, C., and Adams, F., 1992, *ApJ*, **393**, 278.
Loren, R., Sandqvist, S., and Wootten, A., 1983, *ApJ*, **270**, 620.
Meyer, M., *et al.*, 1995, in preparation.
Morgan, W., Abt, H., and Tabscott, J., 1978, *Yerkes Observatory Bulletin.*
Origlia, L., Moorwood, A., and Oliva, E., 1993, *A&A*, **280**, 536.
Simon, M., Ghez, A.M., Leinert, C., Cassar, L., Chen, W.P., Howell, R.R., Jameson, R.F., Mathews, K., Neugebauer, G., Richichi, A., 1995, *ApJ*, **443**, 625.
Skrutskie, M.F., Meyer, M.R., and Coutu, 1995, in preparation.
Strom, K., Kepner, J., and Strom, S., 1995, *ApJ*, **438**, 813.
Taylor, K., and Storey, J., 1984, *MNRAS*, **209**, P5.
Wilking, B., and Lada, C., 1983, *ApJ*, **274**, 698.
Wilking, B., *et al.*, 1995, in preparation.
Wolf, G., Lada, C., and Bally, J., 1990, *AJ*, **100**, 1892.

Multiplicity in the Ophiuchus and Taurus Star Forming Regions

Michal Simon

Astronomy Program, SUNY, Stony Brook, NY, 11794-2100, USA

We describe[1] an IR lunar occultation and direct imaging search for companions in the Ophiuchus star forming region and update a similar search of the Taurus region. The search is sensitive to companions in the angular separation range 0.005 to 10″. In Ophiuchus we surveyed 35 young star targets; this sample contains at least 10 binaries, 2 triples, and one quadruple. Ten of the companion stars are newly discovered. In Taurus, the survey now includes 47 systems among which there are at least 22 binaries and 4 triples. Only 2 companion stars are newly identified because there is strong overlap with prior work. All the triples and quadruple are hierarchical. The observed binary frequency in Ophiuchus, in the 3 to 1400 AU range of separations, is *at least* 1.1 ± 0.3 that of the nearby solarlike stars. This value is a lower bound because we make no corrections for incompleteness. In Taurus, in the same range of separations, the observed binary frequency is *at least* 1.6 ± 0.3 that of the nearby solarlike stars. This value extends Ghez *et al.*'s (AJ, **106**, 2005) and Leinert's *et al.*'s (A&A **278**, 129) determination of an excess binary frequency to 3 AU separation. We used the WT/TT type and the K-L color index to distinguish between systems with and without inner disks. We find no convincing difference in the binary frequency or distribution of separations of the systems with and without inner disks. The 1.3 mm continuum emission of the single systems exceeds that of the multiples suggesting that their extensive outer disks are more massive. The specific angular momenta of the binaries overlap those of molecular cloud cores measured by Goodman *et al.* (ApJ **406**, 528).

[1] The paper presented at the conference is essentially identical in content to the paper "A Lunar Occultation and Direct Imaging Survey of Multiplicity in the Ophiuchus and Taurus Star Forming Regions" by M. Simon *et al.* scheduled for the April 20, 1995 issue of the *Astrophysical Journal*.

A Massive "Core" Associated with the UC HII Region G31.41+0.31

R. Cesaroni[1], L. Olmi[2], C.M. Walmsley[3], E. Churchwell[4], and P. Hofner[4]

[1] Osservatorio Astrofisico di Arcetri, Largo E. Fermi 5, I-50125, Firenze, Italy
[2] I Physicalisches Institut der Universität zu Köln, Zülpicherstrasse 77, D-50937 Köln, Germany
[3] Cornell University, Arecibo Observatory, P.O. Box 995, Arecibo, Puerto Rico 00613
[4] Washburn Observatory, University of Wisconsin-Madison, 475 North Charter St., Madison, WI 53706, USA

Summary. We have used the IRAM Plateau de Bure Interferometer to observe the ground state and vibrationally excited lines of methyl cyanide towards the UC HII region G31.41+0.31. We also obtained a map of the continuum emission at a frequency of 110 GHz. We detect a hot molecular core in the emission both of methyl cyanide and of the 110 GHz continuum. We estimate a temperature of 200 K and a mass of 1000 $M_\odot$ for this compact massive region. We also detect a velocity gradient or shift across the methyl cyanide core whose origin could be due to rotation.

1. Introduction

Ultracompact HII regions (often abbreviated as UC HII's) serve as a tracer for massive star formation throughout our galaxy. Several recent reviews (e.g. Churchwell 1991) summarise our present knowledge of their morphology and physical characteristics. There is some controversy about their lifetime but there seems little doubt that they represent the first easily observable phase in the evolution of an O star. There is also little doubt that they form from extremely dense compact molecular clumps or cores and over the past few years we have been attempting to determine the properties of such molecular cores. This work is described for example in Churchwell et al. (1990), Olmi et al. (1993), and Cesaroni et al. (1994a). One of the results of these studies has been that we have found close to but not coincident with the UC HII's a number of molecular regions with properties analogous to those of the Orion hot core. These are characterised by much higher temperatures than the molecular gas in their surroundings (temperatures of order 100–200 K as compared to 20–30 K). They are also extremely dense and have masses of at least 10 $M_\odot$.

2. The G31.41+0.31 Region

One of the most interesting objects of this type which we discovered was the molecular core associated with the UC HII region G31.41+0.31. NH_3(4,4) VLA observations by Cesaroni et al. (1994a) showed that there was a very hot ammonia clump (temperature 120 K, ammonia column density 10^{19} cm^{-2}, diameter 0.1 pc) associated with this UC HII region. Moreover Churchwell et al. (1992)

found the source to have easily observable lines of methyl cyanide (CH_3CN). Methyl cyanide is a symmetric top with several transitions of differing excitation at neighbouring frequencies. It has thus found applications as a "molecular thermometer" and we therefore decided to observe both ground and vibrationally excited lines of this species using the IRAM Plateau de Bure Interferometer at a frequency of 110 GHz (2.7 mm). Our angular resolution was approximately 2″ which corresponds to 0.08 pc at the distance of G31.41+0.31 (7.9 kpc). In addition to the maps in the various line transitions, we obtained a map of the continuum emission at 110 GHz.

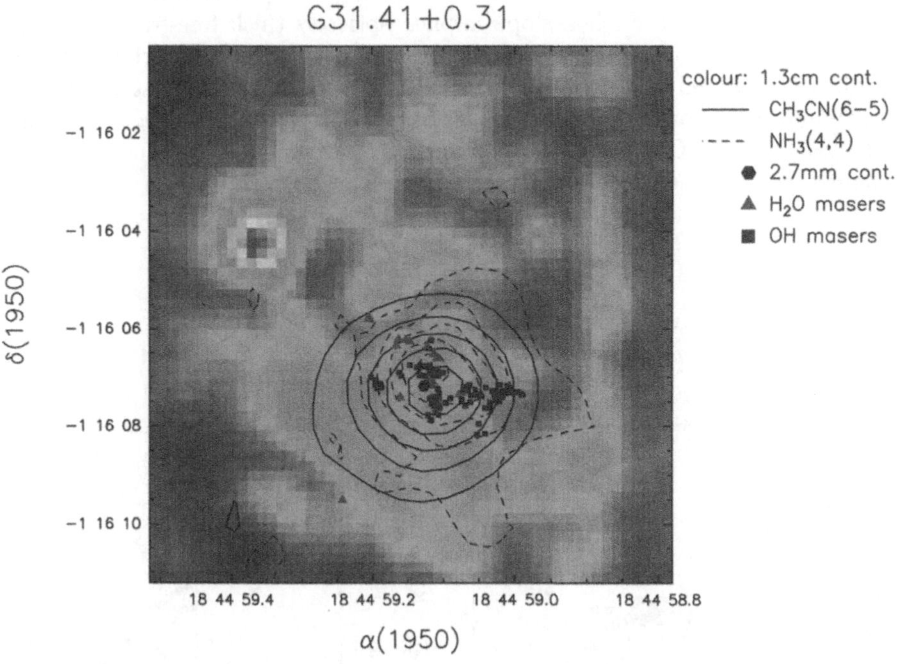

Fig. 1. Map of 1.3 cm continuum emission (colour) towards G31.41+0.31 compared with NH_3(4,4) (dashed contours) and CH_3CN(6–5) (full contours). The position of the 2.7 mm continuum peak is indicated by the black exagon. The red triangles and blue squares represent respectively the H_2O and OH maser spots detected by Hofner & Churchwell (in prep.) and Gaume & Mutel (1987). [This figure is presented in color on CD-ROM.]

In Figure 1, we compare the 1.3 cm continuum map (colour map) of G31.41 with the NH_3(4,4) (dashed contours) and CH_3CN(6–5) emission (full contours); also shown are the H_2O (Hofner & Churchwell in prep.) and OH (Gaume & Mutel 1987) maser spots, represented respectively by the red triangles and blue squares. As we had expected, the methyl cyanide clump coincides with the hot core seen in NH_3(4,4). The methyl cyanide emission appears slightly extended in our beam implying a core diameter of order 0.04 pc. We detect vibrationally excited levels of methyl cyanide more than 500 K above ground and estimate the gas temperature in the line emitting region to be 200 K.

2.1 The Continuum Emission

More surprising was the detection of a 2.7 mm continuum source of approximately the same size as that seen in CH_3CN and which is also coincident with the ammonia hot core: the 2.7 mm continuum peak position is indicated in Figure 1 with a black exagon. There is no corresponding continuum source seen on our VLA 1.3 cm maps above a 3σ limit of 0.04 Jy: this implies a spectral slope of at least 1.9. Such a slope implies optically thick free-free emission at 2.7 mm, but this conclusion contradicts the low deconvolved brightness temperature measured at this frequency ($\sim$ 40 K). We conclude that the 2.7 mm continuum emission is due to heated dust. We derive on this basis a mass of 1000 $M_\odot$ for the G31.41 hot core although this involves "an educated guess" about the 2.7 mm dust emissivity.

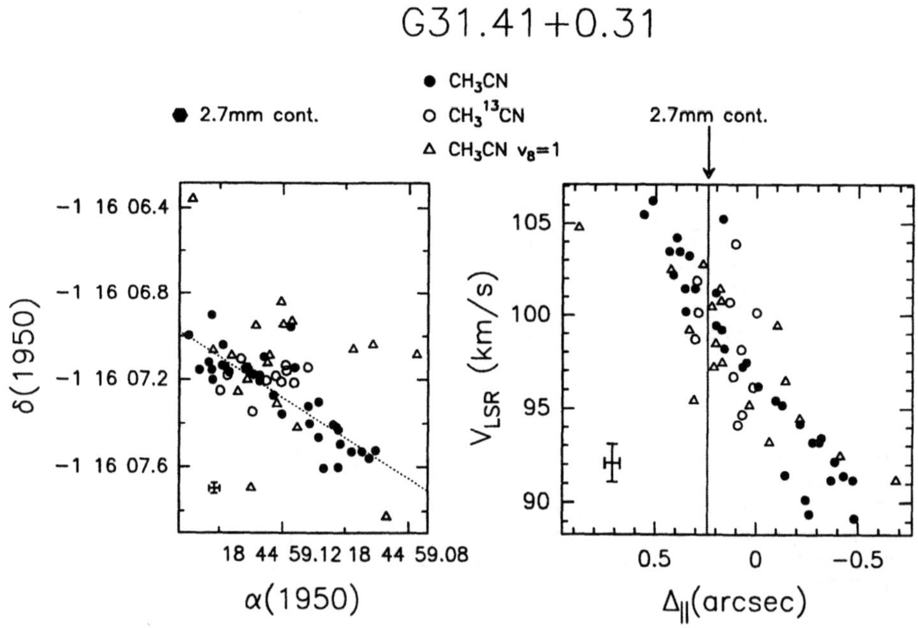

Fig. 2. Left panel shows peak positions measured in different velocity channels for ground state and vibrationally excited methyl cyanide towards G31.41+0.31; the red hexagon indicates the position of the 2.7 mm continuum peak. The right panel shows a plot of channel velocity against the angular offset along the axis denoted by the dotted line in the left panel; the offset corresponding to the 2.7 mm continuum peak is shown by the red line. The crosses in the bottom left of both panels show typical error bars. [This figure is presented in color on CD-ROM.]

2.2 The $CH_3CN(6–5)$ Transition

It is also intriguing that when we map the peak positions determined in different velocity channels and different K-components of methyl cyanide, we see an elongated structure of size approximately 1.2″ (see the left panel of Figure 2). Moreover, when we additionally plot channel velocity against angular offset along the major axis of this structure (right panel of Figure 2), we see evidence for a velocity gradient of 400 $km\,s^{-1}\,pc^{-1}$. The interpretation of this gradient (or velocity shift) is unclear but two explanations seem reasonable: rotation or outflow.

In the former case, the inferred dynamical mass is of the same order as that inferred for the source of the 2.7 mm continuum emission: as a consequence, the disk must be unstable, i.e. less than $\sim 10^3$ yrs old, because a point mass of $\sim 2000\ M_\odot$ in its centre is needed to stabilise it.

Alternatively, if the outflow hypothesis is correct, one can derive a lifetime of ~ 2500 yrs, a mass loss rate of the star powering the outflow of 0.05 $M_\odot\ yr^{-1}$ and a mechanical energy output of 2 $10^5\ L_\odot$, comparable to the IRAS luminosity of the object.

On the basis of the present data it is impossible to decide which of the two interpretations is correct. We can only conclude that we are observing a rather short-lived dynamically unstable massive object.

Finally, we note that the results of this study are given in more detail in an article by Cesaroni et al. (1994b).

References

Cesaroni R., Churchwell E., Hofner P., Walmsley C.M., Kurtz S. (1994a): A&A **288**, 903

Cesaroni R., Olmi L., Walmsley C.M., Churchwell E., Hofner P. (1994b): with the UC HII region G31.41+0.31. ApJL **435**, L137

Churchwell E., Walmsley C.M., Cesaroni R. (1990): A&AS **83**, 119

Churchwell E. (1991): NATO ASI Series, C.J. Lada & N. Kylafis edts., Kluwer

Churchwell E., Walmsley C.M., Wood D.O.S. (1992): A&A **253**, 541

Gaume R.A., Mutel R.L. (1987): ApJS **65**, 193

Olmi L., Cesaroni R., Walmsley C.M. (1993): A&A **276**, 489

Prediction of Stellar Mass in Star Formation: Theory and Its Application to the Orion A Cloud

T. Nakano[1], T. Hasegawa[2], C. Norman[3]

[1] Nobeyama Radio Observatory, National Astronomical Observatory,Nobeyama, Minamisaku, Nagano 384-13, Japan
[2] Institute of Astronomy, University of Tokyo, Osawa, Mitaka, Tokyo 181, Japan
[3] Department of Physics and Astronomy, Johns Hopkins University and Space Telescope Science Institute, Baltimore, MD 21218, U. S. A.

Summary. We have developed a way to estimate the mass of a star from the physical quantities of the cloud core in which it forms. Because matter falls onto a stellar core mainly through an accretion disk, a significant fraction of the mass outflow and ultraviolet radiation escapes from the inner region without interacting with the infalling matter and disturbs the cloud core matter which has not yet contracted much. A high-velocity mass outflow gradually pushes out the surrounding matter forming a thin dense shell. On the other hand, as the central star grows, a compact HII region develops and pushes out the surrounding matter also forming a thin dense shell. A considerable fraction of the initial core matter is blown off when the radius of the bubble or the HII region inside the shell grows to the initial radius of the cloud core, and the remaining matter disperses because it is no longer gravitationally bound. In this way the supply of matter from the cloud core to the disk stops. Consequently accretion onto the star also stops, and the mass of the forming star is fixed. The stellar mass determined in this way is a function of the core density n_c, the mass inflow rate $\dot{M}_I$, $\dot{M}_O/\dot{M}_I$, and M_c/M_J, or a function of n_c, M_c, $\dot{M}_O/\dot{M}_I$, and M_c/M_J, where $\dot{M}_O$ is the mass outflow rate, M_c is the cloud core mass, and M_J is the generalized Jeans mass for the core. When the mass outflow is dominant, we have $M_* \propto M_c^{7/6} n_c^{1/12}$ with the proportionality coefficient dependent on $\dot{M}_O/\dot{M}_I$, and thus the stellar mass is almost independent of n_c. Therefore, the star formation efficiency M_*/M_c is mainly determined by $\dot{M}_O/\dot{M}_I$ and is only weakly dependent on the core parameters M_c and n_c; we obtain $M_*/M_c \sim 0.04$ for $\dot{M}_O/\dot{M}_I = 0.1$ around $M_c \sim 100 M_\odot$. Applying this result to the observed cloud cores in the Orion A molecular cloud by assuming that each core does not contain subclumps we estimate the stellar mass and the initial mass function of stars (IMF) expected in this cloud. As long as $\dot{M}_O/\dot{M}_I \gtrsim 0.02$, the mass outflow is more efficient than the HII region in determining the stellar mass for all the Orion A cores. The IMF at $M_* \gtrsim 4 M_\odot$ can be approximated by a power law $dN_*/d\log M_* \propto M_*^{-1.7}$ for $\dot{M}_O/\dot{M}_I = 0.1$, which is in reasonable agreement with the IMF of field stars $\propto M_*^{-1.5}$ at $M_* \gtrsim 3 M_\odot$.

1. Introduction

Mechanisms of determining the stellar mass and the initial mass function of stars (IMF) are one of the least understood issues in star formation. With a simple assumption that the stellar mass is equal to some constant fraction of the mass of a parent cloud core from which the star forms, some attempts have been made to derive the mass function of cloud cores and the resultant IMF with some assumptions on the process of fragmentation or coagulation. Although the stellar mass should depend on the core mass, not all of the core matter becomes a star in general, and there has been no theoretical background to assume that the fraction of matter transformed into a star should be a universal constant.

Hayashi & Nakano (1965) showed that in the contraction of an interstellar cloud core, a small stellar core nearly in hydrostatic equilibrium forms first and the stellar core (protostar) gradually grows by the inflow of matter. This has been confirmed by computer simulations of cloud contraction (e.g., Nakano, Ohyama, & Hayashi 1968, 1970; Larson 1969, 1972; Narita, Nakano, & Hayashi 1970). The stellar mass must therefore be determined by mechanisms that stop the mass supply to the stellar core. However, little has been understood of such mechanisms and also of mechanisms of determining the star formation efficiency, i.e., the fraction of cloud matter converted into stars.

The mechanisms of stopping mass supply have been discussed in relation to the upper limit to the stellar mass. Attention has been paid mainly to the effects of stellar radiation. Because the luminosity-to-mass ratio is usually an increasing function of the stellar mass, the effect of radiation on the contracting envelope increases relative to the effect of stellar gravity as the stellar core grows. Larson & Starrfield (1971) investigated various effects of stellar radiation on the contracting envelope; radiation pressure, pressure of the gas heated by the radiation, and ionization of the gas. Wolfire & Cassinelli (1987) made numerical simulations assuming spherically symmetric contraction of the envelope by taking into account the effects of thermal radiation of grains and erosion of grains by collision with gas particles in addition to the effects mentioned above. They found that even with a grain abundance about 0.1 times the abundance in the ordinary interstellar clouds a mass inflow rate $\dot{M}_{\rm I}$ greater than 3×10^{-3} and $5 \times 10^{-3} M_\odot\ {\rm yr}^{-1}$ is required for accretion onto a stellar core of 60 and $100 M_\odot$, respectively. Accretion is much more difficult with the ordinary grain abundance. However, contraction of the cloud easily deviates from spherical symmetry by cloud rotation, magnetic fields, and small initial asymmetry in the gas distribution. Nakano (1989) showed that deviation from spherical symmetry makes accretion much easier. For instance, $\dot{M}_{\rm I}$ slightly greater than $10^{-4} M_\odot\ {\rm yr}^{-1}$ is enough for accretion onto a stellar core of $100 M_\odot$ even with the ordinary grain abundance.

In most of these works the stellar radiation is given a role of stopping the inflowing matter. However, because most of the matter flows onto the stellar core through an accretion disk except for very early accretion phases, a significant fraction of the stellar radiation escapes from the inner region without being absorbed by the inflowing matter. The cloud core matter which has not yet contracted much is gradually pushed out by the HII region formed by such stellar ultraviolet radiation.

A moderately collimated mass outflow is often observed around a mass-accreting young star (e.g., the review of Edwards, Ray, & Mundt 1993) and the mass loss rate $\dot{M}_{\rm O}$ as high as $0.1 \dot{M}_{\rm I}$ is expected (Shu et al. 1988). The outflow with this mass loss rate and the velocity $\sim$ 200km s^{-1} observed near its base can supply huge momentum to the surrounding medium and may be able to push out the cloud core matter which has not yet contracted much.

The scenario we develop in this paper is that the growth of the star stops when the mass outflow and the HII region blow off the parent cloud core and as a result the supply of matter to the accretion disk and then to the star ceases.

2. The Model: An Overview

Molecular clouds appear to have a hierarchical structure: high-density clumps in a cloud, and even higher-density subclumps in a clump, etc. We consider here a part of the cloud which is considerably denser than the surrounding, is gravitationally bound, and contains no subclumps, and in which a single star forms. In this paper we call such a part of the cloud a cloud core.

A stellar core forms and grows by mass inflow in a cloud core. Because the mass inflow to the stellar core is mainly through the accretion disk except for very early stages, the mass outflow from (the vicinity of) the central star and the HII region can disturb the surrounding cloud core matter which has not yet contracted much even in the existence of mass inflow. The mass of the accretion disk may be kept small ($\sim$ 0.1) compared with the mass of the central star because instabilities of a massive disk would raise the mass accretion rate onto the star from the disk (Boss 1984; Larson 1984; Lin & Pringle 1987; Anthony & Carlberg 1988).

The mass outflow pushes out the surrounding cloud core matter and forms a thin dense shell. Around the time when the size of the bubble inside the shell grows to the initial size of the cloud core, a considerable fraction of the core matter is blown off. Because the remaining core matter is no longer gravitationally bound, it disperses unless the cloud core is initially far too massive to be in virial equilibrium, and the supply of matter to the accretion disk effectively stops. Because the disk mass is rather small, this isolation of the central star-accretion disk system from the outer cloud core terminates the main accretion phase and virtually fixes the mass of the forming star.

As the stellar mass increases, the luminosity and the surface temperature of the star also increases, and at later stages a part of the cloud core which has not yet contracted begins to be ionized by the stellar ultraviolet radiation. The HII region expands gradually pushing out the surrounding matter and forming a thin dense shell outside it. Around the time when the size of the HII region grows to the initial size of the parent cloud core, the core becomes seriously disturbed and the supply of matter from the cloud core to the disk and then accretion onto the star effectively stops as in the case of the mass outflow. Such an effect of the HII region is significant only when the stellar mass is high, although the mass outflow is efficient even at low stellar masses.

Although the outflow and the stellar ionizing radiation must work simultaneously in general, we estimate the stellar mass at which accretion stops by each mechanism separately. Because the effect of the HII region is very sensitive to the mass of the central star, the range of stellar mass in which both effects are comparable is very narrow. The third condition is added by noting a natural limit, i.e. the stellar mass cannot exceed the mass of the parent cloud core, M_c. The actual stellar mass is therefore given approximately by the smallest of the mass determined by the outflow, the mass determined by the HII region, and the mass of the cloud core. In this way we derive the stellar mass as a function of the density ρ_c of the cloud core, the mass inflow rate $\dot{M}_I$, the core mass relative to the generalized Jeans mass for the core, M_c/M_J, and the mass outflow rate relative to the inflow rate, $\dot{M}_O/\dot{M}_I$.

We describe the model of the cloud core and the effect of the mass outflow in Sect. 3 and the effect of the HII region in Sect. 4. We show the numerical results on the stellar mass in Sect. 5. In Sect. 6 we apply our theory to the observed cloud cores in the Orion A giant molecular cloud and discuss on the expected IMF and the star formation efficiency for this cloud. We discuss some related problems in Sect. 7. Finally in Sect. 8 we give the summary and conclusion of the present work.

3. Effect of the Mass Outflow

Assuming that the cloud core is initially not very far from spherical symmetry, we set a relation among the mass M_c, radius R_c, and mean density ρ_c of the core:

$$M_c \approx \frac{4\pi}{3} R_c^3 \rho_c. \tag{1}$$

When the core begins dynamical contraction from a nearly equilibrium state, M_c is nearly equal to the critical mass for contraction, or the generalized Jeans mass M_J. When the core begins dynamical contraction at a state compressed by an external pressure, M_c can be greater than the critical mass M_J at the compressed state by some factor f_M. Therefore we write in general

$$M_c = f_M M_J \approx f_M \frac{\pi^{3/2} C_{\rm eff}^3}{G^{3/2} \rho_c^{1/2}}, \tag{2}$$

where $C_{\rm eff}$ is the effective sound velocity in the cloud core and is given by

$$C_{\rm eff} = (C_s^2 + V_{\rm turb}^2 + V_A^2)^{1/2}, \tag{3}$$

with the sound velocity C_s, the turbulent velocity $V_{\rm turb}$, and the Alfvén velocity V_A. The mass inflow rate $\dot{M}_I$ is given by the core mass divided by several times the free-fall time, or (Stahler, Shu, & Taam 1980)

$$\dot{M}_I \sim M_c \left(\frac{G\rho_c}{\pi^3}\right)^{1/2} \approx f_M \frac{C_{\rm eff}^3}{G}. \tag{4}$$

Mass outflows are commonly observed around mass-accreting young stars (e.g., Edwards, Ray, & Mundt 1993). Although the outflows exhibit bipolar morphology with various degrees of collimation, they tend to extend their opening angle and clear out material even close to the equator far from the accretion disk as discussed by Heyvaerts & Norman (1989, 1995). The mass loss rate of the high velocity outflow is suggested to be of the order of 0.1 times the mass inflow rate (Shu et al. 1988). Assuming that the mass outflow from the stellar vicinity occurs with a constant rate $\dot{M}_{\rm O}$ and a constant velocity $V_{\rm O} \approx 200$ km s^{-1}, we estimate the influence of this outflow on the parent cloud core.

The outflow pushes out the surrounding cloud core matter forming a thin dense shell. The mass swept up by the shell, $4\pi R_{\rm sh}^3 \rho_{\rm c}/3$, where $R_{\rm sh}$ is the radius of the shell, exceeds the mass ejected from the stellar vicinity, $\dot{M}_{\rm O} R_{\rm sh}/V_{\rm O}$, when

$$R_{\rm sh} > R_{\rm sh}^{\rm (cr)} \approx 0.013 R_{\rm c} f_M^{1/6} \left(\frac{\dot{M}_{\rm O}/\dot{M}_{\rm I}}{0.1} \frac{200 \text{ km s}^{-1}}{V_{\rm O}} \frac{C_{\rm eff}}{1.6 \text{ km s}^{-1}} \right)^{1/2}, \tag{5}$$

as is found from Eqs. (1), (2), and (4). Note that $C_{\rm eff} \lesssim 1.6$ km s^{-1} for $\dot{M}_{\rm I} \lesssim 10^{-3} M_\odot {\rm yr}^{-1}$ and $f_M \geq 1$. The shell enters on this phase in a very early evolutionary stage. Because we consider rather high $\dot{M}_{\rm O}$ and $n_{\rm c}$, the bubble inside the shell is hardly adiabatic for $V_{\rm O} \approx 200$ km s^{-1} (Norman & Silk 1980). Therefore, neglecting the pressure in the bubble compared with the momentum flux of the outflow in the snowplow model, we have the equation of motion of the shell at $R_{\rm sh} > R_{\rm sh}^{\rm (cr)}$:

$$\frac{d}{dt} \left(\frac{4\pi}{3} R_{\rm sh}^3 \rho_{\rm c} V_{\rm sh} \right) = \dot{M}_{\rm O} V_{\rm O}, \tag{6}$$

where $V_{\rm sh} = dR_{\rm sh}/dt$ is the velocity of the shell. In deriving this equation we have neglected the deviation of the solid angle subtended by the outflow from 4π. If this deviation is taken into account, $R_{\rm sh}$ increases faster, and consequently the final stellar mass becomes smaller. However, we have neglected this effect because the deviation may not be very large and also because there are some other uncertainties, e.g., in $\dot{M}_{\rm O} V_{\rm O}$. The momentum input to the shell by the stellar radiation, L_*/c, has been found small compared with the momentum input by the outflow at least for $\dot{M}_{\rm O}/\dot{M}_{\rm I} \gtrsim 0.1$. Integrating Eq. (6) twice assuming $\dot{M}_{\rm O} V_{\rm O}$=constant, we obtain at sufficiently large t (cf. Norman & Silk 1980)

$$R_{\rm sh} \approx \left(\frac{3 \dot{M}_{\rm O} V_{\rm O}}{2\pi \rho_{\rm c}} \right)^{1/4} t^{1/2}. \tag{7}$$

When the shell is at $R_{\rm sh}$, the central star has a mass

$$M_* = \dot{M}_* t \approx \dot{M}_{\rm I} \left(1 - \frac{\dot{M}_{\rm O}}{\dot{M}_{\rm I}} \right) \left(\frac{2\pi \rho_{\rm c}}{3 \dot{M}_{\rm O} V_{\rm O}} \right)^{1/2} R_{\rm sh}^2. \tag{8}$$

Equations (1) and (4) give the radius of the cloud core as

$$R_{\rm c} \approx \left(\frac{3 M_{\rm c}}{4\pi \rho_{\rm c}} \right)^{1/3} \approx \left(\frac{\pi}{G} \right)^{1/6} \left(\frac{3 \dot{M}_{\rm I}}{4} \right)^{1/3} \frac{1}{\rho_{\rm c}^{1/2}}. \tag{9}$$

Around the time when $R_{\rm sh}$ grows to $R_{\rm c}$, a considerable fraction of the cloud core matter is blown off. The remaining cloud core matter disperses because it is no longer gravitationally bound, and accretion onto the forming star virtually stops. In this way the stellar mass is fixed at the value of Eq. (8) with $R_{\rm sh} \approx R_{\rm c}$, which we denote $M_*^{\rm (O)}$ and is given by

$$\begin{aligned} M_*^{\rm (O)} &\approx \left(\frac{\pi}{2}\right)^{5/6} \frac{3^{1/6}}{G^{1/3}} \frac{\dot{M}_{\rm I}^{5/3}}{(\dot{M}_{\rm O} V_{\rm O} \rho_{\rm c})^{1/2}} \left(1 - \frac{\dot{M}_{\rm O}}{\dot{M}_{\rm I}}\right) \\ &\approx 1.7 M_\odot \left(\frac{\dot{M}_{\rm O}/\dot{M}_{\rm I}}{0.1} \frac{V_{\rm O}}{200\ {\rm km\ s^{-1}}}\right)^{-1/2} \left(\frac{1 - \dot{M}_{\rm O}/\dot{M}_{\rm I}}{0.9}\right) \\ &\times \left(\frac{n_{\rm c}}{10^4 {\rm cm^{-3}}}\right)^{-1/2} \left(\frac{\dot{M}_{\rm I}}{10^{-5} M_\odot {\rm yr^{-1}}}\right)^{7/6}, \end{aligned} \tag{10}$$

where $n_{\rm c} = \rho_{\rm c} X / m_{\rm H}$ is the mean density of the cloud core by hydrogen number, $X \approx 0.73$ being the mass fraction of hydrogen. Substituting Eq. (4) for $\dot{M}_{\rm I}$ into Eq. (10) we obtain another expression for $M_*^{\rm (O)}$,

$$\begin{aligned} M_*^{\rm (O)} &\approx 4.3 M_\odot \left(\frac{\dot{M}_{\rm O}/\dot{M}_{\rm I}}{0.1} \frac{V_{\rm O}}{200 {\rm km\ s^{-1}}}\right)^{-1/2} \left(\frac{1 - \dot{M}_{\rm O}/\dot{M}_{\rm I}}{0.9}\right) \left(\frac{M_{\rm c}}{100 M_\odot}\right)^{7/6} \\ &\times \left(\frac{n_{\rm c}}{10^4 {\rm cm^{-3}}}\right)^{1/12}. \end{aligned} \tag{11}$$

The stellar mass due to mass outflow, $M_*^{\rm (O)}$, is determined by the mass of the cloud core $M_{\rm c}$, $\dot{M}_{\rm O}/\dot{M}_{\rm I}$, and $V_{\rm O}$. *The stellar mass is almost independent of the core density $n_{\rm c}$.*

The star formation efficiency, or the fraction of the core mass which will be converted into a star, in the case where the mass outflow determines the stellar mass is obtained from Eq. (11) as

$$\begin{aligned} \frac{M_*}{M_{\rm c}} &\approx 0.043 \left(\frac{\dot{M}_{\rm O}/\dot{M}_{\rm I}}{0.1} \frac{V_{\rm O}}{200\ {\rm km\ s^{-1}}}\right)^{-1/2} \left(\frac{1 - \dot{M}_{\rm O}/\dot{M}_{\rm I}}{0.9}\right) \left(\frac{M_{\rm c}}{100 M_\odot}\right)^{1/6} \\ &\times \left(\frac{n_{\rm c}}{10^4 {\rm cm^{-3}}}\right)^{1/12}. \end{aligned} \tag{12}$$

The star formation efficiency in this case is determined by $\dot{M}_{\rm O}/\dot{M}_{\rm I}$ and $V_{\rm O}$, and is almost independent of the core parameters $M_{\rm c}$ and $n_{\rm c}$.

The time required to form a star in this case is obtained from Eqs. (4) and (10) as

$$t_{\rm SF}^{\rm (O)} = \frac{M_*^{\rm (O)}}{\dot{M}_*} \approx 0.91 f_M^{1/6} \left(\frac{0.1}{\dot{M}_{\rm O}/\dot{M}_{\rm I}} \frac{200\ {\rm km\ s^{-1}}}{V_{\rm O}} \frac{C_{\rm eff}}{1.6\ {\rm km\ s^{-1}}}\right)^{1/2} \left(\frac{32}{3\pi} G \rho_{\rm c}\right)^{-1/2}. \tag{13}$$

Because $C_{\rm eff} \lesssim 1.6$ km s^{-1} for $\dot{M}_{\rm I} \lesssim 10^{-3} M_\odot {\rm yr^{-1}}$ with $f_M \geq 1$, $t_{\rm SF}^{\rm (O)}$ is slightly shorter than the free-fall time of the core, which is considerably shorter than the contraction time of the core because of the supporting forces against the gravity.

4. Effect of the HII Region

An HII region around a protostar pushes out the un-ionized matter forming a thin dense shell. Let $R_{\rm sh}$ be the radius of the shell which is nearly equal to the radius of the HII region and $C_{\rm HII} = 1.0 \times 10^6 {\rm cm\ s^{-1}}$ be the sound velocity in the HII region with a temperature 8000 K and the mean molecular weight 0.66. Because $C_{\rm HII} \gg C_{\rm eff}$, the accretion time scale $M_*/\dot{M}_* \approx M_*/\dot{M}_{\rm I} \sim R_{\rm c}/C_{\rm eff}$ is much greater than the sound crossing time in the HII region, $R_{\rm sh}/C_{\rm HII}$. In addition we have $C_{\rm HII}^2 \gg GM_*/R_{\rm sh}$ except for very early stages of the HII region. Therefore, most of the HII region is in pressure balance and has a nearly uniform density which we denote $\rho_{\rm HII}$. Some density gradient may exist in the HII region because of the stellar radiation pressure acting mainly through dust grains. However, we have confirmed that this effect is not important. The velocity of the shell, $V_{\rm sh}$, satisfies

$$\frac{dR_{\rm sh}}{dt} = V_{\rm sh}, \tag{14}$$

and the equation of motion of the shell reads

$$\frac{1}{R_{\rm sh}^2}\frac{d}{dt}(R_{\rm sh}^2\sigma_{\rm sh}V_{\rm sh}) = P_{\rm HII} - P_{\rm c} + \frac{f_{\rm rec}m_{\rm H}C_{\rm HII}}{3R_{\rm sh}^2}\frac{d}{dt}(R_{\rm sh}^3 n_{\rm HII}), \tag{15}$$

where $\sigma_{\rm sh} = (\rho_{\rm c} - \rho_{\rm HII})R_{\rm sh}/3$ is the column density of the shell, $P_{\rm HII} = C_{\rm HII}^2\rho_{\rm HII}$ and $P_{\rm c} = C_{\rm eff}^2\rho_{\rm c}$ are the pressure of the HII region and of the molecular gas, respectively, and $n_{\rm HII} = \rho_{\rm HII}X/m_{\rm H}$ is the density of the HII region by hydrogen number. The last term in the right-hand side of Eq. (15) represents the recoil effect of the ionized gas ejected from the shell toward the HII region and we take the non-dimensional parameter $f_{\rm rec} = 1$ in the numerical computation.

We regard the HII region as nearly spherical and let $n_{\rm Lym}(r)$ and $f_{\rm Lym}(r)$ be the number density and number flux, respectively, of Lyman continuum photons at distance r from the protostar. Conservation of Lyman continuum photons reads

$$\frac{\partial n_{\rm Lym}}{\partial t} + \frac{1}{r^2}\frac{\partial}{\partial r}(r^2 f_{\rm Lym}) = -\alpha n_{\rm HII}^2 - \kappa\rho_{\rm HII}f_{\rm Lym}, \tag{16}$$

where $\alpha \approx 3 \times 10^{-13}{\rm cm^3\ s^{-1}}$ is the radiative recombination rate coefficient of H^+ to the levels other than the ground state, $\kappa \approx 1 \times 10^3 {\rm cm^2\ g^{-1}}$ is the mass absorption coefficient of dust grains for Lyman continuum photons. Because the sound crossing time of the HII region, $R_{\rm sh}/C_{\rm HII}$, is much shorter than the accretion time scale, $M_*/\dot{M}_*$, the HII region can be regarded as being in quasi-steady state, or $\partial n_{\rm Lym}/\partial t \approx 0$, and Eq. (16) reduces to

$$\frac{dF_{\rm Lym}}{d\tau} + F_{\rm Lym} = -\frac{4\pi\alpha}{(\kappa m_{\rm H}/X)^3 n_{\rm HII}}\tau^2, \tag{17}$$

where $\tau = \kappa\ \rho_{\rm HII}\ r$ is the optical thickness up to r for the dust opacity and $F_{\rm Lym} = 4\pi r^2 f_{\rm Lym}$ is the number of Lyman continuum photons crossing the sphere of radius r per unit time.

With the boundary condition $F_{\rm Lym}(r=0) = N_{\rm Lym}(M_*,\ \dot{M}_*)$, the number of Lyman continuum photons emitted per unit time by a star of mass M_* accreting at a rate $\dot{M}_*$ (see the coming full paper for the details), we obtain analytically from Eq. (17)

$$\begin{aligned} F_{\rm Lym}(\tau) &= N_{\rm Lym}(M_*,\ \dot{M}_*)e^{-\tau} - \frac{8\pi\alpha}{(\kappa m_{\rm H}/X)^3 n_{\rm HII}}(1-e^{-\tau}) \\ &- \frac{4\pi\alpha}{(\kappa m_{\rm H}/X)^3 n_{\rm HII}}\tau(\tau-2). \end{aligned} \tag{18}$$

When dust hardly persists in the HII region, taking a limit of $\tau \ll 1$ in Eq. (18) gives

$$F_{\rm Lym}(r) = N_{\rm Lym}(M_*,\ \dot{M}_*) - \frac{4\pi}{3}r^3\alpha n_{\rm HII}^2, \tag{19}$$

which is the ordinary expression from which the radius of the dust-free Strömgren sphere is derived.

Evolution of the HII region can be pursued in the following way. In a cloud core of density $\rho_{\rm c}$ and mass inflow rate $\dot{M}_{\rm I}$ the central star grows according to $M_* = \dot{M}_* t = \dot{M}_{\rm I}(1 - \dot{M}_{\rm O}/\dot{M}_{\rm I})t$ for some fixed value of $\dot{M}_{\rm O}/\dot{M}_{\rm I}$. At time t the shell is at $R_{\rm sh}$ with a velocity $V_{\rm sh}$. The density of the HII region, $n_{\rm HII}$, is obtained from Eq. (18) or (19) by setting $F_{\rm Lym}(R_{\rm sh}) = 0$ for the given values of $R_{\rm sh}$ and $N_{\rm Lym}$, and the equations of motion (14) and (15) are solved numerically with this $n_{\rm HII}$. When M_* is small, $R_{\rm sh}$, $\sigma_{\rm sh}$, and the recoil effect are very small, and then pressure balance holds between the HII region and the surrounding molecular gas. Therefore the initial position of the shell is determined by this pressure balance at sufficiently small M_* and the initial velocity $V_{\rm sh} \approx 0$.

We determine the final stellar mass as the value at which $R_{\rm sh}$ grows to $R_{\rm c}$, which we denote $M_*^{\rm (HII)}$. This mass is a function of $\rho_{\rm c}$, $\dot{M}_{\rm I}$, f_M, and $\dot{M}_{\rm O}/\dot{M}_{\rm I}$.

5. Numerical Results on Stellar Mass

Although the mass outflow and the HII region must affect the surrounding gas simultaneously, we have investigated their effects separately assuming that only one of them is dominant in a given situation. Because the effect of the HII region is very sensitive to the stellar mass, the stellar mass range in which both effects are comparable is very narrow. Therefore the actual stellar mass at which accretion stops is approximately given by the smallest of the masses determined by the mass outflow, $M_*^{\rm (O)}$, the HII region, $M_*^{\rm (HII)}$, and the mass of the parent cloud core, $M_{\rm c}$.

Figure 1 shows the result for the case of $\dot{M}_{\rm O}/\dot{M}_{\rm I} = 0.1$. The solid lines are the loci of constant stellar mass on the $(\dot{M}_{\rm I},\ n_{\rm c})$ plane for the case of $f_M = 1$. This is the case where the parent cloud core is initially in virial equilibrium. The solid lines are labeled with a value of M_* in units of $M_\odot$. The lower, steeper parts of the lines for $M_* \geq 10 M_\odot$ represent the stellar mass determined by the HII region, $M_* = M_*^{\rm (HII)}$, and the rest (straight lines) represents the stellar mass determined by the outflow, $M_* = M_*^{(\mathrm{O})}$. The dashed lines represent the stellar mass for $f_M = 3$, or for the case where the cloud core begins dynamical contraction at a state where the generalized Jeans mass is 1/3 the core mass, which can occur, for example, as a result of compression by an external pressure: the lines of $M_* = M_*^{(\mathrm{HII})}$ are only slightly shifted from the case of $f_M = 1$, and the lines of $M_* = M_*^{(\mathrm{O})}$ are independent of f_M. There are no regions where the stellar mass is determined by the core mass M_c at least in the region shown in this figure.

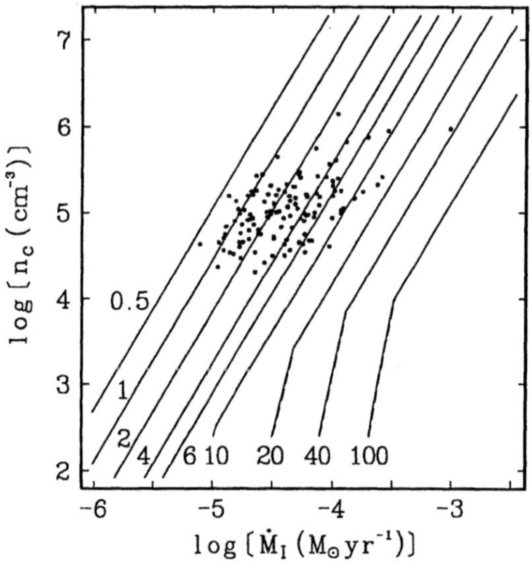

Fig. 1. The mass, M_*, of a star which will form in a cloud core as a function of the hydrogen number density, n_c, and the mass inflow rate, $\dot{M}_\mathrm{I}$, of the cloud core for the case of $\dot{M}_\mathrm{O}/\dot{M}_\mathrm{I} = 0.1$, where $\dot{M}_\mathrm{O}$ is the mass loss rate by the outflow. The solid lines are loci of constant stellar mass for the case of $f_M = 1$, and the dashed lines are for the case of $f_M = 3$. The dots represent the cloud cores in the Orion A giant molecular cloud (Tatematsu et al. 1993). See text for further details.

For $f_M > 1$ the quantities $\dot{M}_\mathrm{I}$ and n_c on the loci of constant stellar mass in Fig. 1 are for the state just after the cloud core has been compressed by an external pressure. The state before compression (nearly in virial equilibrium) is obtained by shifting the point left-downward along the line of constant core mass, $M_\mathrm{c} \propto \dot{M}_\mathrm{I}/n_\mathrm{c}^{1/2}$ = constant, as much as the compression factor. For instance, if the core has been compressed isothermally, the density of the state before compression is lower by a factor of f_M^2 as seen from Eq. (2). Thus the isothermal compression by an external pressure 10 times the initial pressure of the cloud core corresponds to $f_M \approx 3$. When the magnetic force is dominant in supporting the core, $f_M \approx 1$ is maintained during the compression as long as the magnetic fields are nearly frozen in to the gas during the compression. Thus it would be enough to consider the cases with $1 \leq f_M \lesssim 3$.

As seen from Fig. 1, the stellar mass as a function of $\dot{M}_\mathrm{I}$ and n_c is not very sensitive to f_M. This does not necessarily mean that compression of the cloud core by an external pressure is not efficient in increasing the mass of the star to form in the core. For compressed cloud cores the inflow rate $\dot{M}_\mathrm{I}$ shown in Fig. 1 is the inflow rate starting from the compressed state. For example, if the density of a cloud core is enhanced by a factor of 10 by compression, $\dot{M}_{\rm I}$ is increased by a factor of 3 irrespective of the change in f_M as seen from Eq. (4). An increase of this order in $n_{\rm c}$ and $\dot{M}_{\rm I}$ brings about considerable increase in $M_*^{\rm (HII)}$ because the core is compressed along a line of constant $M_{\rm c}$, or $n_{\rm c} \propto \dot{M}_{\rm I}^2$, in Fig. 1, although the effect on $M_*^{\rm (O)}$ is small.

To form a star of $M_* \approx 100 M_\odot$ by spherical accretion a very high inflow rate is required ($\gg 5 \times 10^{-3} M_\odot$ yr^{-1} for the normal dust abundance, Wolfire & Cassinelli 1987). Even with a high mass outflow rate $\dot{M}_{\rm O} \approx 0.1 \dot{M}_{\rm I}$ it is possible to form a star of $M_* \approx 100 M_\odot$, e.g., with a rate $\dot{M}_{\rm I} \approx 3 \times 10^{-4} M_\odot$ yr^{-1} as seen from Fig. 1. This is only slightly higher than the critical accretion rate ($1.1 \times 10^{-4} M_\odot$ yr^{-1}) below which even nonspherical accretion is stopped by radiation pressure (Nakano 1989).

The largest uncertainty in the present work would be in the adopted value of $\dot{M}_{\rm O}/\dot{M}_{\rm I}$. Although $\dot{M}_{\rm O}/\dot{M}_{\rm I} \sim 0.1$ is expected, we have investigated the cases of $\dot{M}_{\rm O}/\dot{M}_{\rm I} = 0.01$ and 0 to see the dependence on this ratio. We have to omit the results of these cases because of the limited pages.

6. Application to the Cloud Cores in Orion A

We have shown above that the stellar mass can be represented as a function of the density $n_{\rm c}$, mass inflow rate $\dot{M}_{\rm I}$, fraction of mass outflow $\dot{M}_{\rm O}/\dot{M}_{\rm I}$, and mass factor f_M of the parent cloud core. Using this result we shall estimate masses of stars which will form in the near future, the IMF, and the star formation efficiency in the Orion A giant molecular cloud which is known as a place of massive star formation.

Using the CS (1-0) emission line survey of the Orion A cloud with the Nobeyama 45 meter telescope, Tatematsu et al. (1993) identified 125 cloud cores and tabulated the line width (FWHM) Δv, radius, mass, and virial mass of each core. Because Δv is much greater than the thermal velocity of a CS molecule at temperature $T \approx 20$K, we can regard this as the turbulent velocity and derive the effective sound velocity according to $C_{\rm eff} = [(\Delta v)^2/8 \ln 2 + C_{\rm s}^2]^{1/2}$, where $C_{\rm s}$ is the sound velocity at $T = 20$K. Because the core mass derived from the CS (1-0) line intensity is rather uncertain due to, for example, the uncertainty in molecular abundance, we regard the virial mass corrected for the sound velocity as the core mass. From this mass and the observed radius we estimate the mean density of the core, and then estimate the mass inflow rate $\dot{M}_{\rm I}$ from the mass and the mean density according to Eq. (4). In this way we have plotted the cores in the Orion A cloud on the ($\dot{M}_{\rm I}$, $n_{\rm c}$) plane in Fig. 1. Assuming that the observed cores do not contain subclumps inside them (they have not been detected by the Nobeyama 45m telescope; see below for the upper mass limit of such clumps), we can estimate the mass of the star which will form in each core with the method developed in the previous sections. A rough estimate can be obtained from the locations of the dots among the theoretical lines in this figure. We should refer to the solid lines (for $f_M = 1$), not to the dashed lines (for $f_M = 3$), because we have assumed virial equilibrium for each core. Even for the case of $\dot{M}_{\rm O}/\dot{M}_{\rm I} = 0.1$ many massive stars are expected to form in this cloud. The core in which the most massive star is expected to form is the one associated with Orion KL. The stellar mass expected in this core for $\dot{M}_{\rm O}/\dot{M}_{\rm I} = 0.1$ is $36 M_\odot$. This is in reasonable agreement with the mass of the protostar IRc2 embedded in this core which has been estimated to be $25 M_\odot$ or somewhat higher from its luminosity (Wynn-Williams et al. 1984; Genzel & Stutzki 1989). It is to be noticed that the mass outflow determines the stellar mass for all the cores as long as $\dot{M}_{\rm O}/\dot{M}_{\rm I} > 0.02$.

The beam size of the Nobeyama 45m telescope for the CS (1-0) line is 40 arcsec, which corresponds to 2.7×10^{17}cm at the distance 450pc of the Orion A cloud. Because the undetected subclumps, if any, must have a radius smaller than this, the upper mass limit of such a clump is $1.3(n_{\rm H}/10^5{\rm cm}^{-3}) M_\odot$, where $n_{\rm H}$ is the mean density of the subclump.

Assuming that each core does not contain subclumps and that a single star will form in each core, we can calculate the initial mass function of stars (IMF) and the star formation efficiency in the Orion A cloud. From the number of stars (cloud cores), dN_*, in each logarithmic stellar mass bin we derive the IMF by

$$\xi_*(M_*) \equiv \frac{dN_*}{d \log M_*}, \tag{20}$$

which is shown by the dots in Fig. 2 for the case of $\dot{M}_{\rm O}/\dot{M}_{\rm I} = 0.1$. The error bars in Fig. 2 have been obtained by assuming that the fluctuation in the star number is equal to the square root of the star number in each stellar mass bin. The IMF at $M_* \gtrsim 4 M_\odot$ can be approximated well by a power law $\xi_* \propto M_*^{-1.7}$ (the dashed line). The hatch in Fig. 2 shows the IMF of field stars, $\xi_{\rm FS}$, including the errors, obtained by Scalo (1986; Fig. 16 therein) assuming that the star formation rate in the galactic disk has been constant for the past 1.2×10^{10} years. He recommends a power-law approximation $\xi_{\rm FS} \propto M_*^{-1.5}$ at $3 M_\odot \lesssim M_* \lesssim 60 M_\odot$. The IMF of the Orion A at $M_* \gtrsim 4 M_\odot$ agrees well with this. On the other hand the IMF of the Orion A has a turnover at $M_* \sim 2 M_\odot$, while the IMF of field stars has one at $M_* \sim 0.3 M_\odot$. Thus the stars of $M_* < 2 M_\odot$ in the IMF of the Orion A are much fewer than in the IMF of field stars. We shall discuss in Sect. 7 the agreement at $M_* \gtrsim 4 M_\odot$ and the disagreement at smaller M_* in the IMFs. The IMF of the Orion OB association agrees pretty well with the field star IMF at $4 M_\odot \lesssim M_* \lesssim 10 M_\odot$ and the flattening at smaller M_* may be due to incompleteness of the survey (Scalo 1986). This IMF is similar to the IMF of the Orion A obtained above.

As seen from Fig. 1, the stellar mass is determined by the mass outflow for all the cores in Orion A at least when $\dot{M}_{\rm O}/\dot{M}_{\rm I} \gtrsim 0.1$. In this case the stellar mass is determined mainly by $M_{\rm c}$ and is quite insensitive to $n_{\rm c}$ as seen from Eq. (11). This suggests that the IMF is closely related to the mass function of the cloud cores in which stars will form. Tatematsu et al. (1993) found that the mass function of the Orion A cores can be fitted by $\xi_{\rm c} \equiv dN_{\rm c}/d \log M_{\rm c} \propto M_{\rm c}^{-0.6}$ at $M_{\rm c} \gtrsim 50 M_\odot$. However, this can be better fitted by $\xi_{\rm c} \propto M_{\rm c}^{-1.6}$ at $M_{\rm c} \gtrsim 300 M_\odot$.

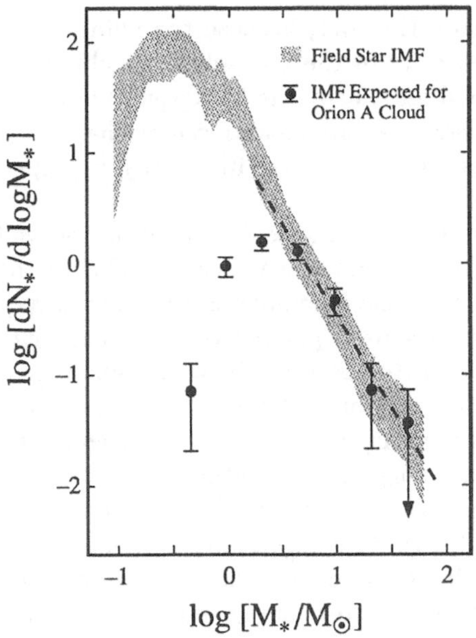

Fig. 2. The dots represent the initial mass function of stars (IMF), $\xi_*(M_*) \equiv dN_*/d\log M_*$, estimated for the Orion A giant molecular cloud with $\dot{M}_{\rm O}/\dot{M}_{\rm I} = 0.1$ assuming that each core does not contain subclumps and a single star forms therein. The dashed line shows the power-law approximation $\xi_* \propto M_*^{-1.7}$ at high stellar masses ($M_* \gtrsim 4M_\odot$). The hatch shows the IMF of field stars including the errors obtained by Scalo (1986). The unit of the ordinate is stars per unit $\log M_*$ per $\rm pc^2$ of the galactic disk appropriate for the field star IMF. The IMF of the Orion A has been shifted in the vertical direction by an arbitrary amount.

In addition we have to notice that their mass function is for the core masses estimated by the intensities of the CS (1-0) line. The mass function for the virial masses corrected for the sound velocity can be fitted by a power law $\xi_c \propto M_c^{-2}$ at $M_c \gtrsim 70M_\odot$. When the mass function of cloud cores is approximated by $\xi_c \propto M_c^{-\alpha}$ and the stellar mass M_* is related to the core mass M_c by Eq. (11), we have the IMF $\xi_* \propto M_*^{-6\alpha/7}$. For $\alpha \approx 2$ we have $\xi_* \propto M_*^{-1.7}$, which is, as expected, in good agreement with the IMF for $M_* \gtrsim 4M_\odot$ obtained above.

7. Discussion

Molecular clouds seem to have a hierarchical structure. What effect does star formation in the subclumps of the lowest level of the hierarchy (which we have called cloud cores) have on the structure of the larger system containing many cores (which we call a cloud here)? As can be confirmed from Eq. (7), the shell velocity $V_{\rm sh}$ decreases to a value not much greater than $C_{\rm eff}$ when $R_{\rm sh}$ increases to R_c. Because a condensation of greater size usually has a greater velocity dispersion (Larson 1981), the mass outflows from the forming stars in the cores can hardly enhance the velocity dispersion of the cloud. In this turbulent cloud, new cloud cores may form by, e.g., shock compression during the collision of the shells as discussed by Norman & Silk (1980), and stars will form in these new cores. Because the contraction time of the core is considerably shorter than the contraction time of the cloud as a whole, the cloud must experience several cycles of this core formation and star formation. Finally the most massive star in the cloud will form by the contraction of the cloud as a whole. In this case the shell expands in a medium of clumpy structure. However, because the filling factor of the cores is low and also because the mass of the unresolved subclumps is very small, most of the matter in the cloud must be in the background medium. Therefore, the shell motion in this case may be approximated well by the motion in a uniform medium and the stellar mass can be estimated in the way developed in Sects. 3 and 4.

The IMF in Fig. 2 estimated for the Orion A molecular cloud agrees well with the IMF of field stars at $M_* \gtrsim 4M_\odot$ but not at smaller M_*. One of the causes for this discrepancy at small M_* is in the incompleteness of the survey at low core masses. This incompleteness has been pointed out by Tatematsu et al. (1993). There may be much more small-mass cores in which small-mass stars will form. Another possible cause is that some of the cloud cores in the Orion A may contain some subclumps though they have not been resolved by the Nobeyama 45m telescope. Recent infrared imaging observations have revealed that a cluster of stars, not a single star, forms in a clump of the molecular cloud (e.g., Carpenter et al. 1993; Chen et al. 1993). This suggests that smaller-mass stars have formed from smaller-scale subclumps which have not been resolved by millimeter-wave observations. If some observed cores contain many subclumps, many stars will form in each of these cores, whose masses are much smaller than the mass obtained in Sect. 7 by assuming that there are no subclumps in the core. After several cycles of subclump formation and star formation, the core will finally contract as a whole and form the most massive star in the core as discussed above. Because the energy input into the core from the subclumps by the outflow will not affect much the random velocity in the core as mentioned above, and also because the mean density of the core may not change greatly by star formation in subclumps because of low star formation efficiency, the mass of the star which forms at the final stage must be not much different from the mass estimated in Sect. 6 from the mean physical quantities of the observed core by assuming that the core does not contain subclumps. Because the mass of the subclump must be much smaller than the mass of the core (see Sect. 6), the subclumps form only small-mass stars. Therefore the IMF in Fig. 2 at high M_* must approximate the true IMF pretty well. Because many stars of small mass may form in unresolved small cores and in unresolved subclumps in the cores, the true IMF at small M_* must be much greater than that in Fig. 2. Therefore, Fig. 2 does not necessarily mean that the IMF at $M_* < 4M_\odot$ in the Orion A region is different from that of the field stars.

Although the star formation efficiency in each subclump is very low (~ 0.04) as seen from Eq. (12), the total star formation efficiency in the clump may be several times higher because the formation of subclumps and stars may be repeated several times.

Because the stellar mass is determined by the mass outflow for most of the realistic cases, and the stellar mass in this case is determined by the mass of the cloud core almost independent of the core density as shown in Eq. (11), the agreement between the IMF estimated for the Orion A region and the IMF of field stars at $M_* \gtrsim 4M_\odot$ may suggest that the mass spectrum of cloud cores is universal and there may be some mechanisms to make up the universal mass spectrum, e.g., fragmentation of the cloud and coagulation of smaller fragments.

There are a number of studies which indicate that the mass spectra of cloud cores in some molecular clouds are approximated by a power law $dN_c \propto M_c^{-s} d\log M_c$ with the index s between 0.4 and 0.7 (e.g., Lada et al. 1991; Nozawa et al. 1991; see also the review of Blitz 1993). This is significantly flatter than the IMF of field stars at $M_* \gtrsim 1M_\odot$. One of the causes for this difference must lie in the large beam sizes of the telescopes used in their observations with which many small cores remain to be resolved.

8. Summary and Conclusion

We have developed a way to estimate the mass of a star from the physical quantities of the cloud core in which it forms. Because the cloud core matter falls to the stellar core (protostar) mostly through an accretion disk, growth of the stellar core virtually stops and the mass of the forming star is fixed when the supply of matter to the disk is stopped. As the physical mechanisms that stop the supply of matter to the disk, we have considered the development of a cavity due to mass outflow from the vicinity of the forming star and also the expansion of an HII region around the forming star. The stellar mass determined in this way is a function of the density of the cloud core, n_c, mass inflow rate in the cloud core, $\dot{M}_I$, the ratio of the cloud core mass M_c to the generalized Jeans mass M_J for the cloud core, $f_M \equiv M_c/M_J$, and the ratio of the mass outflow rate to the inflow rate, $\dot{M}_O/\dot{M}_I$. We have calculated the stellar mass as a function of $\dot{M}_I$ and n_c for the cases of $\dot{M}_O/\dot{M}_I = 0.1$, 0.01, and 0, though the results have been shown only for the case of $\dot{M}_O/\dot{M}_I = 0.1$.

For $\dot{M}_O/\dot{M}_I \gtrsim 0.1$ most cloud cores seem to locate in the part of the $(\dot{M}_I, n_c)$ plane where the mass outflow is dominant in determining the stellar mass. When the mass outflow is dominant, we have $M_* \propto M_c^{7/6} n_c^{1/12}$ with the proportionality coefficient dependent on $\dot{M}_O/\dot{M}_I$. Then the stellar mass is almost independent of the core density n_c. Therefore, the star formation efficiency, M_*/M_c, is mainly determined by $\dot{M}_O/\dot{M}_I$ and is only weakly dependent on the core parameters M_c and n_c. We have typically obtained $M_*/M_c \sim 0.04$ for $\dot{M}_O/\dot{M}_I = 0.1$.

We have applied our method to 125 cloud cores identified in the Orion A giant molecular cloud and estimated the mass of the star which will form in each core assuming that each core does not contain subclumps. We have obtained the following results.

1) As long as $\dot{M}_O/\dot{M}_I > 0.02$ the stellar mass is determined by the mass outflow in all the cores of the Orion A.

2) In the case of $\dot{M}_O/\dot{M}_I = 0.1$ we have found a fairly good agreement with the observations. For instance, the stellar mass estimated for the core associated with the Orion KL is $36M_\odot$, while the protostar IRc2 embedded in it is estimated to have a mass $\sim 25M_\odot$ or somewhat higher. The initial mass function (IMF) obtained for the Orion A is shown in Fig. 2 and is well fitted by a power law

$dN_*/d\log M_* \propto M_*^{-1.7}$ at $M_* \gtrsim 4M_\odot$, which is in reasonable agreement with the IMF for field stars $\propto M_*^{-1.5}$ at $M_* \gtrsim 3M_\odot$ (Scalo 1986). At $M_* < 2M_\odot$, however, the IMF of the Orion A contains much fewer stars than the IMF of field stars.

Molecular clouds seem to have a hierarchical structure: high-density clumps in the cloud, and even higher-density subclumps in the clump, etc. On the basis of the above results we have proposed a scenario of star formation in such a cloud. A star forming in a subclump expels the remaining matter of the subclump by the outflow. This delays somewhat the decay of the turbulence in the clump. New subclumps are formed in the turbulent clump, e.g., by shock compression, and new star formation occurs in these subclumps. After repeating this process several times the clump finally contracts as a whole and forms the most massive star in the clump. Because the mass of the star formed finally by the contraction of the whole clump is not much different from the stellar mass estimated from the mean physical quantities of the clump, the IMF estimated for the Orion A region by assuming that each core does not contain subclumps is a good approximation for the true IMF at high stellar masses, but may be much smaller than the true IMF at smaller stellar masses because many small-mass stars will form in subclumps. Although the star formation efficiency in each subclump is very low (~ 0.04), the total star formation efficiency in the clump must be considerably higher because formation of subclumps and stars must be repeated several times in the clump.

This work was supported in part by the Grant-in-Aid for Scientific Research on Priority Areas (Interstellar Matter) of the Ministry of Education, Science, and Culture, Japan (04233106, 04233208).

References

Anthony, D. M., & Carlberg, R. G. 1988, ApJ, **332**, 637
Blitz, L. 1993, Protostars & Planets III, eds. E. H. Levy & J. I. Lunine (Univ. Arizona Press, Tucson), p. 125
Boss, A. P. 1984, MNRAS, **209**, 543
Carpenter, J. M., Snell, R. L., Schloerb, F.P., & Skrutskie, M. F. 1993, ApJ, **407**, 657
Chen, H., Tokunaga, A. T., Strom, K. M., & Hodapp, K.-W. 1993, ApJ, **407**, 639
Edwards, S., Ray, T., & Mundt, R. 1993, Protostars & Planets III, eds. E. H. Levy & J. I. Lunine (Univ. Arizona Press, Tucson), p. 567
Genzel, R., & Stutzki, J. 1989, Ann. Rev. A. Ap., **27**, 41
Hayashi, C., & Nakano, T. 1965, Prog. Theor. Phys., **34**, 754
Heyvaerts, J., & Norman, C. 1989, ApJ, **347**, 1055
. 1995, in preparation
Lada, E.A., Bally, J., & Stark, A.A. 1991, ApJ, **368**, 432
Larson, R.B. 1969, MNRAS, **145**, 271
. 1972, MNRAS, **157**, 121
. 1981, MNRAS, **194**, 809
. 1984, MNRAS, **206**, 197

Larson, R.B., & Starrfield, S. 1971, A&A, **13**, 190
Lin, D. N. C., & Pringle, J. E. 1987, MNRAS, **225**, 607
Nakano, T. 1989, ApJ, **345**, 464
Nakano, T., Ohyama, N., & Hayashi, C. 1968, Prog. Theor. Phys., **39**, 1448
. 1970, Prog. Theor. Phys., **43**, 672
Narita, S., Nakano, T., & Hayashi, C. 1970, Prog. Theor. Phys., **43**, 942
Norman, C., & Silk, J. 1980, ApJ, **238**, 158
Nozawa, S., Mizuno, A., Teshima, Y., Ogawa, H., & Fukui, Y. 1991, ApJ Suppl, **77**, 647
Scalo, J.M. 1986, Fundam. Cosmic Phys., **11**, 1
Shu, F. H., Lizano, S., Ruden, S. P., & Najita, J. 1988, ApJ, **328**, L19
Stahler, S.W., Shu, F.H., & Taam, R.E. 1980, ApJ, **241**, 637
Tatematsu, K., Umemoto, T., Kameya, O., Hirano, N., Hasegawa, T., Hayashi, M., Iwata, T., Kaifu, N., Mikami, H., Murata, Y., Nakano, M., Nakano, T., Ohashi, N., Sunada, K., Takaba, H., & Yamamoto, S. 1993, ApJ, **404**, 643
Wolfire, M.G., & Cassinelli, J.P. 1987, ApJ, **319**, 850
Wynn-Williams, C. G., Genzel, R., Becklin, E. E., & Downes, D. 1984, ApJ, **281**, 172

Star-Disk Interactions in Small N Clusters: How to Form Binary Stars

J.M. McDonald and C.J. Clarke

Institute of Astronomy, Madingley Road, Cambridge, CB3 0HA, England

Summary. We investigate the effect of circumstellar disks on the dynamical evolution of a small cluster of protostars formed by 'prompt initial fragmentation'. In particular we study how the presence of disks affects the resultant mass components of binaries formed in the cluster. We find that when the stars are assigned circumstellar disks the occurrence of lower mass stars in binaries is greatly increased compared with diskless simulations. This is due to the fact that disks both increase the number of binaries formed and also randomise the selection of secondary mass companions to each primary. For a cluster of 10 stars with massive disks the predicted binary fraction is in good agreement with observations. We also find that disks boost the number of triple and quadruple systems formed, and suggest that the eventual disruption of the less hierarchical multiples might account for the excess of binaries among pre-main sequence stars.

1. Introduction

The recognition in recent years that most stars are in binaries, both during the main sequence and pre-main sequence stage (Duquennoy & Mayor 1991, Ghez et al 1993, Simon et al 1994, Leinert et al 1993, Reipurth & Zinnecker 1993) has made the question of binary star formation one of the central unresolved problems of star formation theory. Despite the intensive efforts devoted to tackling this problem through numerical simulation over the last decade, there has been a disappointing lack of progress in developing models with clear predictive properties that can be compared with observational data. This failure can readily be traced to the small mass fraction contained in the binaries (as opposed to distributed gas) at the stage when the simulations are terminated due to computational expense. As a result, not only the ultimate binary parameters but their very survival remains in doubt.

These problems are considerably less serious, however, in the case of binary formation scenarios in which distributed gas is a relatively minor component of the system. Models in this category may be termed 'prompt initial fragmentation', implying that the break-up of a molecular cloud core into a number of distinct self-gravitating units occurs before the parent core has collapsed by a large factor (Pringle 1989). In such a picture the initial configuration is essentially a small N cluster, and the problem of binary formation becomes one of achieving the necessary energy loss and angular momentum redistribution required for the formation of bound pairs.

We limit our investigation to small N clusters since in large N clusters star-disk capture can only form a small fraction of the observed binary population (Clarke & Pringle 1991). In the case of a large N virialised cluster the disk is more likely to undergo an interaction with a fast passing star, resulting in its destruction, before capture of a slower moving star can occur. However the velocity dispersion in a small N cluster is considerably smaller reducing the frequency of the destructive disk encounters without capture. The small N systems that we consider here are likely to be formed with interstellar separations of order the Jeans length in clumps of dense star forming gas (i.e of order 10^4 A.U.) and represent substructures in a larger complex of ongoing star formation. Such small N non-hierarchical clusters, although initially bound, dissolve after several internal crossing times (a few time 10^5 years), dispersing hierarchical multiples and single stars into the surrounding complex. The short lifetimes of these clusters in their initial compact non-hierarchical state, compared with the lifetime of T Tauri stars argues against their detection in this initial state. The more diffuse clustering observed in the spatial distribution of T Tauri stars may, however, contain fossil evidence of an origin of stars in the compact clusters envisaged here.

In this talk we examine the processes by which stars pair up in small N clusters and link these processes with the mass ratio distributions of the resultant binaries.

In a previous paper (McDonald & Clarke 1993) we considered the limit in which the role of distributed gas is negligible and in which the system therefore evolves essentially as a small N-body ensemble. Such a situation has been extensively investigated by a number of authors (van Albada 1968, Harrington 1974,1975, Anosova 1968, 1991). In each case it is found that the cluster dissolves on the order of a few tens of crossing times, the energy required for this process being released by the formation of one central binary star per cluster. Furthermore, this binary generally contains the two most massive stars in the cluster, thus allowing McDonald and Clarke to use this prescription to generate analytically a number of general properties of the resulting binary population. Whilst such an analysis provides a useful tool for assessing the role of point mass gravitational interactions in binary formation, it is unlikely that gas dynamical forces can be altogether neglected: distributed gas is likely to be present not only throughout the cluster as a relatively tenuous background but also in the form of centrifugally supported disks around the individual stars. It is not then at all clear how the pairing properties of the stars are modified by such a dissipative medium and one cannot in this case call upon existing N-body numerical results. Thus in the present paper we undertake a numerical investigation of small N-body cluster dynamics with the inclusion of dissipative encounters between stars and circumstellar disks.

The structure of the talk is as follows. In Sect. 2 we describe the model small N cluster employed in our simulations and offer a brief description of how the dissipative effects of the star-disk interactions are included. In Sect. 3 we summarise the numerical results and use them as inputs to an analytical calculation of the properties of the resultant binary population, comparing these with corresponding quantities calculated for the diskless case. We stress that though we have reasonable confidence in the order of magnitude estimates for the star-disk energy dissipation on a dynamical timescale, which determines the binary pairings, these calculations cannot be used to assess the secular evolution of binary systems. As a result we cannot comment on the orbital periods or eccentricities. Finally we summarise our conclusions in Sect. 4.

2. Calculations

A direct N-body code (Aarseth 1985) is used to study the dynamical evolution of a small cluster of point masses. The code employs a predictor-corrector type integrator which implements individual time steps and which conserves energy to better than one part in 10^4 per cluster crossing time. In these simulations, and also in the simulations where disks are included, we use a softening length of 0.01 (in units of the initial cluster size) to allow direct comparison between both data sets.

Initially the N particles are assigned random positions and velocities, which are then scaled so that Q, the ratio of the total kinetic to potential energies is some specified fraction of its virial value (0.5). Different random seeds are used for the positions and velocities for each simulation. In order to investigate the influence of the distribution of stellar mass, we assign stellar masses at random from each of the following mass functions in turn, where in each case $f(m)dm$ is the fraction of stars in the mass range m to $m + dm$:

$$f_A(m) \propto m^{-2.35} \tag{1}$$
$$f_B(m) \propto m^{-1.5} \tag{2}$$
$$f_C(m) = \delta(m - m_0) \tag{3}$$

For each mass function we used the same mass selection in every simulation. The reader should assume that the stellar masses are selected from a Salpeter mass function (equation (1)) unless otherwise stated.

Table 1. Table of star numbers and corresponding masses, disk masses and radii

star number	M_*	R_d(in units of $R_{\rm cluster}$)	M_d
1	0.373	0.580	0.280
2	0.152	0.228	0.114
3	0.102	0.153	0.077
4	0.079	0.119	0.059
5	0.065	0.098	0.049
6	0.056	0.084	0.042
7	0.050	0.075	0.040
8	0.045	0.068	0.034
9	0.041	0.062	0.031
10	0.037	0.058	0.028

The simulations are run for 25 crossing times of the initial cluster. After this time the cluster generally consists of one relatively tightly bound binary (termed the 'central binary') with respect to which the remaining stars are either weakly bound or unbound. In the present study we focus on the question of the pairing of the central binary and thus, in the interests of computational economy and of following the initial interactions with high accuracy (ie, employing short timesteps) we terminate the calculations after 25 crossing times, by which time the pairing of the central binary is well established.

2.1 The Inclusion of Disks

At time $t = 0$ each star is assigned a disk with initial mass, M_{d0}, and initial radius R_{d0}. It should be noted that the disks are taken to have zero mass as far as the dynamics of the cluster is concerned as their only purpose is to tell whether an interaction should take place and to determine the resulting outcome. The majority of the calculations are performed with disk mass and radius directly proportional to stellar mass (for a diskussion of the plausibility of this prescription see Murray & Clarke 1993) where the constants of proportionality are adjusted so that the resulting disks are less than the initial cluster radius and greater than the softening length. Table 1 lists the disk masses and radii employed in all simulations unless otherwise stated. Note that since we interact any star lying within a spherical volume of radius R_d around each star, the most massive star typically has 1–2 stars initially in its disk for the parameters in Table 1. We also perform calculations where the disk parameters are equal and also where the masses and radii are swapped around so that the most massive star has the disk parameters previously assigned to the lightest star and vice versa. By using the above prescriptions we are able to investigate extensively the influence of disk mass and radius on the star disk interactions. The disks are viscously evolved throughout the simulation using the viscous timescale prescription of Lin & Pringle(1987, 1990). Furthermore, once the disk mass falls below $10^{-3}M_*$ we assume that the disk plays no further role (for more details the reader is referred McDonald & Clarke 1995).

2.2 The Star-Disk Interaction

The appropriate prescription for energy loss in star-disk interactions has still to be established in the nonlinear regime appropriate to encounters that penetrate the disk (Hall, Clarke & Pringle in prep.). As a result the prescriptions employed here are 'educated guesses', based partly on the results of star-disk encounter calculations of disks and massless test particles (Clarke & Pringle 1993). The uncertainties in these estimates are sufficiently large for us to check carefully which of our results are sensitive to the details of the interaction prescription employed. The variations that we consider are whether the star-disk drag effect is concentrated at periastron, what is the response of the disk to the encounter and various treatments of the situation, not uncommon at early times in the case of large disks, of there being more than one star in a disk at a time. To include all the various possibilities of interaction mechanism the code was implemented with a series of options that determined whether the interactions occurred continuously or only at periastron, involved all stars in a particular disk or only the closest star to the disk center and finally how the disk responded to the interactions.

As an example we will now consider an encounter between two stars A and B in which B currently lies within the spherical volume of A's disk. Following an interaction, A's disk is shrunk to a new radius, R_t, which is determined by option 2. We assume that the energy lost in the interaction is equal to the energy required to unbind all of A's disk exterior to R_t. Thus for an assumed power law disk surface density profile

$$\Sigma = \Sigma_0 r^{-\frac{3}{2}}, \tag{4}$$

(where Σ_0 is a function of time for disks subject to viscous evolution) the energy loss is given by,

$$\Delta E = \frac{G M_A M_d}{2 R_d} \left(\left(\frac{R_d}{R_t} \right)^{\frac{1}{2}} - 1 \right). \tag{5}$$

In order to obtain the total amount of energy lost in the interaction of A and B we also have to consider whether A is also in the disk of B and if so perform a similar calculation for the energy lost in unbinding part of B's disk. The total energy lost in unbinding the disk material is taken from the stellar orbits instantaneously at interaction and it is furthermore assumed that the net momentum removed by the unbound disk material is zero. Hence the velocities of the stars, in the center of mass frame of the two stars, V_{AC} and V_{BC} respectively, are reduced by a factor λ, where,

$$\Delta E = \frac{1}{2}(M_A V_{AC}^2 + M_B V_{BC}^2)(1 - \lambda^2) \tag{6}$$

and λ can be calculated by equating equations (10) and (11). In the case where the two stars do not have sufficient energy to unbind the required amount of disk material then no interaction is deemed to take place until they do. This assumption cuts down on the computational expensive situation where both stars fall directly towards one another. After such an interaction the disk radii are updated to R_t and the corresponding disk mass is then given by,

$$M_d = M_{db} \left(\frac{R_t}{R_{db}} \right)^{\frac{1}{2}}, \tag{7}$$

where M_{db} and R_{db} are the disk mass and radius prior to the interaction.

If the interaction is chosen to be a multiple type encounter then all the stars in the disk around A have to be taken into account. We assume that each star contributes to unbinding the disk material which lies exterior to it. Thus in Figure 1 we assume that all stars in the disk of A contribute to unbinding the outermost segment (darkest section), only C and D contribute in unbinding the middle segment (dark grey) and D alone unbinds the innermost segment (light grey). We furthermore assume that the energy lost in unbinding a particular disk segment is shared equally among all the stars contributing to the unbinding of that segment.

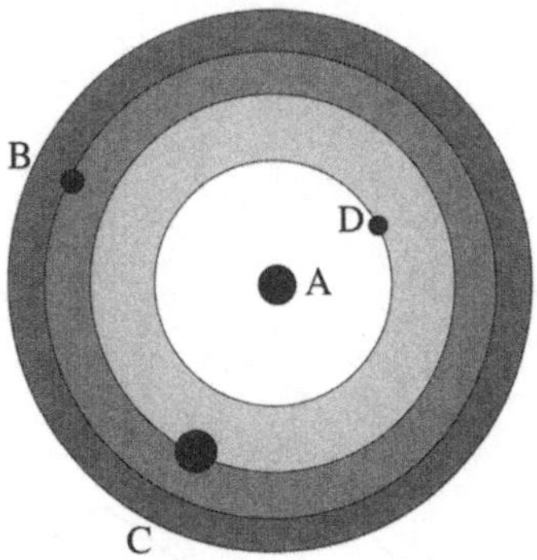

Fig. 1. A schematic picture of a multiple type encounter

3. Results

The simulations indicate that the resultant binary pairing statistics in the presence of disks are intermediate between those of the diskless case and those of the 'saturated' disk models, such as the standard model. [We term models 'saturated' if increasing the star-disk interactive strength has a negligible effect on the binary pairings]. In 'saturated' models the disks are massive (i.e, comparable with the stellar masses), large (a significant fraction of the cluster size), do not lose a substantial fraction of their mass by accretion onto their central stars over a few cluster crossing times and are interacted continuously with stars that intersect them. It should be noted that although these disk masses are much higher than those observed around T Tauri stars, the systems considered in these simulations are very young ($\leq 10^5$ years), and are thus likely to be surrounded by massive infalling envelopes at this stage rather than ordered, centrifugally balanced disks. Furthermore these massive disks should evolve on a relatively short timescale and hence be very difficult to observe.

In such 'saturated' models the mean number of binaries in a cluster of ten star is increased from 1.2 to 1.9, implying a mean binary fraction (defined as the ratio of the number of binaries to the number of binaries plus single stars) of 24%. (Note that this is a mass averaged binary fraction: we consider the binary fraction as a function of mass later in this section). In Figure 2(a) we have plotted the number of times each star ends up in a binary and then in Figure 2(b) we plot the number of times each star ends up in a binary with the heaviest star in the cluster. It is apparent that the first binary always contains the most massive star in the cluster, paired at random with the remaining stars;

additional binaries also show a bias in primary selection towards more massive stars although, again, the secondary is selected at random.

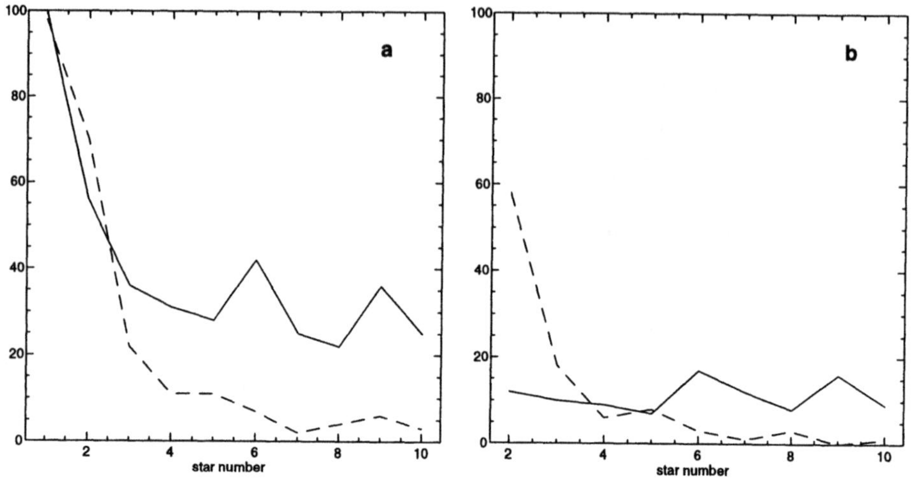

Fig. 2. (a) The number of times each star ends up in a binary for models with (solid) and without (dashed) disks. (b) The number of times each star ends up in a binary with the heaviest star for models with (solid) and without (dashed) disks.

As is evident from Figure 2(b) we find that the presence of disks leads to the formation of a higher number of binary and higher order multiple systems. This is due in part to the increase in the cross-sectional area for interaction and in part due to the hardening of temporary binaries that form in the diskless case so that they are able to withstand destructive encounters with more massive stars in the cluster. Thus the presence of disks weakens the dynamical bias in the diskless model and allows lower mass stars to be incorporated into binaries. We find that in order for this disk hardening to be fully effective in assuring the survival of temporary binaries, the orbits need to be shrunk to pericentres smaller than those they would attain in the absence of dissipation: thus a star-disk interaction prescription that is implemented only at periastron (and which thus keeps periastron fixed) gives results intermediate between the saturated and diskless models. This result underlines the necessity of obtaining reliable prescriptions for star-disk interactions in the non-linear regime (Hall, Clarke & Pringle in prep.).

We find that the incidence of higher order multiple systems (i.e. triples/quadruples) is significantly increased by the presence of disks. Throughout this talk we quote the number of binaries by including also triples and quadruples as providing two binaries per system, and evaluate this number after 25 cluster crossing times. In recognition of the fact that not all these multiples are stable hierarchical systems we have performed crude stability checks by demanding if the pericentre of the outer binary is greater than five times the apocentre of the inner binary: we list the resultant statistics in Table 2 as 'stable (?)'. We stress that such a criterion, although probably conservative, is very approximate since in reality, the stability borderlines in hierarchical systems are a complicated function of the masses, eccentricities and relative orientations in the system (Kiseleva & Anosova in prep., Kiseleva, Eggleton & Orlov in prep.). We do not feel that a more detailed stability analysis is warranted at this stage since, as we stressed in the introduction, our star-disk interaction prescription is completely unreliable on a greater than dynamical timescale so that the orbital parameters of the resultant binaries are not well known. The results in the 'stable (?)' column of Table 2 are therefore highly uncertain, but do qualitatively illustrate the expectation that the binary fraction should decrease with time, owing to the large numbers of higher order multiples that are generated in the presence of disks. This effect therefore provides a promising mechanism for explaining the claimed over-abundance of binaries amongst pre-main-sequence systems (Ghez et al 1993; Simon et al 1994; Leinert et al 1993). [Note that 25 cluster crossing times corresponds to a few million years in the case of a cluster of 10 stars with a radius of 0.1 parsecs]. The predicted estimates of the frequency of higher order multiple systems, for our standard disk model, (quadruple and triple systems) are 18% and 2% for pre-main-sequence and main-sequence systems respectively. These values are within a factor two to three of the observed values, (35% for pre-main-sequence (Ghez et al. 1993, Simon et al. 1994) and only 5% for main-sequence (Duquennoy & Mayor 1991)which, given the uncertainties in the binary orbital parameters, suggests that this may be a viable m for explaining the excess of pre-main-sequence binaries compared with the main-sequence population.

Table 2. Number of binary, triple and quadruple systems formed per cluster

	Diskless Model		'Standard' Model	
	short lived	stable(?)	short lived	stable(?)
binary	1.17	1.12	1.91	1.51
triple	0.04	0.02	0.35	0.03
quadruple	0.07	0.04	0.24	0.11

We now turn to the statistical properties of the resultant binary population that can be derived from our numerical simulations. Rather than using the simulations (whose results are subject to $\sqrt{N}$ noise) to construct a mass ratio distribution directly, we instead use the simulations to extract the fractional occurrences of each of the more common (ie, occurring greater than 5% of the time) binary mass pairings and make a general overview of the rest. We may then analytically generate the statistical properties of the resultant ensemble, in a manner analogous to McDonald & Clarke (1993), who performed such an exercise in the diskless case. For a derivation of the expressions for the resultant mass ratio distribution and binary fractions as a function of mass the reader is referred to McDonald & Clarke (1995). The results of this exercise (assuming that the single stars are assembled in clusters of either 10 or 4 by random selection from a Salpeter mass function) are shown in Figures 3(a) and (b) and 5(a) and (b).

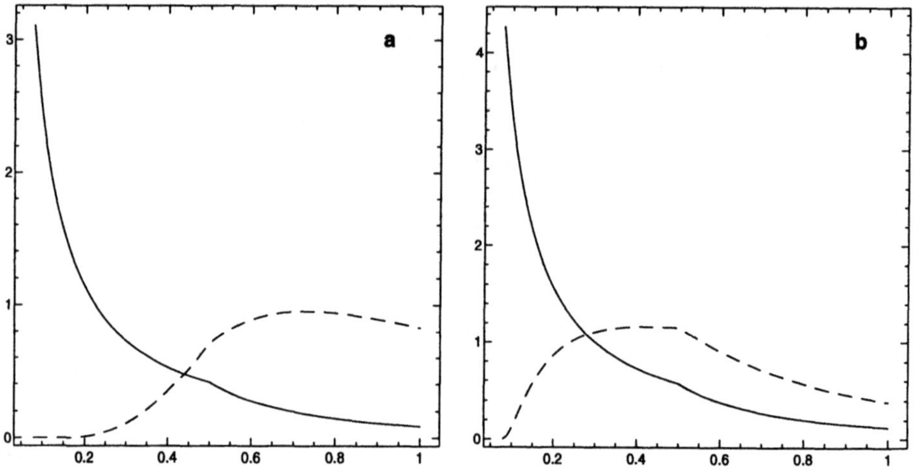

Fig. 3. Secondary mass distributions for a 1 solar mass primary for models with (solid) and without (dashed) disks for clusters of (a) $N = 10$ and (b) $N = 4$ (normalised to give the same area).

In Figure 3 we plot $F_2(M_2, M)$, this being the fraction of one solar mass binary primaries with companions in the mass range M_2 to $M_2 + dM_2$. For the standard disk model (solid), $F_2(M_2, M)$ coincides with the initial mass function, irrespective of cluster size: this illustrates the random sampling of secondary masses with the inclusion of disks. In contrast, the secondary mass distribution for the diskless model (dashed) is depressed at lower masses and elevated at higher masses relative to the initial mass function, illustrating a dynamical bias in favour of the heavier stars which is particularly marked in higher N clusters. We note that the prediction in the presence of disks of a companion mass function proportional to the single star mass function is broadly compatible with the results of surveys of stars of solar mass and below (Duquennoy & Mayor 1991, Fischer & Marcy 1992), but is incompatible with the non-random pairing inferred in the case of OB stars (Clarke & Pringle 1992, Garmany, Conti and Massey 1980, Eggleton, Fitchet and Tout 1989). We also note that the removal of the dynamical bias in secondary selection implies that the presence of brown dwarfs in binaries should trace their frequency in the field, in contrast to the diskless results.

Another difference between the results with and without disks is related to the dependence of the companion mass function on primary mass. In the diskless case, $f_2(M_2, M)$, [which is the fraction of stars of mass M which have companions in the mass range M_2 to $M_2 + dM_2$] is independent of M (McDonald & Clarke 1993). Such a universal secondary mass function implies, for example, that one would expect equal fractions of G stars and M stars to have brown dwarf companions. In the presence of disks, however, $f_2(M_2, M)$ can be expressed in the form:

$$f_2(M_2, M) = g(M)f(M_2), \tag{8}$$

$f(M_2)$ is the single star and $g(M)$ is a function of primary mass only. Such an expression implies that the companion mass function is always proportional to the single star mass function, but that the normalisation depends on the primary mass. In Figure 4 we have plotted $g(M)$ for clusters of 10 and 4 stars. In the $N = 4$ star case (dashed lines) $g(M)$ rises monotonically with M; in this case the greater probability of a more massive star being a binary primary outweighs the fact that a smaller fraction of companions is to be found in a particular mass range. The $N = 10$ case is somewhat more complex since $g(M)$ rises with decreasing mass for masses that approach the lower cutoff in the mass function (taken to be $0.08M_\odot$): this is because, whilst a high fraction of stars are in a given mass range, there is a non-negligible probability that a low mass star is a primary in a second binary in a large N cluster. For masses above $\sim 0.2M_\odot$, $g(M)$ rises with mass in both cases, implying that there is a higher probability of there being companions, in any given mass range, about higher mass stars. We note, however, that the greater observational difficulty of detecting binaries with extreme mass ratios may offset this predicted effect when it comes to devising strategies for detecting, for example, brown dwarfs in binaries.

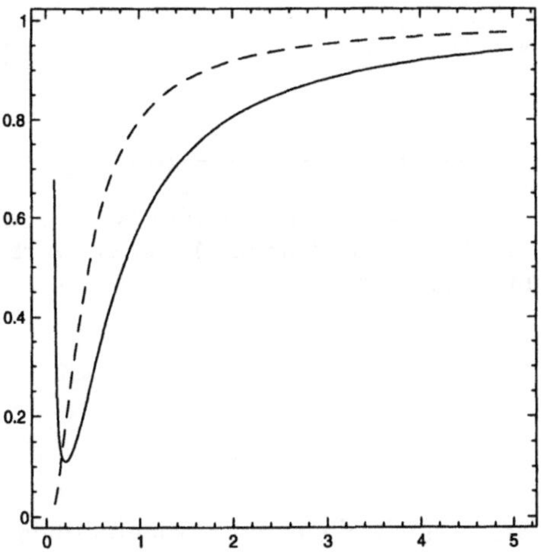

Fig. 4. Analytic form of dependence of $f_2(M_2, M)$ on primary mass, M, for clusters of $N = 4$ (dashed) and $N = 10$ (solid).

In Figures 5a, b we plot the binary fraction, *fbin*(M), which is the fraction of apparently single stars with binary companions. As in the diskless case, *fbin*(M) rises monotonically with M for all cluster sizes. The rise is considerably less steep, however, when disks are included because a star can now be a binary primary without having to be the most massive star in the cluster. The inclusion of disks actually reduces the binary fraction amongst higher mass stars since stars ranked 2 in the cluster, which without disks would always be secondaries, can sometimes escape as singles when disks are included. This effect is particularly marked in low N clusters, where stars have a high probability of being ranked 2 even at low masses. Thus, for $N = 4$, the crossover with the diskless results occurs at around $0.5 M_{\odot}$. It should be noted that although Figure 5(b) might suggest to a cruising glance that disks reduce binary formation, the greater incidence of lower mass stars (for which binarity is increased) means that the overall binary fraction is increased by disks. We note that the $N = 10$ and $N = 4$ disk models bracket the observational datapoints for the binary fractions among M and G dwarfs (Fischer & Marcy 1992, Duquennoy & Mayor 1991), implying that a mixture of clusters containing 4-10 stars would provide a very acceptable fit to the observations. The uncertainties in the disk models do not encourage us to constrain the cluster membership number more precisely at this stage.

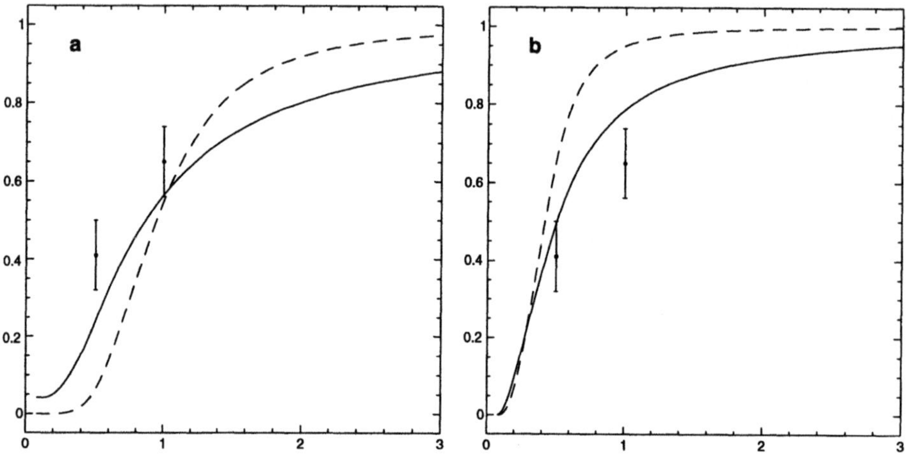

Fig. 5. Binary fraction as a function of primary mass from models with (solid) and without (dashed) disks with the observational binary fractions of M and G dwarfs marked as points for cluster of (a) $N = 10$ and (b) $N = 4$.

4. Conclusions

Disks promote the formation of binary stars in small N clusters through providing a greater interaction cross-sectional area and hardening temporary binaries sufficiently that they can survive encounters with massive stars in the cluster. As a results, the dynamical bias towards more massive stars in binaries found in diskless simulations is substantially reduced. Although the binary fraction increases with primary mass this rise is shallower than in the diskless case: for clusters of 4-10 stars the agreement with observational estimates of this quantity are good when star-disk dissipation is included.

Another observational signature of disk mediated binary formation is that the companion mass function is proportional to the mass function for single stars. This means that such a model cannot be applied to the formation of OB binaries (whose components are somewhat correlated in mass) unless, as is likely, the mass function is different in small clusters containing massive stars (see, eg, Clarke & Pringle 1992, Garmany et al. 1980, Eggleton et al. 1989). For stars of around $1M_\odot$ and below the predicted companion mass function agrees well with observations. It is noteworthy that, unlike in the diskless case, there is no turnover in the companion mass function, relative to the single star mass function, at low masses, an effect implying that brown dwarfs should be commonly found in binaries. We find, however, that they are more likely to be found around more massive stars, where the extreme mass ratios make their detection problematic.

Another consequence of the presence of dynamically significant disks is the formation of a substantial number of multiples (i.e, triple, quadruples), compared with the diskless case. Many of these systems survive over 25 cluster crossing times at least: the probability that some are unstable on considerably larger timescales offers a promising mechanism for explaining the reported excess of binaries among pre-main-sequence stars (Ghez et al. 1993; Simon et al. 1994; Leinert et al. 1993).

Finally we note that the 'disk' results described above refer to the case of large and massive disks (see Sect. 4) where the interaction prescription allows stars to spiral into separations that are roughly an order of magnitude less than those attained in the absence of disks. For the case of weaker disks, the predicted binary statistics are intermediate between the strong disk and those obtained without disks.

Acknowledgement. The authors would like to thank Sverre Aarseth, Jeremy Heyl, Derek Richardson, Andrea Ghez and Jim Pringle for all their help and support throughout this project.

References

Aarseth, S.J., 1985, in: Multiple Time Scales, eds, T.U Bradibill and B.I Cohen, Academic Press, New York, p. 377
Anosova, J.P., 1968, Trans. Astron. Obs. Leningr. Univ. **25**, 100
Anosova, J.P., 1991, Comm. Astophys. **15**, 283
Clarke, C.J. & Pringle, J.E., 1991, MNRAS **249**, 588
Clarke, C.J. & Pringle, J.E., 1992, MNRAS **255**, 423
Clarke, C.J. & Pringle, J.E., 1993, MNRAS **261**, 192
Duquennoy, A. & Mayor, M., 1991, A & A **248**, 48
Eggleton, P.P., Fitchet, M.J. & Tout, C.A., 1989 Astrophys.J. **347**, 998
Fischer, D.A & Marcy, G.W, 1992, ApJ **396**, 178
Garmany,C.D., Conti, P.S. & Massey, P., 1980, Astrophys.J. **347**, 998
Ghez, A.M., Neugebauer, G. & Matthews, K., 1993, A.J **166**, 2005
Hall, S., Clarke, C.J. & Pringle, J.E., in prep.
Harrington, R.S., 1974, Celest. Mech. **9**, 465
Harrington, R.S., 1975, AJ **80**, 1081

Heller, C.H., 1993, Ap.J. **408**, 337
Kiseleva, L.G., Eggleton, P.P. & Orlov, V.V., 1994, MNRAS, **267**, 161
Kiseleva, L. & Anosova J.P., in prep.
Leinert, Ch., Zinnecker, H., Weitzel, N., Christou, J., Ridgway, S.T., Jameson, R., Hass, M. & Lenzen, R., 1993, Astron & Astrophys **278**, 129
Lin, D.N.C. & Pringle, J.E., 1987, Astrophys. J. **358**, 515
McDonald, J.M & Clarke, C.J, 1993, MNRAS **262**, 800
McDonald, J.M & Clarke, C.J., 1995, in prep.
Murray, S. & Clarke, C.J., 1993, MNRAS **265**, 169
Ostriker, E., 1994, Phd thesis, University of California, Berkeley
Pringle, J.E., 1989, MNRAS **239**, 361
Reipurth, B. & Zinnecker, H., 1993, A & A **273**, 81
Simon, M., Ghez, A.M., Leinert, Ch., Cassar, L., Chen, W.P., Howell, R.R., Jameson, R.F., Matthews, K., Neugebauer, G., & Richichi, A., 1995, Ap.J., **443**, 625
van Albada, T.S., 1968, Bull. Astron. Neth. **19**, 479

Proper Motion Measurements in Jets from Young Stars

Jochen Eislöffel

Laboratoire d'Astrophysique, Observatoire de Grenoble, F-38041 Grenoble Cedex, France

1. Introduction

Recently, proper motion measurements of the knots in jets from young stars and of the condensations in their bow shocks have been carried out on deep, high-quality CCD images. The availability of such CCD images allows us to measure the apparent motion of many more knots in the jets and bow shocks than was possible in the earlier works using photographic plates by G. Herbig and co-workers more than a decade ago (Cudworth & Herbig 1979; Herbig & Jones 1981, 1983; Jones & Herbig 1982). Much better sampling of the tangential velocity field in these objects has now become feasible, opening up the possibility of a variety of tests of jet and bow shock models. Some of these tests and their results will be described in the following sections.

2. Knot Motion and Pattern Speed

The first example of a jet system studied in great detail was HH 34 (Eislöffel, Mundt & Ray 1989; Eislöffel & Mundt 1992; Heathcote & Reipurth 1992). In Figure 1 the proper motions in the jet of this object are shown. We see that all the knots move and that the directions of their motions are quite well aligned to the jet axis. The tangential velocity increases from about $100\,\mathrm{km\,s^{-1}}$ for the knots close to the source to more than $300\,\mathrm{km\,s^{-1}}$ for knots near the bow shock. These results are contrary to models of steady jets in which the knots are interpreted as the recombination regions of oblique stationary crossing shock waves, which form naturally when an overpressured jet bores its way through the surrounding interstellar medium (e.g., Königl 1982; Falle, Innes & Wilson 1987; Cantó, Raga & Binette 1989; Raga, Binette & Cantó 1990). The observed proper motions are obviously in direct contradiction to interpretating the knots as being the cooling region of standing shock waves.

It is interesting to ask if the knots themselves move at the speed of the flow or if they propagate at a different velocity, i.e., do they show some kind of pattern motion relative to the flow? Once the angle of an outflow with respect to the plane of the sky is known (measured, e.g., from the knots in the bow shock, see Eislöffel & Mundt 1992), its measured radial velocities, together with this angle, are sufficient to determine the spatial velocity of the flow, while the measured tangential velocities, together with this angle, permit us to determine the spatial

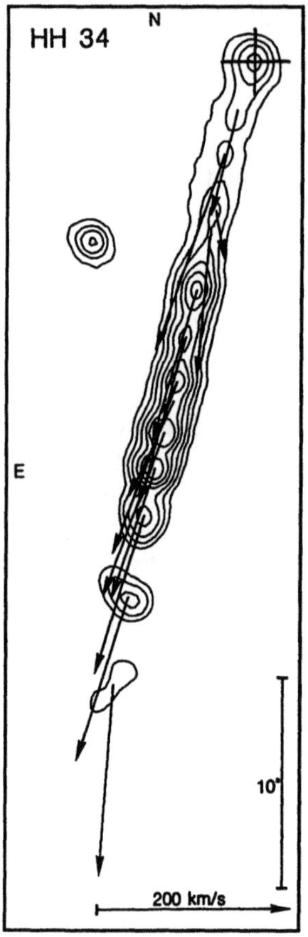

Fig. 1. Proper motions in the HH 34 jet. The tangential velocities amount to about 100 to 200 $\mathrm{km\,s^{-1}}$.

velocity of the knots. The ratio of the spatial velocities of the knots and of the flow then parameterizes how closely the knots move at the flow speed. Even if the angle of the outflow with respect to the plane of the sky is not known well enough to measure the absolute value of the knot pattern motion, one can still detect relative changes in the pattern motion of the knots along the flow. Such changes were indeed found in HH 34 and a number of other outflows. In HH 34 these changes of the knot pattern speed along the flow amount to a factor of about two, while in HH46/47 and the jet from FS Tau B (see Figure 2) to a factor of about three and in the jet from DG Tau B to a factor of about five (Eislöffel & Mundt 1992, 1994, 1995). In those cases where the absolute value of the knot pattern speed can be measured, it seems to be somewhat slower than the flow speed. This knot pattern motion turns out to be a serious constraint on models of jets with internal working surfaces. In these models variability of the jet source is thought to cause variations in the jet velocity. In such a case faster moving jet gas catches up with slower moving gas ahead of it and creates a shock wave – a small bow shock in the flow. This bow shock is interpreted as a jet knot, which should then move along at close to the flow speed (Rees 1978; Raga et al.

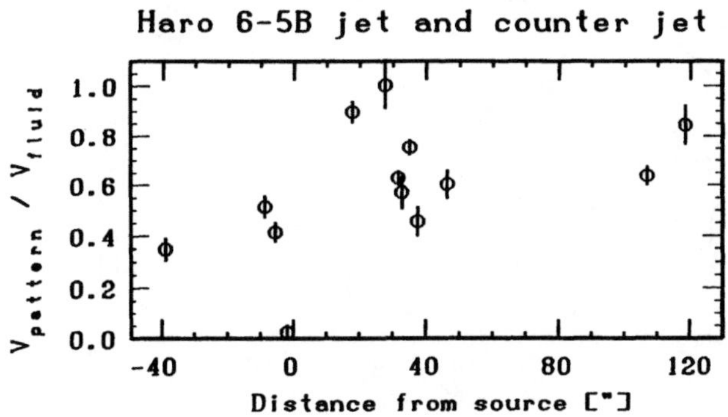

Fig. 2. Ratio of the pattern motion of the knots relative to the flow speed in the Haro 6-5B jet and counterjet. A variation of about a factor of three is found. This diagram is based on the tangential velocities of the jet knots as measured from Hα-images.

1990). However, this seems to be contrary to the observed pattern motions in some jets.

3. Entrainment of Ambient Material into Jets

Proper motion measurements might also allow us some insights into the entrainment of ambient material into the flow. The jet in HH 46/47 seems to offer the opportunity to study such an interaction. This wiggly and highly structured jet splits up into two strands for a distance of about 15 arcsec. Proper motion measurements have shown that the average tangential velocity of the knots in the southern on-axis strand of knots is clearly higher at about 255 $\mathrm{km\,s^{-1}}$ than in the northern off-axis strand, where the knots move at 140 $\mathrm{km\,s^{-1}}$ on average (Eislöffel & Mundt 1994). From their Fabry–Perot imaging of the HH 46/47 jet, Hartigan et al. (1993) also found higher radial velocities on-axis than off-axis in that part of the jet. At the same time they found a higher Hα/[SII] ratio, i.e. higher-excitation shocks, for the off-axis part. Both facts, higher tangential and radial velocities on-axis, and the higher excitation off-axis, have been interpreted as a sign of the interaction of the flow with the wall of the jet channel and the entrainment of ambient material into the flow in a turbulent mixing layer (Raymond et al. 1994). Evidence for such entrainment of ambient material has also been found in HH 24 (Solf 1987) and in HH 1 (Eislöffel, Mundt & Böhm 1994).

4. Internal Motion in Bow Shocks and Comparison with Other Physical Parameters

Proper motion measurements not only tell us something about processes in the jets themselves, but also in their bow shocks. HH 47D, one of the bow shocks in the HH 46/47 jet system, serves as an example to demonstrate the internal motion in a bow shock. From the measured proper motions of the condensations

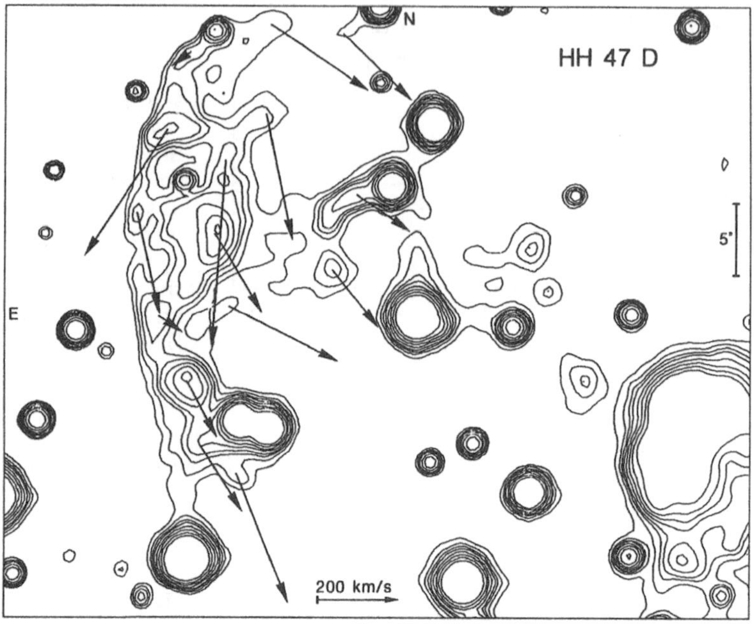

Fig. 3. Internal motion of the condensations in HH 47D. From the measured proper motions a plausible vector for the motion of HH 47D was subtracted.

in this bow shock we subtract a vector that has the direction of the outflow axis and the magnitude of the motion of the bow's apex. That way we will get the internal flow pattern in the bow shock relative to its apex (Figure 3). Now we can compare this result to the pattern expected from simple bow shock models. In these models a spherical obstacle is assume to plough through the ambient medium thereby creating a parabola-like bow (see, e.g., Hartigan, Raymond & Hartmann 1987). The motion of the condensations in this bow should be directed backwards along the wing and their relative velocity should increase further away from the apex. If we compare that to the pattern found in the observations (Figure 3) they are indeed in good agreement, at least when we exclude the condensations in the region where the jet enters the bow shock. This region cannot be described by such a simple model. Good agreement between the observed internal motions and the model are also found for other bow shocks like like HH 1 (Eislöffel, Mundt & Böhm 1994) and HH 47C (Eislöffel & Mundt 1994).

Once the tangential velocities of the condensations are known, other tests of bow shock models can be carried out as well, for example the comparison of the tangential velocity field with other astrophysically meaningful parameters. In HH 2 detailed spectroscopic observations have been carried out of a number of knots (e.g., Solf & Böhm 1991; Böhm & Solf 1992) and so it is possible to compare their tangential velocities with their excitation and radial velocity dispersion. HH 2 is thought to be a bow shock which broke up into clumps by instabilities. These clumps move individually now and are the condensations

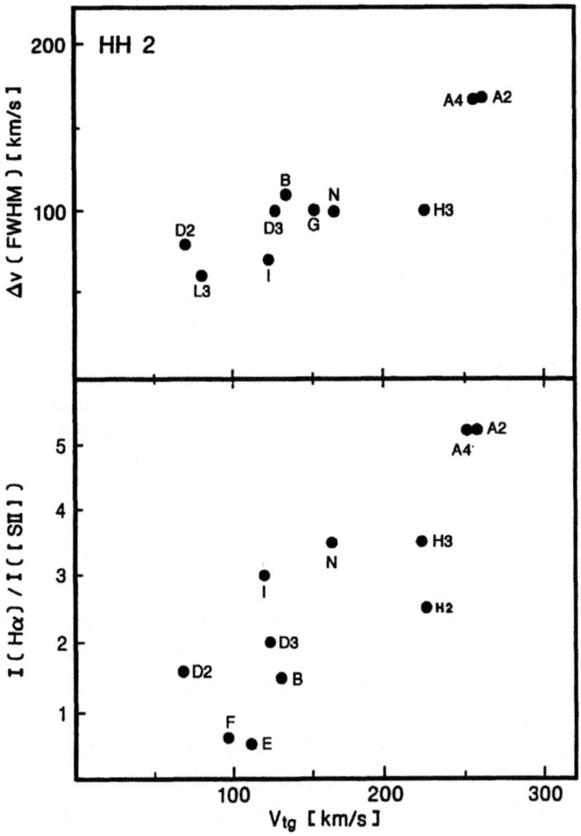

Fig. 4. a) Tangential velocities versus Hα / [SII] λλ6716, 6731 ratio in HH 2. b) Tangential velocities versus radial velocity dispersion in HH 2. A linear correlation is found in both cases.

that we see. Since their bow shock structure is not really resolved, they should have stronger shocks, and so show higher excitation spectra, the faster they are moving. Exactly this behaviour is found in a nice correlation between tangential velocity and excitation measured as the ratio between Hα and the [SII] lines in Figure 4a. In the same model one expects a higher radial velocity dispersion, i.e. a higher line width, the faster the knots move, because they will cover a larger range of velocities from the apex to the wings. This relationship is seen in a good correlation between the tangential velocities and radial velocity dispersion shown in Figure 4b. Both presented correlations (from Eislöffel, Mundt & Böhm 1994) nicely agree with the model of HH 2 being a bow shock that has now broken up into individually moving clumps.

5. A Relation Between Source Variability and Knot Production?

VLBI observations of extragalactic jets show that knot production close to the source is accompanied by photometric outburst of the jet source beforehand

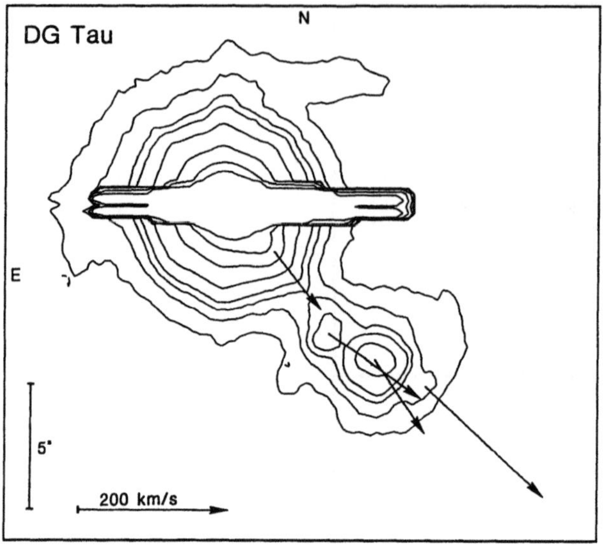

Fig. 5. Proper motions of the knots in the DG Tau jet. For the three knots closest to the source ejection dates of 1970 (±6 years), 1958 (±4 years) and 1936 (±4 years) are derived from their distance to the source and their motion.

(Babadzhanyants & Belonkon' 1984; Belokon' 1991). So in these sources there seems to be some direct connection between the source activity and knot production. On the other hand in stellar jets the appearance of only one new knot has been reported in the 12 years since Mundt & Fried (1983) discovered the first jets: when comparing a sequence of images of the L1551 IRS5 jet taken between 1983 and 1987 Neckel & Staude (1987) discovered a new knot that had appeared at the beginning of the optically visible jet. But since the source IRS5 is deeply embedded, that knot was first seen when it was already several arcseconds away from the source. So the actual time of its ejection is not known. Is it possible to learn something about source activity and knot production in spite of such a poor detection rate of newly ejected knots? For DG Tau the proper motion measurements in the jet (Figure 5) show that the knots were ejected within the last 60 years (assuming that they moved at constant velocities throughout that entire timespan). Since DG Tau is bright, of 12 mag in V, it should be visible even on old photographic plates, and since the advent of photoelectric photometry it has certainly been measured again and again in various colours. So by searching through plate archives and by collecting the available data published in the literature it should be possible to put together a complete photometric light curve of this source and to compare it with the ejection dates of the knots from the proper motion measurements. That way we might be able to learn something about a possible connection between source activity and knot production without having to monitor a large number of jet sources over a rather long time.

6. Conclusions

We have discussed some aspects of proper motion measurements of the knots in jets from young stars and of the condensations in their bow shocks for our understanding of the outflow phenomenon. The measured proper motions permit some decisive tests of the various jet models and point out problems for both the steady jet model and the presently favoured model of internal working surfaces. The fact that the knots move at all is in contradiction with the principal prediction of the steady jet models, which explain the knots as stationary shock waves, while the observation that the knots show motions relative to the flow cannot (yet) be accounted for by the model of internal working surfaces. On the other hand bow shocks are generally well explained even by simple models. The observations of the internal velocity field and comparison of the tangential velocities of the knots with other physical parameters, like excitation or radial velocity dispersion, reproduce well the predictions of these models.

Acknowledgements. It is a pleasure to thank Reinhard Mundt, together with whom the work presented in this paper has been carried out. I am grateful to Tom Ray for many discussions on stellar jets and for his critical reading of the manuscript. This work was partially supported by the the Human Capital and Mobility Programme of the European Community under grant ERBCH-BGCT920205.

References

Babadzhanyants, M.K., Belonkon', E.F. 1984, Astrofizika **21**, 217
Belokon', E.T. 1991, Sov. Astr. **35**, 1
Böhm, K.H., Solf, J. 1992, AJ **104**, 1193
Cantó, J., Raga, A.C., Binette, L. 1989, Rev.Mex.Astr.Astrof. **17**, 65
Cudworth, K.M., Herbig, G.H. 1979, AJ **84**, 548
Eislöffel, J., Mundt, R. 1992, A&A **263**, 292
Eislöffel, J., Mundt, R. 1994, A&A **284**, 530
Eislöffel, J., Mundt, R. 1995, in preparation
Eislöffel, J., Mundt, R., Ray, T.P. 1989, Astr. Ges. Abstr. Ser. **3**, 35
Eislöffel, J., Mundt, R., Böhm, K.H. 1994, AJ **108**, 1042
Falle, S.A.E.G., Innes, D.E., Wilson, M.J. 1987, MNRAS **225**, 741
Hartigan, P., Raymond. J., Hartmann, L.W. 1987, ApJ **316**, 323
Hartigan, P., Morse, J.A., Heathcote, S., Cecil, G. 1993, ApJ **414**, L121
Heathcote, S., Reipurth, B. 1992, AJ **104**, 2193
Herbig, G.H., Jones, B.F. 1981, AJ **86**, 1232
Herbig, G.H., Jones, B.F. 1983, AJ **88**, 1040
Jones, B.F., Herbig, G.H. 1982, AJ **87**, 1223
Königl, A. 1982, ApJ **261**, 115
Mundt, R., Fried, J.W. 1983, ApJ **274**, L83
Neckel, Th., Staude, H.J. 1987, ApJ **322**, L27
Raga, A.C., Binette, L., Cantó, J. 1990, ApJ **360**, 612
Raga, A.C., Cantó, J., Binette, L., Calvet, N. 1990, ApJ **364**, 601

Raymond, J.C., Morse, J.A., Hartigan, P., Curiel, S., Heathcote, S. 1994, ApJ **434**, 232
Rees, M.J. 1978, MNRAS **184**, 61p
Solf, J. 1987, A&A **184**, 322
Solf, J., Böhm, K.H. 1991, ApJ **375**, 618

Molecular Entrainment by Optical Jets

Stephen Bence and Rachael Padman

Mullard Radio Astronomy Observatory, Cavendish Laboratory, Madingley Road, Cambridge CB3 0HE, United Kingdom

Summary. Molecular outflows around young stellar objects (YSOs) have been studied extensively. Many models have tried to explain their appearance, and currently the most popular are those which invoke momentum-conserving collimated jets, emanating from the YSO. Two competing models describe the mechanism by which the jet sets the molecular cloud in motion: 'prompt' entrainment and 'turbulent' entrainment. We present a method for distinguishing between the two cases by examining the velocity structure and the morphology of outflows. We also examine the effects on the observations of viewing angle and show how two well-studied outflows fit into the global picture of jet-molecular cloud interaction. Finally we show evidence for time-dependent ejection of the jet.

1. Introduction

The idea that optical jets from YSOs drive molecular outflows has recently gained acceptance. Initial objections, that the total momentum supplied by the jets over their lifetime was insufficient to drive the outflow, have now been answered. This has been brought about by a re-examination of the lifetime of the outflow [Parker et al. (1991) and Fukui et al. (1993)] showing that the dynamical lifetime of outflows (i.e. the total spatial extent of the outflow divided by the highest velocities) underestimates the true value by an order of magnitude. Recent authors have considered the nature of this æsthetically pleasing direct interaction.

Molecular entrainment may be convenient divided into two categories [De Young (1986)]: prompt entrainment at the jet head (a bowshock) and steady-state entrainment along the sides of the jet due to Kelvin-Helmholtz instabilities. The two mechanisms predict radically different shapes and velocity structures. Prompt entrainment will produce high velocity emission at the jet head and an expanding parabolic shell away from the head, whereas turbulent entrainment will show an outflow with the lowest velocities at the head of the jet and an outflow with spatial extent increasing away from the source. Stahler (1993) and Masson & Chernin (1993) have discussed the cases in detail and the two senarios are summarized in Figure 1.

2. RNO 43 – A Case Study

The molecular outflow RNO 43 (L 1582B) shows both optical and molecular emission and so is an ideal source in which to observe the direct interaction between the two. The global picture is shown in Figure 2, where the integrated in-

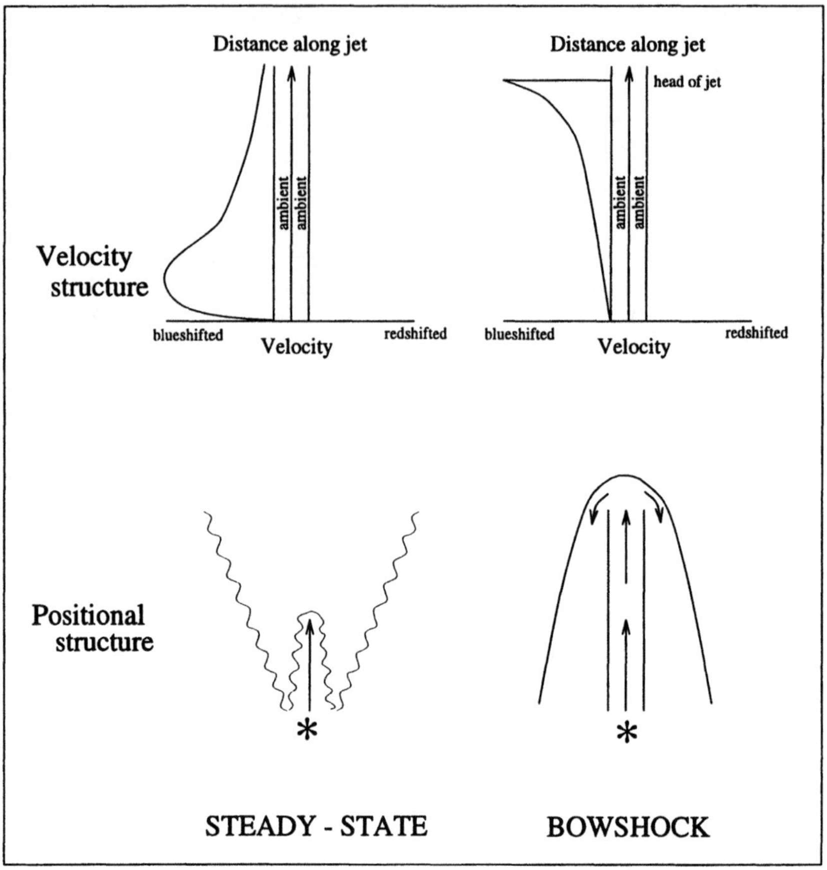

Fig. 1. The positional and velocity structure of the two competing models of jet–cloud interaction.

tensity of the CO emission is overlaid with optical emission [Mundt et al. (1987)]. Recent work [Bence et al, in preparation] shows that all the molecular and optical emission is associated with one, central, FIR driving source and also that the region is more extended than shown. Although each region is of interest, we will limit our discussion to the D region which has fully sampled CO spectra and thus affords the greatest detail. Figure 3 shows the velocity structure along a cut joining the head of the D region to the FIR source. This velocity structure is the classical bowshock structure shown in Figure 1 and we also note that the spatial structure is parabolic as expected for a bowshock.

2.1 Geometric Effect

The theoretical velocity profile shown in Figure 1 relates to the velocity of CO moving in a direction along the axis of the jet. For each individual side of a bipolar outflow this profile is naturally asymmetric in velocity and we can see

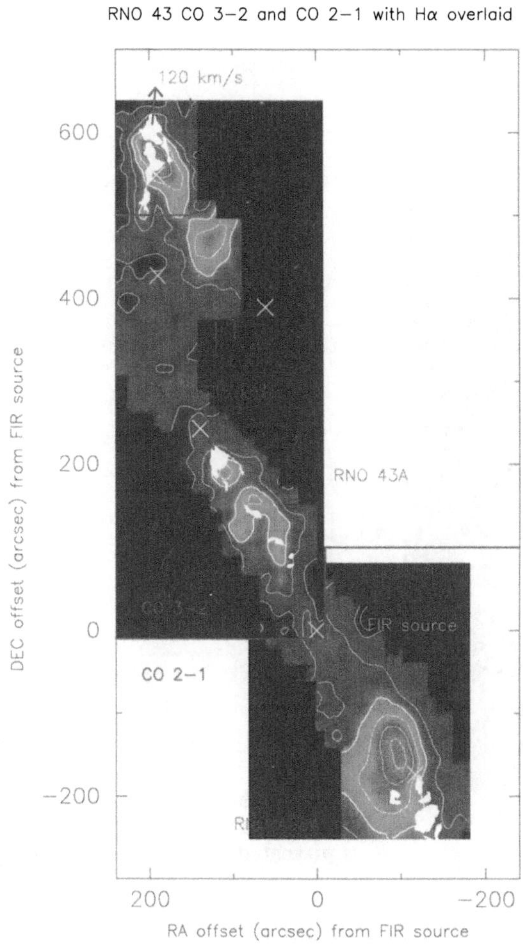

Fig. 2. The global picture of the RNO 43 molecular outflow. Yellow crosses mark 3.6 cm sources. [This figure is presented in color on CD-ROM.]

this asymmetry in the velocity structure of the blue-shifted RNO 43D (Figure 3). However, in the case of RNO 43, the outflow is almost in the plane of the sky [Cabrit et al. (1988)] and so we also expect to see emission associated with CO moving transverse to the the jet axis (for outflows which have a significant angle away from the plane of the sky this component becomes so small that it is dominated by the longitudinal component). Assuming axisymmetry, which is globally reasonable by conservation of momentum, the transverse emission is symmetrical in blue- and red-shifted components and so we expect it to show up as a symmetrical component in the averaged spectrum. This is shown in Figure 4.

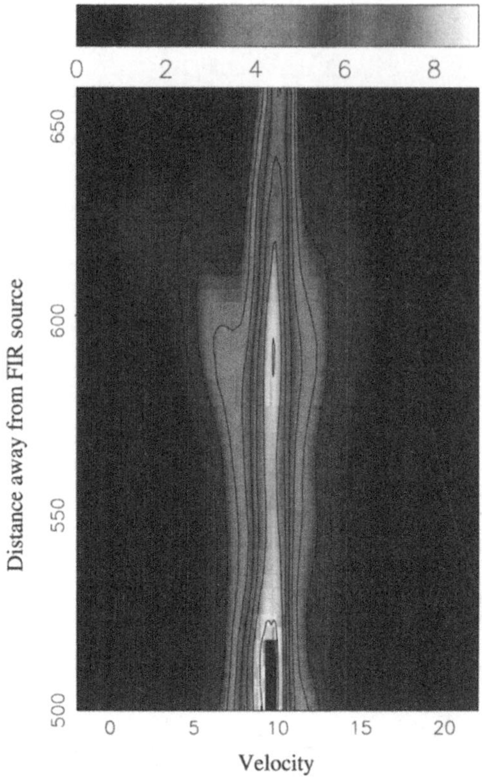

Fig. 3. The velocity structure of the RNO 43D region. [This figure is presented in color on CD-ROM.]

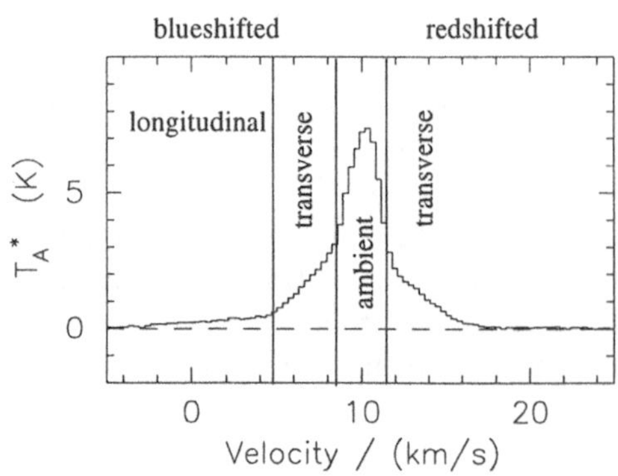

Fig. 4. The average spectrum of the RNO 43D region showing three distinct components: the emission associated with CO longitudinal to the jet direction appears as an asymmetrical component, and that associated with the CO perpendicular to the jet direction is symmetrical.

3. The Multiple Bowshocks of NGC 2024

Unlike RNO 43, the molecular outflow NGC 2024 (Figure 5) does not show optical emission although it shows shocked H_2 S(1) [Richer et al. (1992)]. Since the visual extinction is estimated to be about 50 magnitudes we expect any 'optical' jet to be obscured, but nevertheless we assume that the outflow is driven by an unseen jet. Figure 6 shows the velocity along the jet's axis and just as in the case of RNO 43, we see classical bowshock behaviour. Five distinct

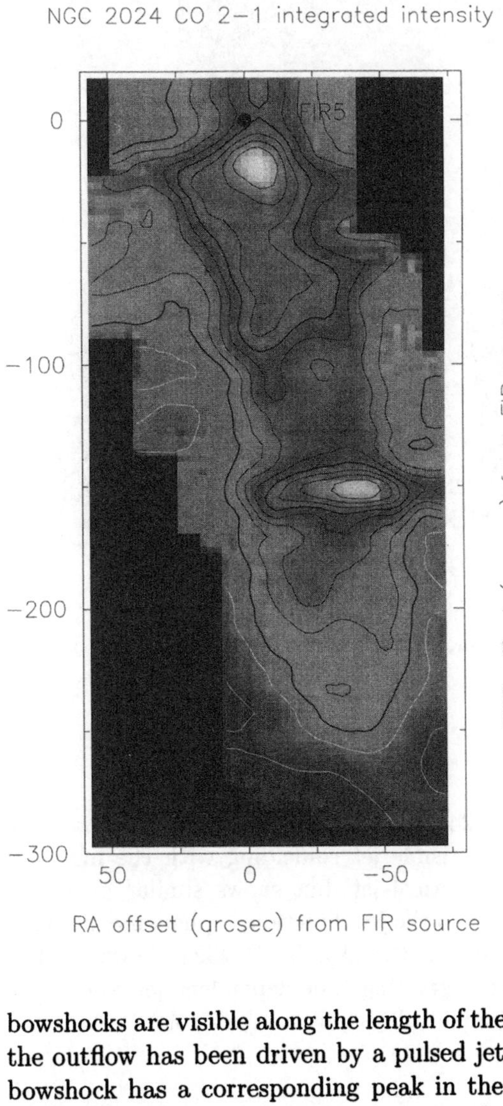

Fig. 5. The CO integrated intensity emission for the molecular outflow NGC 2024. [This figure is presented in color on CD-ROM.]

bowshocks are visible along the length of the outflow in this case, suggesting that the outflow has been driven by a pulsed jet. The high velocity material of each bowshock has a corresponding peak in the integrated CO emission (Figure 6) which also shows a parabolic shape at the end of the outflow. Although this parabola is broad compared to theoretical models its morphology is that of a classical bowshock.

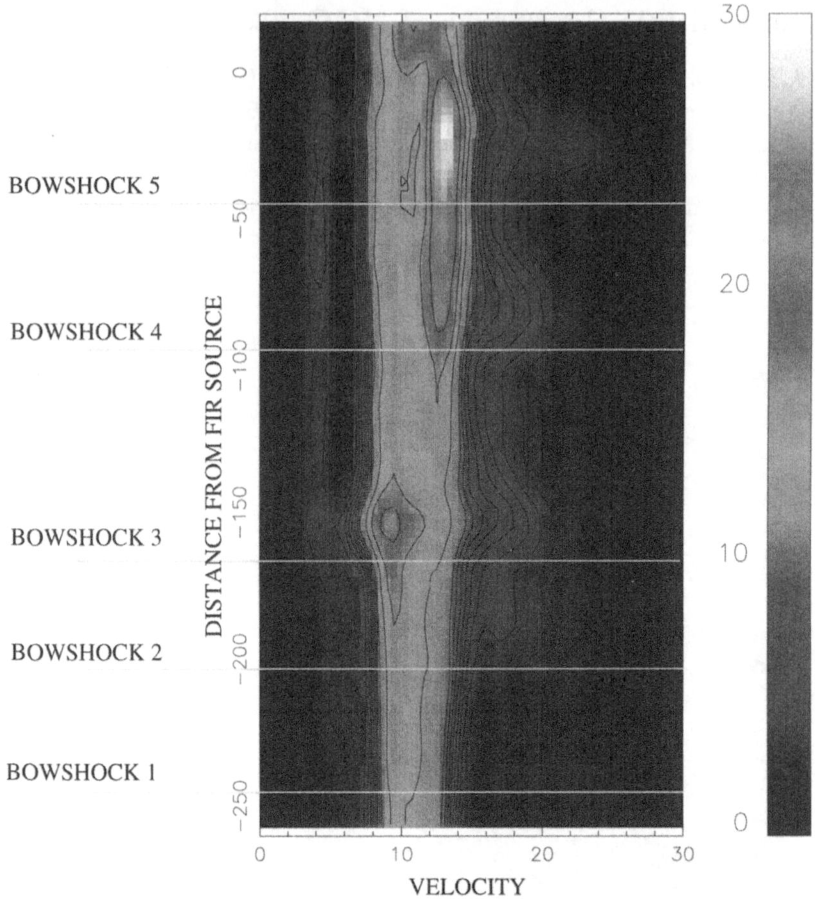

Fig. 6. The velocity structure of the molecular outflow NGC 2024. [This figure is presented in color on CD-ROM.]

4. Conclusion

The outflows RNO 43 and NGC 2024 show prompt entrainment (bowshock) behaviour. RNO 43 has an optically-visible jet coinciding with the molecular emission; NGC 2024 has no visible 'optical jet' but shows similar behaviour. RNO 43 shows CO moving transverse to the jet direction, a phenomenon which relies on it lying almost in the plane of the sky. NGC 2024 shows multiple bowshocks along the axis of the jet suggesting time-dependent jet ejection. In both these sources turbulent entrainment plays only a minor role.

References

Bence S. J., Richer J. S., Padman R., 1995, in prep
Cabrit S., Goldsmith P. F., Snell R. L., 1988, ApJ, **334**, 196
De Young D. S., 1986, ApJ, **307**, 62
Fukui Y., Iwata T., Mizuno A., Bally J., Lane A.P., 1993, Protostars & Planets III, eds. Levy E. H., Lunine J. I., Univ. Arizona: Tuscon
Masson C.R., Chernin L.M., ApJ, **414**, 230
Mundt R., Brugel E. W., Bührke T., 1987, ApJ, **319**, 275
Parker N. D., Padman R., Scott P. F., 1991, MNRAS, **252**, 442
Richer J. S., Hills R. E., Padman R., 1992, MNRAS, **254**, 525
Stahler S. W., 1993, in Astrophysical Jets, ed. D. Burgarella, M. Livio & C. O'Dea, (Cambridge University Press), 183.

HH 212: The Most Beautiful Protostellar Jet Known to Date

Hans Zinnecker[1], Mark McCaughrean[2], and John Rayner[3]

[1] Institut für Astronomie und Astrophysik, Universität Würzburg, Am Hubland, D-97074 Würzburg, Germany
[2] Max-Planck-Institute für Astronomie, Königstuhl 17, D-69117 Heidelberg, Germany
[3] Institute for Astronomy, University of Hawaii, 2680 Woodlawn Drive, Honolulu, HI 96822, USA

Summary. We report the discovery of the most symmetric embedded twin-exhaust jet known to date. It is located in a dense molecular star forming core in Orion, not far from the famous Horsehead Nebula, and is about 0.3 pc long on either side. Each side contains an inner series of spatially resolved knots (with inter-knot emission) and an outer series of giant bow-shocks, all seen in the v=1–0 S(1) line of shock excited H_2 at 2.12 μm. Each pair of bow-shocks represents a distinct ejection event, lasting of the order of 300 yr. The regular spacing of the inner knots (of order 0.01 pc) which are almost perfectly matched on opposite sides of this bipolar jet, provide the strongest evidence yet for a physical model of a time-variable pulsed jet, with a period of (small) velocity variations as short as 30 yr. The high degree of symmetry also allows us to see that the two opposite halves of the jet are not completely co-linear and that there is a 1–2 degree asymmetry angle. The jet (named HH 212) originates from a very cold infrared and mm-continuum source, *i.e.,* a very young embedded stellar object with a luminosity about 15 $L_\odot$, likely to be powered by accretion from an edge-on disk. Furthermore, the position of the exciting source as inferred from the symmetry of the jet coincides very well with the position of a compact radio H_20 maser, this maser being the origin of our initial interest in this cold IRAS source.

1. Infrared Observations

The birth of a low-mass, solar-type star is often first signalled by a powerful jet and outflow emanating from an invisible object still deeply embedded in its parent gas and dust cloud. At the earliest ages, the jet itself may also not be visible at optical wavelengths due to the same extinction, and we must resort to longer wavelengths in order to find and study them.

As part of a continuing near-infrared imaging search for embedded young binary and multiple systems in cold protostellar IRAS sources, we observed the source IRAS 05413−0104 with the 256x256 pixel infrared camera NSFCAM at the 3-m IRTF on Mauna Kea on December 18, 1993.

We first observed the source in the continuum K band (2.2 μm); these images showed some faint structures. Therefore we took images through narrow-band filters centered on the v=1–0 S(1) line of molecular hydrogen at 2.12 μm and H^+ Brγ line at 2.17 μm, both with effective integration times of 4 minutes per pixel. The seeing was 1″ FWHM. While the Brγ images showed no emission, the H_2 images did. Although it was immediately clear that we had discovered a spectacular jet, the full size of the two-sided object was not revealed until we had obtained a 2×4 arcmin mosaic in the H_2 line.

The H_2 mosaic is shown in Figure 1. We can see a highly symmetric bipolar jet with a series of knots and bowshocks, and as the data have not been continuum subtracted, there are also a number of stellar point sources visible. This jet was subsequently named HH 212 (Reipurth 1994), where HH stands for Herbig–Haro and the number (212) reminds us of the wavelength at which the object was discovered (2.12 μm). The total length of the jet is about 240″ or 0.6 pc at the distance of the Orion B molecular cloud (500 pc) where the object is located, about 90 arcmin north-east of the Horsehead Nebula. Thus the double-sided jet extends about 0.3 pc on either side of the IRAS position. The mirror symmetry is obvious and striking, and it is this high degree of symmetry which makes this molecular hydrogen jet arguably the most beautiful jet discovered so far.

2. Observed Facts

Next we list the observed facts and features contained in the H_2 images, together with some basic numbers derived from these data.

1. The jet is apparently bipolar and highly symmetric with respect to a position between the innermost two bright H_2 knots, a position that coincides with the location of the totally obscured driving source, as discussed below. There is a system of 3 almost regularly spaced inner knots (spacing 4″ or a projected separation of 2000 AU) on both sides, each of which seems to end in a small double bow-shock at a distance of about 20,000 AU from the central position[1]. The symmetry extends further: there is a larger bowshock on either side at about 40,000 AU. Finally, the symmetry is broken with a final large bowshock to the SW, at about 60–70,000 AU from the centre: it has no counterpart on the NE side.
   If the bulk velocity of the two-sided jet were 300 km/s, the kinematic age of the jet (time to reach the outermost bow-shock) would be ~1000 yr; the regular spacing of the bowshocks then suggests a large-scale periodicity of about 300 yr. If the jet itself is molecular, the assumed high speed and acceleration of the jet raises questions as to the survival of H_2, as the H_2 molecules should not survive shocks more than 20–40 km/s. If, on the other hand, the jet consists of ionized material, the H_2 emission must arise in the ambient medium entrained by the jet and the existence of low-velocity shocks would be no problem.
2. The inner H_2 emission is more or less equally bright on either side, indicative of equally strong intrinsic emission and equal extinction on either side. Thus, we may infer that the jet axis lies close to an axis of symmetry of the parent cloud and also close to the plane of the sky. Radial velocity measurements

[1] While the brightness of the inner knots decreases with distance from the driving source, it should be noticed that the emission from the double bow-shocks is relatively strong again.

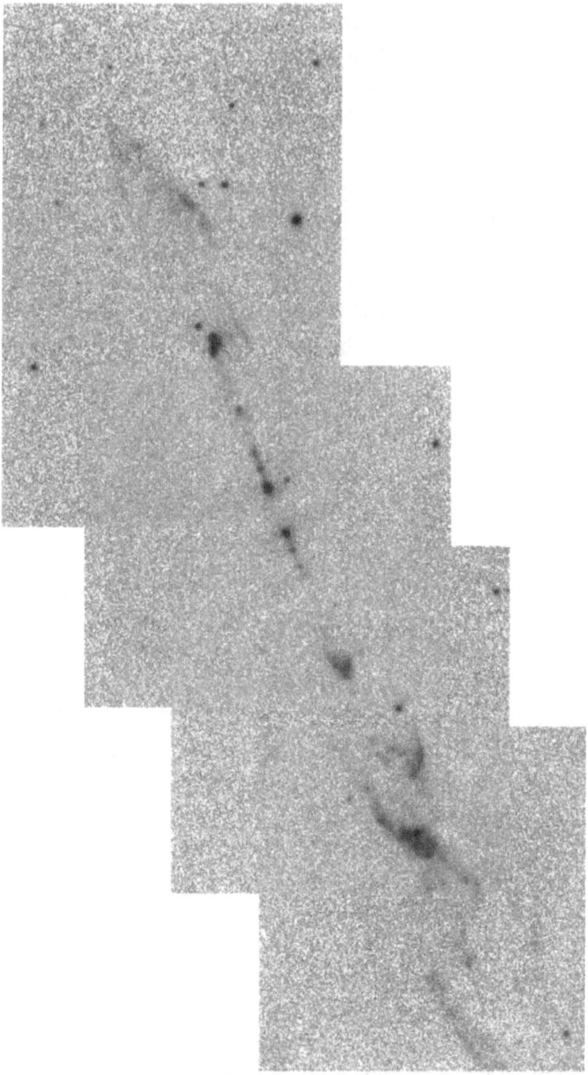

Fig. 1. Image of the shock-heated H_2 gas (ro-vibrational v=1–0 S(1) line emission at 2.12μm) in the bipolar protostellar jet HH 212 associated with the very cold infrared/mm source IRAS 05413–0104 in Orion. A number of stellar point sources are also visible, as this image has not been continuum subtracted. The brightest H_2 knots have a surface brightness of 6×10^{-4} erg/s/cm^2/sr and the total S(1) line luminosity is 0.007 $L_\odot$, not corrected for extinction. Note the high spatial symmetry between the two sides of the jet, especially in the inner part. The exciting source is located in between the two innermost knots and coincides with an H_2O maser. The image is $\sim 2\times4$ arcmin in size and is a mosaic of several infrared array images obtained in December 1993, with NSFCAM at the 3-m NASA IRTF on Mauna Kea.

of the knots at 2.12 μm, kindly provided by R. Gredel and J. Bally, confirm this geometry. Further out, the H_2 emission in the giant outer bow-shocks is no longer equally bright on the two sides of the jet, presumably because there the jet is interacting with the external interstellar medium, which is likely to be distributed unevenly.

3. The system of inner knots (up to the small double bow-shock on either side) defines the direction of the northern and southern parts of the jet. These two directions are not completely colinear, and there is an "asymmetry angle" of 1–2 degrees. This obliquity could be indicative of relative motion of the central driving source (on the order of a few km/s) with respect to the cloud core in which it is embedded, or of relative motion (ram pressure) of the core with respect to the ambient cloud medium; the latter is supported by the presence of multiple velocity structure at the position of the HH 212 exciting source (Wouterloot *et al.* 1989; J. Bally, personal communication). Furthermore, the apices of the outer bow-shocks appear to be displaced from the directions defined by the inner jet by small amounts in opposite directions, as if the outer jets were guided along the two sides of some mildly S-shaped curve, hinting at precession of the source (*e.g.*, Fukui & Yokoo 1986). Precession could be due to the initial settling of the orbital plane of a circumstellar disk (free precession) or could be caused by torques in a young binary system (forced precession). If this S-shape were due to rotation of the ambient medium, the rotation axis of the medium would not be colinear with the jet.
4. The innermost knots are clearly spatially resolved in our images, with a seeing-deconvolved intrinsic size of about 1″ or 500 AU. They are slightly larger across the jet axis, suggestive of small bow shocks. There is also faint continuous interknot H_2 emission. The ratio of knot spacing to knot diameter is between 3:1 and 5:1. Such ratios could, in principle, be consistent with the knots arising from Kelvin–Helmholtz instabilities (m=0 pinch mode) in a MHD jet with small (*i.e.*, 2–3) magnetosonic Mach number (Appl & Camenzind 1992; M. Camenzind, personal communication), but the bow-shock nature of the knots, if confirmed, would argue against this explanation and would favour a pulsed jet origin. Note that the actual width of the jet must be smaller than the knot diameter, unless the knots represent compact ejecta.
5. The bows of the outer bow-shocks are rather open and paraboloidal in shape, similar to those modelled by Smith (1991). H_2 emission in the wings of the bows is weak and does not disappear near the working surfaces (apices). This is somewhat surprising, as H_2 dissociation caps might be expected for fast shocks associated with jet heads advancing with almost the bulk speed. We note that the opening angle and the H_2 emission of the bows is also a function of density contrast between the jet and ambient medium and the cooling of hot gas squirted out from the working surface where shock dissipation occurs. The orientation and strength of the magnetic field in the ambient medium may also play a role (Smith 1991). Last but not

least, it is worth noting that the two outermost southern bows appear to be interconnected, exhibiting a twisted morphology reminiscent of 3D helical motion. This might be evidence for a directional variability of the jet early on before it settled for the straight course witnessed by the inner knots (L. Kofman, personal communication).

6. The fact that the outer bow-shocks and the inner knots exhibit H_2 emission at the same time, despite a LTE cooling time of less than a year for shock excited H_2 (at T=2000 K), implies that the emission must be powered all the time, hence that the jet fluid flow cannot be strongly interrupted. On the other hand, the presence of at least 2 outer bow-shock systems argues in favour of multiple ejection events with time intervals of around 300 yr for an assumed jet bulk velocity of 300 km/s, as discussed above. It is also noteworthy that there are no knots between the outer bow-shocks.
7. There is some near-infrared reflection nebulosity around both of the innermost knots, with dark lane in between. As we know the jet lies close to the plane of the sky, it seems likely that we are seeing light reflected from the central source above and below an edge-on circumstellar disk, which obscures the object from direct sight in the near-infrared.
8. Deep optical CCD images in Hα and [SII] at ESO (courtesy of W. Brandner and B. Reipurth, respectively) have failed to detect the jet[2]; it is therefore fair to say that HH 212 is an embedded pure infrared molecular hydrogen jet, the second of its kind after HH 211 (McCaughrean, Rayner, & Zinnecker 1994).

3. The Nature of the HH 212 Exciting Source

Next we discuss the nature and the position of the exciting source. This source (IRAS 05413−0104) is listed in the IRAS point source catalog, and is detected only at 25, 60, and 100 μm, with flux densities of 0.31, 17.3, and 59.5 Jy, respectively. It is too cold to be detected at 12 μm. Sub-mm and mm continuum measurements have been carried out (Zinnecker *et al.* 1992) and the source was detected at 1.1 mm at the JCMT at a level of about 0.2 Jy[3].

The IRAS and JCMT data points can be fit by a simple, spherically symmetric model with an $r^{-1.5}$ density distribution inside R=0.01 pc, containing about 0.4 $M_\odot$ of gas and dust. It is worth noting that this radius is about equal to the location of the first bright inner knot of H_2 emission. The luminosity is 15 $L_\odot$ at an assumed distance of 500 pc. We note that disks observed edge-on can also be fitted by spherically symmetric models, but the derived luminosity

[2] Only two very faint diffuse spots of [SII] emission were detected in a 30 min ESO-NTT exposure, located near the apices of the outer bow shocks some 170″ apart (B. Reipurth, personal communication)

[3] IRAS maximum entropy maps clearly resolve a second source at 60 μm some 2.5′ towards the southwest at PA=30 degrees (R. Assendorp, personal communication). This coincides with a very red point source and an associated reflection nebula seen in broad-band near-infrared images, and seems to be unrelated to the jet.

is then underestimated by the amount of near-infrared flux escaping unobserved in polar directions (*cf.* Sonnhalter *et al.* 1995).

The reason we originally singled out this cold IRAS source for study was that it showed rather strong H_20 maser emission (Wouterloot & Walmsley 1986), which itself was shown to lie near the center of an ammonia core (Wouterloot *et al.* 1988). The FWHM of that ammonia core is about 40″ or 0.1 pc, and assuming that the core is roughly twice the FWHM in size, we arrive at a core diameter of ~0.2 pc. It is interesting to note that it is at *this* diameter that the highly symmetric system of inner knots ends in a pair of bow-shocks, and where the transition to the system of outer bow shocks occurs, as if the jet were breaking out of its parent protostellar core.

A small offset between the maser position and the IRAS position was observed (Wouterloot *et al.* 1988). However, the lack of IRAS 12 μm data made the IRAS position too uncertain to associate any significance to the small positional mismatch, and we can assume that the peak of the IRAS emission coincides with the maser position. The maser position lies very close (within 1″) to the position of the exciting source, as inferred from the symmetry of the jet and near-infrared astrometry of stars surrounding it.

The positional coincidence of the maser and central source has been confirmed recently by 1.3 cm continuum observations with the VLA (Claussen, Wilking, & Wootten 1995, in preparation). These measurements have found the maser to be compact at a synthesized spatial resolution of 0.3″ or 150 AU. The maser has two velocity components at 2 km/s and 4.5 km/s, to be compared with the systemic velocity of 1.5 km/s of the NH_3/CO cloud core (Wouterloot *et al.* 1989). If the two velocity components correspond to receding and approaching motion of dense gas in a circumstellar disk or in a highly inclined bipolar disk wind, the protostellar system velocity would be in between (3.25 km/s) and hence somewhat higher than the cloud velocity, consistent with differential motion between cloud and embedded protostar (see below). It is also possible that the two velocity components of the maser correspond to two stellar components, consistent with a binary system with projected separation less than 150 AU and a projected orbital velocity difference of 1.75 km/s. These numbers would suggest component masses of order 1 $M_\odot$ or less.

Note that no continuum source at 3.6 cm could be detected with the VLA, consistent with our non-detection of near-infrared Brγ emission. This does not however place a strong limit on the absence of ionized gas, as calculations show. Nevertheless we suspect that there is probably little ionized gas and that the source drives a neutral, most likely molecular jet. This is in contrast to most numerical simulations of protostellar jets (*e.g.*, Stone & Norman 1993) which assume a jet plasma of temperature $T = 10^4$ K, suggested by observations of optical jets (*cf.* Mundt 1986).

4. Comparison with Other Jets from Young Stars

The beauty of the HH 212 jet invites comparisons to other spectacular jets, such as HH 34 (Bührke *et al.* 1988; Heathcote & Reipurth 1992) HH 111 (Reipurth 1989; Reipurth *et al.* 1992; Gredel & Reipurth 1994; Davis *et al.* 1994c), and HH 211 (McCaughrean *et al.* 1994). All of these are two-sided, but less obviously symmetric relative to HH 212.

HH 34 is a very long and very straight two-sided jet (total length $\sim$6 pc), terminated by bow-shocks at each end, and has been dubbed a 'superjet' (Bally & Devine 1994). However, HH 34 is largely unembedded, and its optical emission does not have a molecular hydrogen counterpart. By contrast, only one lobe of the HH 111 jet is optically revealed, with the other lobe embedded in a molecular cloud. Both lobes are detected in H_2 testifying to the underlying symmetry, although the optically revealed lobe is also the brighter lobe in H_2 emission (Gredel & Reipurth 1994; Davis *et al.* 1994c). It is also noteworthy that both HH 34 and HH 111 have been observed with mm-interferometry, revealing gaseous structures around the exciting stars, oriented perpendicular to the jet direction (Stapelfeldt & Scoville 1993).

Other optical jets and HH-objects which have also recently been imaged in H_2 include HH 1/2 and HH 46/47 (Davis *et al.* 1994b; Eislöffel *et al.* 1994). Also, the highly collimated CO outflow in L 1448 exhibits bipolar H_2 emission (Bally *et al.* 1993, Davis *et al.* 1994a).

HH 211 is another kind of H_2 jet altogether: like HH 212, it is completely embedded and undetected in the optical, and it is very young, with an inferred kinematic age of <1000 yr (McCaughrean *et al.* 1994). HH 211 is not as extended as HH 212, and appears more irregular and asymmetric, with a broken limb-brightened structure. This structure may reflect slow shocks where the invisible inner jet sweeps up external gas into the the observed coincident molecular outflow (Raga & Cabrit 1993; Masson & Chernin 1993). The broken structure of HH 211 may well be due to local density variations in the ambient medium, not to velocity or mass outflow variations in the driving source.

In contrast, the regularity and strong two-sided symmetry of HH 212 appears to imply that it is driven by an intrinsically time-variable source. The symmetric inner knots may be related to some form of young stellar magnetic cycle as described by Clarke *et al.* (1995), while the symmetric outer bow-shocks may be due to periodic FU Orionis-type thermal instabilities discussed in detail by Bell & Lin (1994)[4]. Thus, the HH 212 jet may have a key role to play in unraveling protostellar jet physics (see, *e.g.*, Kofman & Raga 1992; Biro & Raga 1994) and knot physics (Falle & Raga 1995).

Finally, it is worth noting that HH 212 is at least morphologically reminiscent of some extragalactic jets (*e.g.*, 3C 449; Miley 1980), and we therefore speculate

[4] Interestingly enough, the FU Ori ionisation instabilities occur at disk radii similar to those where the MHD wind is expected to be launched in the models of Camenzind (1990, 1993) and/or Shu *et al.* (1994), *i.e.*, near 0.25 AU. Collimation, on the other hand, would not occur that far in but rather at radii comparable to the light cylinder, *i.e.*, at 10–100 AU (Camenzind 1990, 1993)

that protostellar jets may simply be scaled-down versions of basically similar extragalactic jets, with similar periodic ejection mechanisms and collimation processes at work underneath (*cf.* Rees 1978; Camenzind 1990).

5. What Next?

In the future, a number of additional crucial observations need to be carried out. First, we should monitor the inner knots, looking for intensity changes and movements, in order to disentangle changes due to rapid cooling and rapid proper motion.

At an assumed jet bulk velocity of 300 km/s, we would expect proper motions of $\sim 0.1''$ per year, which should be detectable in just a few years using modern infrared imaging cameras. Such proper motion measurements will, in principle, enable us to discriminate between the different types of shocks in which the H_2 emission could arise: stationary oblique shocks, which would have zero proper motion; quasi-ballistic bullet-like bow-shocks, which would have the maximum proper motion, corresponding to the full jet speed; and shocks due to the shear interaction between the jet flow and the external medium (Kelvin-Helmholtz instabilities), which would have proper motions intermediate between zero and the full jet velocity (see Ray & Mundt 1993; Reipurth & Heathcote 1993). Of course, we do not know what the full jet speed is in HH 212, as the flow is nearly in the plane of the sky. Thus, we cannot easily relate the measured proper motions of the knots to the bulk speed of the jet, unless we measure very slow or very fast proper motions. If we measure a proper motion corresponding to intermediate velocities at (say) 150 km/s, then it will not be possible to know whether this is a pattern speed of the knots riding on a fast bulk motion, or if we are in the ballistic limit where the knots ride at full speed on a relatively slow jet fluid.

Another important experiment will be to measure the velocity dispersion and velocity profile in each knot through high resolution infrared spectroscopy of the v=1–0 S(1) H_2 line (*cf.* Zinnecker *et al.* 1989). If the dispersion turns out to be large (*i.e.*, >20–25 km/s, the speed above which H_2 is dissociated in purely hydrodynamic shocks), then we will have evidence that MHD shocks with magnetic percursors are at work, *i.e.*, C-shocks as opposed to J-shocks. In C-shocks, the velocity dispersion should be as high as 40–50 km/s (Smith & Brand 1990; Draine & McKee 1993).

Another set of measurements would probe the interaction of the jet with its surroundings, to test the entrainment picture and see if and where the bipolar jet accelerates the ambient cloud material. This will be traced in the high velocity wings of various molecular tracers such as CO and SiO at millimetre wavelengths (Stahler 1993; Masson 1995). However, since the jet lies close to the plane of the sky, the geometry is not favourable for such radial velocity observations. On the other hand, the geometry is very favourably disposed to the possible detection of a circumstellar disk around the central source seen edge-on: millimetre interferometry in CO, SiO, and continuum is already underway at Owens Valley to test this hypothesis. Finally, looking further ahead, HH 212 should be a prime target for the NICMOS infrared instrument on the Hubble Space Telescope, with its 0.2″ angular resolution at 2.12 μm. A particular focus will be to resolve the structure of the inner knots, in order to prove that they truly are bow-shocks. Ground-based adaptive optics on 4–10 m telescopes offer higher resolution in principle, but in practice, the HH 212 field contains few if any stars bright enough for even tip-tilt correction.

Acknowledgement. We would like to thank John Bally, Wolfgang Brandner, Max Camenzind, Dave Devine, Roland Gredel, Lev Kofman, Andreas Quirrenbach, Bo Reipurth, Frank Shu, Mark Shure, Mike Smith, Jim Stone, and Harold Yorke for help at various stages of this work.

References

Appl, S. & Camenzind, M. 1992, *Astron. Astrophys.*, **256**, 354
Bell, K.R. & Lin, D. N. C. 1994, *Ap. J.*, **427**, 987
Biro, S. & Raga, A.C. 1994, *Ap. J.*, **434**, 221
Bally, J. & Devine, D. 1994, *Ap. J.*, **428**, L65
Bally, J., Lada, E.A., & Lane, A. P. 1993, *Ap. J.*, **418**, 322
Bührke, Th., Mundt, R., Ray, T. 1988, A&A **200**, 99
Camenzind, M. 1990, *Rev. Mod. Astronomy*, **3**, 234
Camenzind, M. 1993, in *Stellar Jets and Bipolar Outflows*, eds. L. Errico & A.A. Vittone (Dordrecht: Kluwer), p. 289
Clarke, C. J., Armitage, P.J., Smith, K. W., & Pringle, J.E. 1995, *Mon. Not. R. astr. Soc.*, **273**, 639
Davis, C.J., Dent, W.R.F., Matthews, H.E., Aspin, C. & Lightfoot, J.F. 1994a, *Mon. Not. R. astr. Soc.*, **266**, 933
Davis, C.J., Eislöffel, J. & Ray, T.P. 1994b, *Ap. J.*, **426**, L93
Davis, C.J., Mundt, R., & Eislöffel, J. 1994c, *Ap. J.*, **437**, L55
Draine, B.T. & McKee, C.F. 1993, *Ann. Rev. Astron. Astrophys.*, **31**, 373
Eislöffel, J., Davis, C.J., Ray, T.P. & Mundt, R. 1994, *Ap. J.*, **422**, L91
Falle, S.A.E.G. & Raga, A.C. 1995, *Mon. Not. R. astr. Soc.*, **272**, 785
Fukui, J. & Yokoo, T. 1986, *Nature*, **321**, 841
Gredel, R. & Reipurth, B. 1993, *Ap. J.*, **407**, L29
Gredel, R. & Reipurth, B. 1994, *Astron. Astrophys.*, **289**, L19
Gredel, R., Reipurth, B. & Heathcote, S. 1992, *Astron. Astrophys.*, **266**, 439
Heathcote, S., Reipurt, B. 1992, *A. J.*, **104**, 2193
Kofman, L. & Raga, A.C. 1992, *Ap J.*, **390**, 359
Masson, C. 1995, in *Circumstellar Matter*, eds. G. Watt & P.M. Williams, in press
Masson, C. & Chernin, L.M. 1993, *Ap. J.* **414**, 230
McCaughrean, M.J., Rayner, J.T. & Zinnecker, H. 1994, *Ap. J.*, **436**, L189
Miley, G. 1980, *Ann. Rev. Astron. Astrophys.*, **18**, 165
Mundt, R. 1986, *Canadian J. Physics*, **64**, 407
Norman, M.L. 1993, in *Astrophysical Jets*, eds. M. Livio, C. O'Dea & D. Burgarella (Cambridge: Cambridge Univ. Press), p. 211
Raga, A.C., Cabrit, S. 1993, *Astron. Astrophys.*, **278**, 367
Ray, T., Mundt, R. 1993, in *Astrophysical Jets*, eds. M. Livio, C. O'Dea & D. Burgarella (Cambridge: Cambridge Univ. Press), p. 145
Rees, M. 1978, *Mon. Not. R. astr. Soc.*, **184**, 61P
Reipurth, B. 1989, *Nature*, **340**, 42

Reipurth, B. 1994, *A general catalogue of Herbig-Haro objects*, electronically published via anonymous ftp from `ftp.hq.eso.org`, directory `/pub/Catalogs/Herbig-Haro`
Reipurth, B. & Heathcote, S. 1993, in *Astrophysical Jets*, eds. M. Livio, C. O'Dea & D. Burgarella (Cambridge: Cambridge Univ. Press), p. 35
Reipurth, B., Raga, A.C. & Heathcote, S. 1992, *Ap. J.*, **392**, 145
Shu, F., Najita, J., Ostriker, E., Wilkin, F., Ruden, S., & Lizano, S. 1994, *Ap. J.*, **429**, 781
Smith, M.D. & Brand, P.W.J.L. 1990, *Mon. Not. R. astr. Soc.*, **245**, 108
Smith, M.D. 1991, *Mon. Not. R. astr. Soc.*, **252**, 378
Sonnhalter, C., Preibisch, T., Yorke, H.M. 1995, *Astron. Astrophys.*, **299**, 545
Stahler, S.W. 1993, in *Astrophysical Jets*, eds. M. Livio, C. O'Dea & D. Burgarella (Cambridge: Cambridge Univ. Press), p. 183
Stapelfeldt, K.R. & Scoville, N.Z. 1993, *Ap. J.*, **408**, 239
Stone, J.M. & Norman, M.L. 1993, *Ap. J.*, **413**, 210
Wouterloot, J.G.A. & Walmsley, C.M. 1986, *Astron. Astrophys.*, **168**, 237
Wouterloot, J.G.A., Walmsley, C.M., & Henkel, C. 1988, *Astron. Astrophys.*, **203**, 367
Wouterloot, J.G.A., Henkel, C., & Walmsley, C.M. 1989, *Astron. Astrophys.*, **215**, 131
Zinnecker, H., Mundt, R., Geballe, T.R. & Zealey, W.J. 1989, *Ap. J.*, **342**, 337
Zinnecker, H., Bastien, P., Arcoragi, J.P., & Yorke, H.W. 1992, *Astron. Astrophys.*, **265**, 726

Dust Scattering as a Diagnostic Tool for Neutral Winds in Bipolar Outflows – The Case of IRAS 08159–3543

H.J. Staude and Th. Neckel

Max-Planck-Institut für Astronomie, Heidelberg-Königstuhl

1. Abstract

The gas density and velocity field of the neutral outflow through a young bipolar reflection nebula can be determined by longslit spectroscopy, if the (unseen) driving source is a strong Hα emitter of high luminosity. The method makes use of the Hα and stellar continuum radiation scattered by the dust carried along by the outflowing gas; its application to the Class I point source IRAS 08159-3543 led to the detection of a strong neutral wind, which carries enough momentum to drive molecular outflows typical of high luminosity YSOs.

2. Structure of bipolar winds

The first young stellar object (YSO) showing clearly the essential morphological elements of the outflow phase was the bipolar nebula S 106 (Eiroa et al. 1979). In S 106, the structure and velocity field of the gas flowing outward through the lobes can be analyzed by observing its optical emission lines and radio continuum emission (Hippelein and Münch 1981, Solf and Carsenty 1982, Staude et al. 1982, Felli et al. 1984). S 106 remains an outstanding exception and a key object for the understanding of young bipolar outflows: it owes this property to the fact that its central source is an O9-B0 star which ionizes (and by this makes readily observable) its fast bipolar wind flowing out in the lobes. OB stars are rare, and their bipolar outflow phase does not last for more than few thousand years: the presence of even a single object like S 106 within 5 kpc from the Sun must be considered as a lucky chance. In all other known young bipolar nebulae with similar morphological characteristic as S 106 (deeply embedded central star, thick equatorial dust belt, bright polar lobes of sharply defined parabolic shape, which delimitate essentially 'empty' cavities), but whose central stars are of later spectral type, the existence and physical properties of the primary wind flowing away from the YSOs at large distances must be inferred either from optical spectroscopy and radio continuum observations of their central stars (e.g., bueshifted NaID and CaII H+K absorption lines, collisionally excited expanding envelopes), or from the effects of the winds on the material of the surrounding molecular cloud: the shock-excitation of Herbig-Haro objects, the acceleration of secondary, massive bipolar CO flows, and the formation of the 'empty cavities' observed as bright bipolar reflection nebulae near the YSOs (see the review by Staude and Elsässer, 1993).

During their bipolar phase, young stellar objects of later spectral type frequently drive highly collimated, partially ionized Herbig-Haro jets flowing outward at high velocity. A matter of debate is whether these jets provide an adequate visualization of the entire outflowing wind: there are strong arguments indicating that the momentum carried by the jets fails to account for the acceleration of the secondary CO outflows by up to at least two orders of magnitude (Sarcander et al. 1985; Mundt et al. 1987). Also, the characteristic parabolic shape of the widely open 'empty cavities' in the dense surrounding clouds seems unlikely to be due to the ram pressure exerted by the narrow jets alone. These findings suggest the existence of a broad high-velocity component of the primary wind, which pervades the entire bipolar 'cavities'. In the case of S 106, this component is manifest, since it is ionized. Near stars of later spectral type, it should be neutral; where collimated jets appear, they may be embedded in this broad component. It was the aim of the work presented below to test this hypothesis by optical observations.

3. Optical detection of the neutral wind

Under very special conditions, a neutral outflow can be detected by optical means. Consider the situation illustrated in Figure 1: the central YSO is hidden by its thick equatorial dust belt, since the polar axis of the bipolar nebula lies nearly in the plane of the sky; in this situation the bright lobes are clearly separated. At a given position within the nebula the surface brightness due to reflected light is determined by the luminosity of the YSO, the internal extinction, the number of scattering grains within the column behind the surface element considered, and the extinction in front of the nebula.

If the postulated neutral wind pervading the lobes carries along the right amount of dust, the light scattered by this dust dominates over the light scattered by the walls of the 'cavity' (this case is then a rare exception, since usually the light scattered by the walls is the prevailing component). Then, the surface brightness and color of the stellar continuum, which is scattered by the outflowing grains, contains information about their number. From this, with standard assumptions about the optical properties of the grains and about the dust-to-gas ratio, we obtain the total amount of outflowing neutral gas in the lobes. In order to prove that the light is indeed scattered by the outflowing grains (and not by the walls of the 'cavity'), we have to determine their velocity relative to the central source. If the YSO is a strong Hα line emitter, high-resolution spectroscopy of the Hα line emitted by the YSO and scattered by the dust grains flowing through the lobes can be performed. From the Doppler shift of the scattered Hα the velocity field of the scattering dust grains (and hence of their invisible carrier, the neutral wind) can be derived. Having determined the amount of gas in the lobes and its outward velocity, we can estimate mass loss and momentum transfer (force) of the neutral wind.

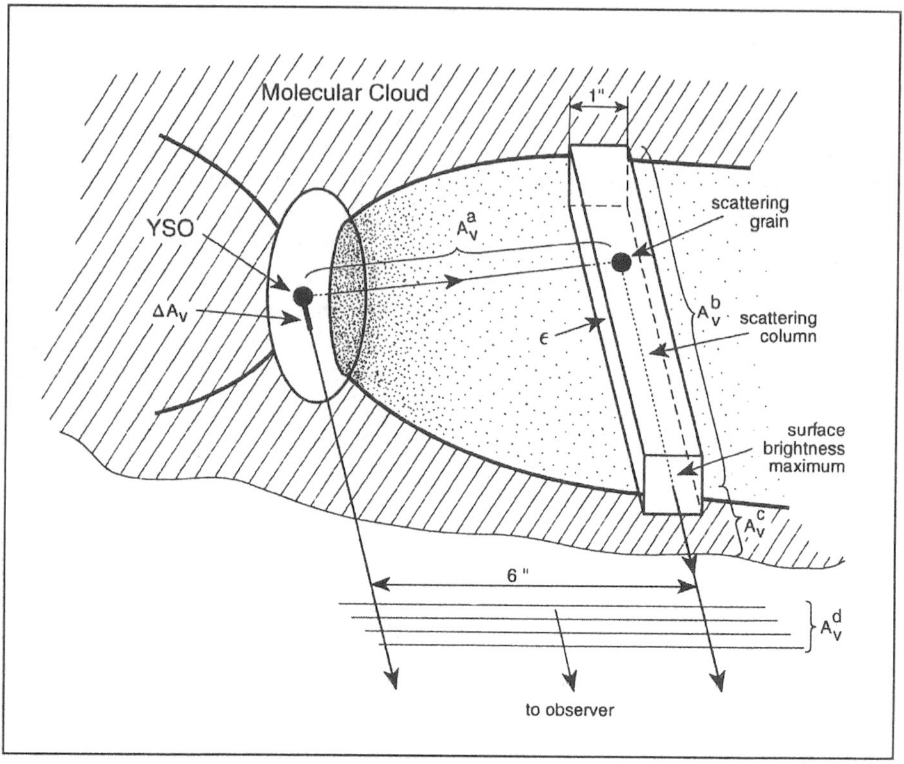

Fig. 1. This orientation of a bipolar nebula is favourable for the observations described here. The components of extinction along the path of scattered light are indicated. This geometry holds for the nebula associated with IRAS 08159-3543, to which the indicated distance (6″) between the YSO and the nebular brightness maximum refers.

Thus, in searching for an object suitable for this kind of observations, we have used the following criteria:

- it should be a Class I IRAS source of high luminosity,
- it should have no optical counterpart (orientation!),
- it should be associated with a bipolar nebula, which should be bright in the red continuum as well as in Hα.

4. Observations of IRAS 08159-3543

After a long lasting search, we found that the point source IRAS 08159-3543 fulfills these criteria. At the 2.2 m telescope on La Silla, we obtained broad and narrow band images from R to K and a set of longslit spectrograms covering the entire associated nebula. A detailed analysis of this comprehensive set of data has appeared recently (Neckel and Staude, 1995): here we will describe the essential results.

The YSO is located at the distance $d \approx 4.3$ kpc, its bolometric luminosity is $L \approx 2 \times 10^{-4}\ L_{\odot}$. It has no optical counterpart, but we could identify the central star on our deep K image: it is hidden behind a dust belt causing $A_v \geq 43$ mag. The associated bipolar nebula extends over about 30″, with a bright core of about 10″. Figure 2 shows the K image, and also the $K - H$ color map, on which the central source appears as a highly reddened spot near the IRAS position. The reddening of the nebula increases from SW to NE, indicating that the SW lobe lies near the surface, while the NE lobe is embedded deeper inside the associated molecular cloud. This is in accordance with our spectroscopic data, which show that the material in the SW lobe is approaching, while in the NE lobe it is receding at high velocity.

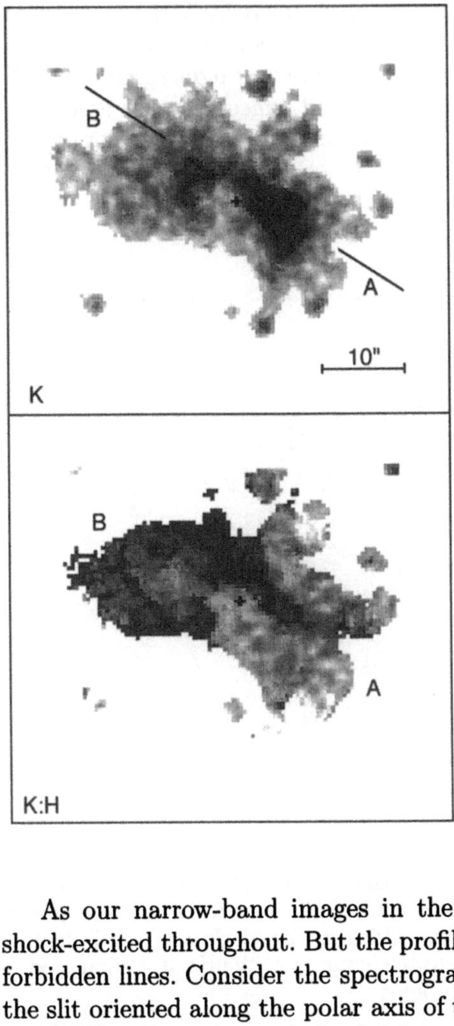

Fig. 2. K image and $K : H$ color map of the bipolar nebula associated with IRAS 08159-3543. A and B denote the approaching and the receding lobe respectively. The bright red spot west of the cross marking the formal IRAS position (best seen in the color map) is the central YSO. The slit position of the spectrogram III discussed below (see Figure 3) is indicated.

As our narrow-band images in the [SII] lines demonstrate, the nebula is shock-excited throughout. But the profile of Hα strongly differs from that of the forbidden lines. Consider the spectrogram shown in Figure 3: it was taken with the slit oriented along the polar axis of the nebula, as indicated in Figure 2. Hα and [NII] emission is seen in both lobes, above and below the dark lane hiding the central YSO, and the lines are redshifted in the NE lobe and blueshifted in the

SE lobe. Hα consists of a broad and a narrow component. The decomposition, row by row, of the Hα line into two Gaussians (see Figure 4) shows that the two components are of comparable strength. The narrow Hα component has the same FWHM and shows the same variation with position as the forbidden lines: consequently it is emitted by the shock-excited gas in the lobes together with the forbidden lines. In contrast, the broad component shows a much larger jump in velocity when passing from the receding to the approaching lobe. Radial velocity of the broad and narrow Hα components, as well as of the forbidden lines, are shown in Figure 5 as a function of position along the polar axis of the nebula. Since the broad Hα component has no counterpart in other lines emitted in the lobes, we conclude that it is emitted by an envelope of the central YSO and scattered by dust in the lobes. This result has been confirmed by the spectropolarimetry of the nebula, which was performed by Gary D. Schmidt (note added in proof to Neckel and Staude 1995): it shows that the nebular continuum is strongly polarized ($p \approx 20\%$), as expected for light scattered in the given geometry; the forbidden lines are unpolarized, since they are emitted in the lobes; Hα is polarized to about 10%, as expected if half of it is emitted in the lobes, and half scattered like the continuum.

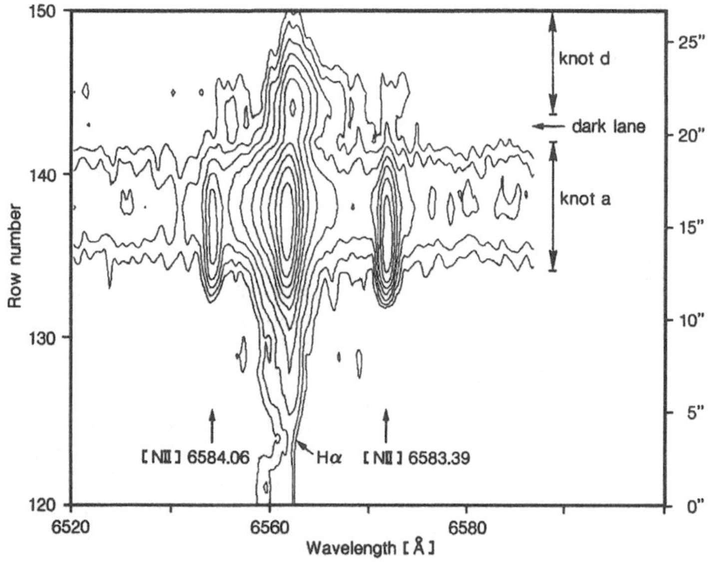

Fig. 3. Section of spectrogram III, which was taken with the slit along the polar axis of the nebula (see Figure 2), including Hα and the [NII] lines. 'Knot a' is the brightest part of the approaching lobe, the YSO is embedded in the 'dark lane', and 'knot d' belongs to the receding lobe. Note the composite character of the Hα line compared to [NII], and the strong scattered continuum in knot a.

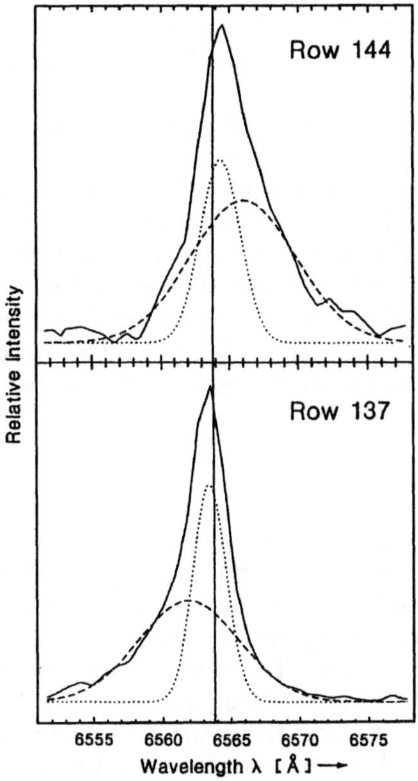

Fig. 4. Representation of the Hα profiles by the sum of two Gaussians (two examples). *Continuous lines:* the Hα profiles in row 144 (receding lobe) and in row 137 (approaching lobe). *Dotted lines:* Gaussians obtained by a two-component fit to the observed profiles. The central vertical line marks the wavelength of Hα in the rest frame of the source. The sum of the two Gaussians reproduces the observed profiles almost exactly.

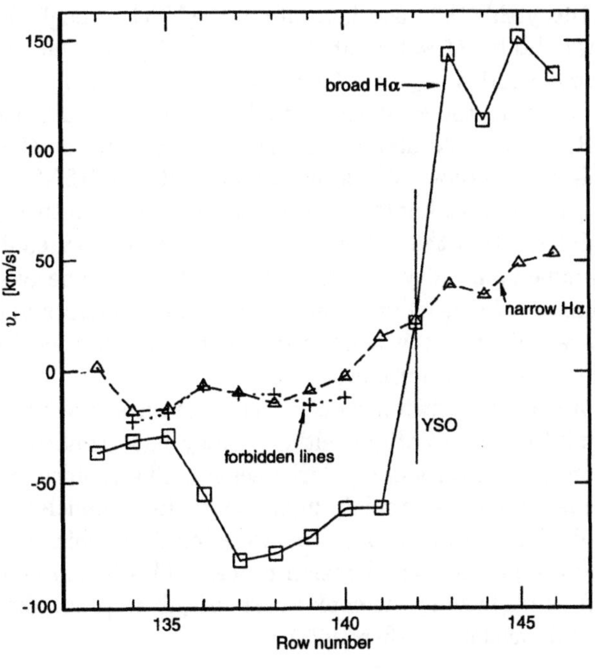

Fig. 5. Radial velocity of the broad and narrow Hα and of the forbidden lines as a function of position along the slit in spectrogram III. The position nearest to the YSO is marked. The radial velocities are corrected for the systemic velocity, $v_{sys} = +35 \text{ km s}^{-1}$.

The detailed analysis of all spectrographic data shows:

- The envelope emitting the broad Hα component near the YSO is expanding at ≈ 570 km/s.
- The axis of the bipolar nebula is inclined by about 80° against the line of sight.
- The dust grains scattering the broad Hα component in the lobes are flowing outward at about the same speed as the Hα envelope expanding away from the YSO. We assume that they are carried along by a neutral wind which, since no limb-brightening is seen in the nebula, pervades the entire lobes.
- Each dust grain receives only blueshifted Hα radiation. This means, that from each lobe only the approaching part of the emitting envelope is visible. Thus, the envelope is bisected, as expected if it arises above the inner, hottest part of a massive accretion disk.

5. Discussion

The fact that the velocity v_w of the neutral wind pervading the lobes agrees with the expansion velocity of the Hα emitting envelope near the YSO, implies that this wind does not consist of swept up material of the associated molecular cloud, but rather of material, which has been accelerated by the primary mechanism at work near the YSO (momentum conservation!). In addition to the velocity of this primary wind, we can estimate the number of the scattering dust grains in the lobes from surface brightness and color of the scattered stellar continuum: for this we have to assume standard grain properties. With the normal gas-to-dust ratio, the amount of dust in the lobes leads to the mass of the neutral outflowing wind. Together with its velocity, this yields the mass loss rate of the YSO through the neutral wind. In this way we obtain: $\dot{M} \approx 6 \times 10^{-5} \ldots 2 \times 10^{-4} \mathrm{M}_\odot/\mathrm{yr}$. The momentum carried by the neutral wind is $M_w v_w = 0.03 \ldots 0.1 (\mathrm{M}_\odot/\mathrm{yr})(\mathrm{km/s})$. This high mass loss rate agrees well with the mass loss rate derived by McMuldroch et al. (1995) for the FUor RNO 1C in its high state, under the assumption that it drives the associated molecular outflow. We do not know if IRAS 08159-3543 drives a molecular outflow; but our estimate of its mass loss and momentum carried by the neutral wind lies within the range of values derived by Levreault (1988) for sources of comparable luminosity and obeys the $\dot{M}/L$ relation derived by Levreault for a large number of molecular outflow sources of all masses and luminosities. On this basis we identify the neutral dust-carrying wind as the main accelerator of secondary molecular outflows.

Since the dust consists of material which has been accelerated near the YSO, we are left with two possibilities for its origin: either it may have condensed out 'in flight' from the primary neutral wind (which seems unlikely, since the gas density in the wind probably does not remain high enough for a sufficiently long flight time), or we are dealing with dust from the accretion disk, which was entrained by the gas during the primary acceleration process. This is expected for centrifugally accelerated winds leaving the cooler parts of the accretion disk outside the hot ionized central envelope (Safier 1993).

The fast neutral wind detected in the bipolar lobes causes shock-excitation within almost the entire nebula. However, the filling factor of the shock-excited material, estimated from our data, is only $f \approx 0.04$. Since for the forbidden line emission, as for the scattered continuum, no limb brightening is observed, we conclude that the shock-excited material is confined to small clumps, which are evenly dispersed throughout the lobes. They move outward at about one-fourth the speed of the neutral outflow. These clumps are probably preexisting cloudlets hit by the wind.

6. Conclusions

We have observed a broad neutral wind emanating from the luminous Class I source IRAS 08159-3543 ($L = 2 \times 10^4 L_\odot$) and pervading the entire lobes of its associated bipolar nebula. The wind carries enough momentum to accelerate the secondary molecular outflows typically driven by YSOs of this high luminosity, and obeys the $\dot{M}/L$ relation derived by Levreault (1988) for molecular outflow sources of all masses and luminosities.

Three independent findings favour the notion that the observed wind is driven by an active accretion disk rather than by the star. First, our spectroscopic observations indicate that the envelope emitting Hα is bisected by an optically thick layer. Second, the fact that the fast dust in the lobes has probably been accelerated together with the primary wind, also speeks in favour of a disk wind, since the dust could not survive an acceleration in the immediate vicinity of the star. Third, the 'chromospheric' Fe II emission lines observed in YSOs of all masses, which are thougt to arise above the cooler regions of dense, active accretion disks, are prominent also in the scattered spectrum of the central source.

The spectral type of the central YSO, estimated from the color and spectrum of the stellar continuum scattered in the lobes, is late A or F. Thus, in the H-R diagram the YSO is located far above the theoretical stellar birthline derived by Palla & Stahler (1990). This fact may be related to its high mass loss rate, which, together with its high luminosity, is a necessary condition to allow the observations described above: possibly we observe the YSO in a transient overluminous stage due to enhanced mass accretion, similar to the FU Orionis phenomenon. Note that the mass loss rate derived for the FUor RNO1C by McMuldroch et al. (1995) from its bipolar CO outflow is close the what we found here for IRAS 08159-3543.

References

Eiroa, C., Elsässer, H., Lahulla, J.F.:1979, A&A 74,89

Felli, M., Staude, H.J., Reddmann, T., Massi, M., Eiroa, C., Hefele, H., Neckel, T., Panagia, N.: 1984, A&A 135,261

Hippelein, H.H., Münch, G.: 1981, A&A 99,248
McMuldroch, S., Blake, G.A., Sargent, A.J.: 1995, AJ 110,354
Mundt, R., Brugel, E.W., Bührke, T.: 1987, ApJ 319,275
Levreault, R.M.: 1988, ApJ 330,897
Neckel, Th., Staude, H.J.: 1995, ApJ 448,832
Palla, F., Stahler, S.S.: 1990, ApJ(Lett.) 360,L47
Safier, P.N.: 1993, ApJ 408,115
Sarcander, M., Neckel, Th., Elsässer, H.: 1985, ApJ(Lett.) 239, L17
Solf, J., Carsenty, U.: 1982, A&A 113,142
Staude, H.J., Elsässer, H.: 1993, A&A Rev. 5,165
Staude, H.J., Lenzen, R., Dyck, H.M., and Schmidt, G.D.: 1982, ApJ 255,95

A Long-Slit Spectroscopic Survey of Forbidden Emission Lines in T Tauri Stars

Gerhard A. Hirth[1], Reinhard Mundt[1], and Josef Solf[2]

[1] Max-Planck-Institut für Astronomie, Königstuhl 17, D-69117, Heidelberg, Germany
[2] Thüringer Landessternwarte, Sternwarte 5, D-07778 Tautenburg, Germany

1. Introduction

In the past ten years the optical forbidden emission lines (FELs) of T Tauri stars (TTSs), e.g. the [OI]$\lambda\lambda$ 6300, 6363 and the [SII]$\lambda\lambda$ 6716, 6731 lines, have been used widely as a probe for the outflows from young stellar objects (YSOs) since they occur relatively frequently in many YSOs, e.g. about 30% of all TTSs show strong FELs with W_λ([OI]6300) > 1 Å. These lines are expected to be formed by internal shocks within the outflow.

In general the FELs show blueshifted, double-peaked profiles with a high-velocity component (HVC) having typical radial velocities between -50 and -150 $\mathrm{km\,s^{-1}}$ and a low-velocity component (LVC) with values around -5 to -20 $\mathrm{km\,s^{-1}}$. Already Jankovics, Appenzeller & Krautter (1983) concluded from the similar velocity distributions of Herbig-Haro (HH)- objects and the blueshifted HVC of TTSs that the FELs of TTSs originate in spatially unresolved anisotropic outflows. In this context the blueshifted profile is attributed to the existence of an optically thick circumstellar disk, obscuring the receding part of the bipolar outflow.

Spectroscopic data of the FELs from about 45 TTSs are presently available in the literature (e.g. Appenzeller, Jankovics & Östreicher 1984; Edwards *et al.* 1987; Hamann 1994). However all these surveys deal mainly with the spectral properties of the FELs only. It was Solf (1989) who showed that not only the kinematic but also the spatial properties of the FELs are important for a better understanding of the physical processes in the outflow, especially in the immediate vicinity of the source.

Motivated by the first results concerning HL Tau, XZ Tau and DG Tau (Solf 1989; Solf & Böhm 1993) we conducted a long-slit spectroscopic survey of the FELs of TTSs in order (1) to investigate the spatial properties of the FELs (e.g. detection of yet unresolved jets and search for the outflow direction), (2) to study in detail the physical properties of the outflowing material in the immediate vicinity of the source (e.g. kinematics, excitation, collimation, variability) and (3) to look for evolutionary effects in outflows by comparing jets from TTSs with those from embedded YSOs of similar luminosity.

2. Observations and Data Reduction

Long-slit spectra of 38 classical TTSs were obtained at various position angles during several observing runs between 1988 and 1993 with the coudé spectrograph at the 2.2-m telescope and the Cassegrain twin spectrograph at the 3.5-m telescope of the Calar Alto Observatory in Spain. The spectral resolution was 20 kms^{-1} and 50 kms^{-1} (FWHM), respectively. The spatial resolution, which was determined by the spatial width (FWHM) of the stellar continuum, varied between 1.4″ and 2.0″. Additionally in 1993 we used the ISIS spectrograph on the 4.2-m William Herschel Telescope providing a spectral and spatial resolution of 35 kms^{-1} and 1.3″, respectively.

The spectra have been reduced by a program developed by one of us (J.S.). Following the standard data reduction, a special continuum subtraction procedure was performed in order to reveal the relatively weak FEL regions near the source. This step is essential for deriving information about the spatial and kinematic properties of the outflow in the immediate vicinity of the YSO and is discussed in more detail elsewhere (e.g. Hirth, Mundt & Solf 1994)

The direction of any bipolar outflow was found by changing the position angle from exposure to exposure in steps of 20° – 40° and by measuring the spatial extent of the FEL regions and their positional offset with respect to the stellar continuum. The differential nature of our method enabled us to determine spatial offsets of at least 0.1″ from the stellar position. The detection limit of a spatial extent in the FEL region was about 0.3″.

3. Results of the Survey and Interpretation

3.1 Spatial Properties

One of the most surprising results of our survey was the discovery that all but one of the 14 nearby TTSs (d = 140 – 200 pc) which have been sampled by at least 4 – 6 slit positions show spatially extended FEL regions. These objects are: AS 353 A, V536 Aql, CW Tau, DD Tau, FS Tau B, FS Tau, Haro 6-10, UZ Tau E, HN Tau, DO Tau, DP Tau, UY Aur and RW Aur. The values for the spatial extent range from the detection limit up to several arcsec and are in general about 10 to 100 times smaller than the *typical* jets from YSOs (e.g. Ray & Mundt 1993). The most spectacular bipolar small-scaled jets are the ones of RW Aur (length up to 20″), DP Tau, DO Tau (Hirth *et al.* 1994) and CW Tau (Hirth, Mundt & Solf 1994). *This implies that spatially extended FEL regions are a common phenomenon among all classical TTSs with strong FELs (e.g.* W_λ *([OI]6300) > 1 Å).* Note that only the bulk part of the forbidden emission originates in a very compact region (d < 30 AU) as suggested by the emission measure (cf. Edwards *et al.* 1987).

The extended forbidden emission arizes presumably in a bipolar emission region with typical dimensions of 100 – 500 AU. One nice example of a small-scaled bipolar jet emanating from a TTS is shown in Figure 1.

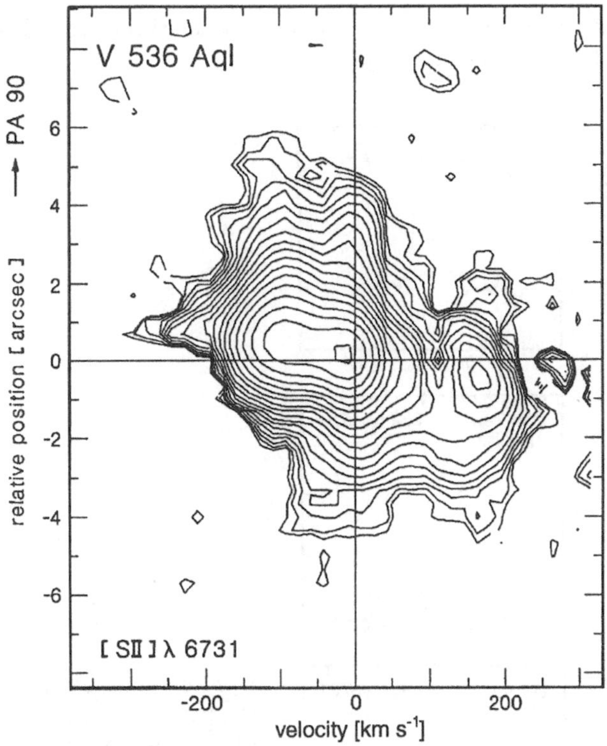

Fig. 1. Continuum-subtracted position-velocity map of the [SII]λ 6731 line of V536 Aql along the outflow direction (PA = 90°). Note that the redshifted part of the outflow shows about 60% higher radial velocities in the HVC compared to the blueshifted part. The spatial offset is given relative to the stellar position. The velocity is referred to the rest frame of the TTS. The spacing of the contours is logarithmic corresponding to a factor $2^{0.5}$.

Interestingly more than half of the TTSs with extended FEL regions are associated with a reflection nebula. This result is unusual compared to typical jet sources which in nearly all cases have a reflection nebula. Presumably the objects studied in this survey represent a more advanced evolutionary stage where the circumstellar material has been dispersed so effectively that its optical depth is too low to be detected as a reflection nebula.

3.2 Spectral Properties

Figure 2 shows some FEL profiles of selected TTSs. About 65% of the TTSs studied in our survey show a double-peaked FEL profile. In about 40% of the objects we also detected the [NII]λ 6583 line which shows similar spatial properties as the [SII] lines but has only a HVC. This suggests a different excitation for the HVC and LVC. The observed different intensities of the HVC and LVC from line to line within the same object (e.g. CW Tau) and their individual temporal variation (see separate Poster by Hirth in this volume) give additional support for the distinct nature of the two velocity components. According to the qualitative model of Kwan & Tademaru (1988) the HVC originates in a well collimated jet, whereas the LVC is formed in a disk wind and/or disk atmosphere (see Hirth, Mundt & Solf 1994 for more details). A strong argument that the

HVC is formed in a jet is provided by the similar radial velocities of the HVC at the stellar position and further out in the jet.

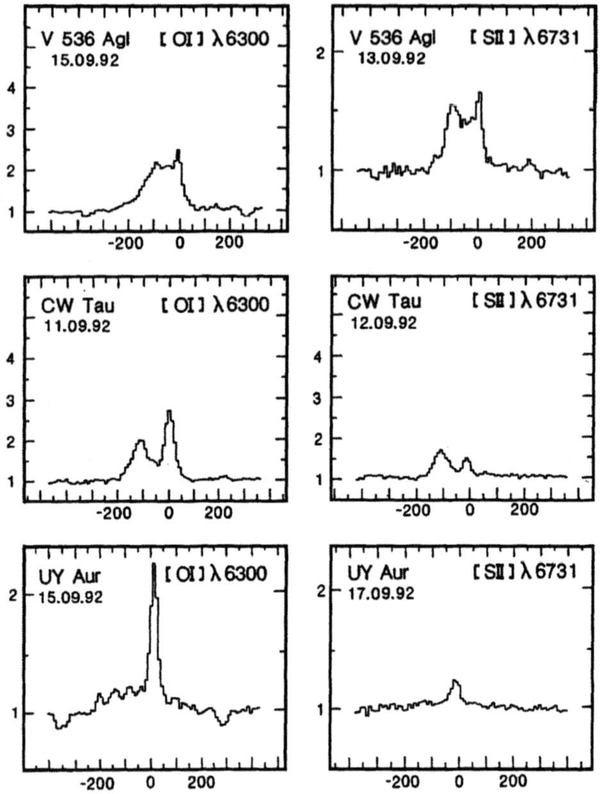

Fig. 2. Line profiles of the [OI]λ 6300 and [SII]λ 6731 lines of selected TTSs from our survey. The radial velocities (in kms^{-1}) are given relative to the star. The line intensity is normalized to the continuum.

Note that only eight TTSs show one velocity component. It is possible that in these cases we do not always observe a LVC but a HVC which lies practically in the plane of the sky. This is the case for FS Tau B and FS Tau. The existence of the [NII]λ 6583 at similar low velocities gives additional support for this idea in these two cases.

3.3 Collimation of the Jet Component

Long-slit spectroscopy provides a unique opportunity to determine the collimation properties of outflows from YSOs at scales below 500 AU which are not accessible by direct imaging techniques due to the strong continuum contribution of the TTS.

Using the average radial velocity of the HVC derived for the TTSs of our sample (85 $\mathrm{km\,s^{-1}}$) we estimated the jet velocity to be 170 $\mathrm{km\,s^{-1}}$ assuming that all jets have the same velocity and are oriented randomly in space ($v_{jet} = 2\ v_{rad}$). Assuming further that the emission in the jet can be approximated by a filled cone of emission with the flow vectors having all the same length we can roughly estimate the full opening angle from the radial velocity and velocity dispersion (for details see Hirth, Mundt & Solf 1994). The values for the TTSs of our sample range between 10 and 20 degrees within distances of 100 to 1000 AU. At distances below 200 AU from the source the opening angles are about a factor 2 – 3 higher.

3.4 Asymmetries in Bipolar Jets

The most interesting and surprising result of our long-slit spectroscopic survey are the quite different flow velocities found for the opposite sides of the bipolar jets of RW Aur, DO Tau and V536 Aql (cf. Figure 1). In these cases the observed radial velocities with respect to the star indicate that the velocity of the outflowing gas on the one side of the star is about as twice as high as the velocity on the opposite side. Including all radial velocity data on bipolar HH-flows and HH-jets available in the literature and in unpublished spectra of our institutes (neglecting flows located close to the plane of the sky, like the jet from HH 30) we found that in 9 out of 16 cases (RW Aur, DO Tau, V536 Aql, HL Tau, DG Tau B, R Mon, Th 28, HH 46/47 and RNO 91) the observed radial velocities with respect to the star indicate velocity ratios between the two sides of the bipolar outflow of 1.4 to 2.6, while only in 7 stars (DP Tau, XZ Tau, CW Tau, FS Tau B, AS 353 A, HH 24 and DD Tau) this ratio is between 1.0 and 1.2. These results imply that beside the asymmetries in the spatial and excitation properties of bipolar jets known so far, different outflow velocities seem to be a common phenomenon too. *We suggest that in these cases the bipolar outflow is accelerated to intrinsically different velocities on either side of the star.* (For a detailed discussion see Hirth *et al.* 1994).

3.5 Evolutionary Aspects

Comparing the typical jet velocity of TTSs (170 $\mathrm{km\,s^{-1}}$) with the jet velocity derived from a sample of 30 embedded and therefore presumably younger outflow sources (200 $\mathrm{km\,s^{-1}}$) we conclude that at least for the outflow velocity no significant evolutionary changes are evident.

References

Appenzeller, I., Jankovics, I., Östreicher, R. 1984 **A&A**, 141, 108
Edwards, S., Cabrit, S., Strom, S.E., Heyer, I., Strom, K.M., Anderson, E. 1987, ApJ **321**, 473
Hamann, F. 1994, ApJS **93**, 485
Hirth, G.A., Mundt, R., Solf, J. 1994, A&A **285**, 929
Hirth, G.A., Mundt, R., Solf, J., Ray, T.P. 1994, ApJ **427**, L99
Jankovics, S., Appenzeller, I., Krautter, J. 1983, PASP **95**, 883

Kwan, J., Tademaru E. 1988, ApJ **332**, L41
Ray, T.P., Mundt, R. 1993, in *STScI Symposium on Astrophysical Jets*, eds. L. Errico, A. Vittone, Kluwer, Dordrecht, p. 241
Solf, J. 1989, in *ESO Conf. Proc. 33, Low Mass Star Formation and Pre-Main Sequence Objects*, editor: B. Reipurth, ESO, Garching/München, p. 399
Solf, J., Böhm, K.-H. 1993, ApJ **410**, L31

Aperture Synthesis Observations of CO Outflows in Low-Mass Star Forming Regions

N. Hirano[1], O. Kameya[2], T. Umemoto[3], and M. Suzuki[3]

[1] Laboratory of Astronomy and Geophysics, Hitotsubashi University
[2] Mizusawa Astrogeodynamics Observatory
[3] Nobeyama Radio Observatory

Summary. We have made aperture synthesis observations of the CO outflows associated with low-mass young stellar objects using the Nobeyama Millimeter Array. Our angular resolution is 5–8″, which corresponds to 850 - 2100 AU.

We have detected both blueshifted and redshifted CO emission in B335, L483, L723, L1527, TMC1A and VLA1623, while only blueshifted one in B1. The CO emission generally peaks at 1000–3000 AU offset from the central source and is rather weak at the position of the central source. The only exception is the CO outflow in VLA1623, which shows strong blueshifted emission at the position of the mm-continuum source. Such centrally peaked distribution of the CO emission suggests that the outflow in VLA1623 is in the earliest stage of its evolution.

Morphological characteristics such as biconical shape in B335 and in L1527 and quadrupolar-like structure in L723 are clearly seen in our interferometer maps. This suggests that the morphology of the CO outflows has been settled within a few thousand AU of their origin.

1. Introduction

Molecular outflows are frequently associated with young stellar objects (YSOs). Recent surveys of CO outflow with high spatial resolution (e.g. Terebey, Vogel, & Myers 1989, Moriarty–Schieven et al. 1993, and Hirano et al. 1995) show that almost all stars experience a phase of molecular outflow in their embedded stage. However, the mechanism by which molecular outflows are driven and collimated is not yet understood. To answer this question, it is important to elucidate the small scale structure of the outflows adjacent to their origin.

In this paper we present aperture synthesis observations of the CO outflows associated with the nearby low-mass star forming regions, B335, L483, B1, L723, L1527, TMC1A and VLA1623. The central sources are the low luminosity far-infrared sources classified Class 0 or Class I.

Aperture synthesis ^{12}CO (J=1–0) observations were made with the Nobeyama Millimeter Array (NMA). The synthesized beam size is 5–8″, which corresponds to 850–2100 AU. In Table 1, we list the source name, the IRAS name of the central source, the distance, the luminosity of the central source, the size of synthesized beam, and the characteristics of the outflow.

Table 1. List of Observed Sources

Object	IRAS Name	Distance [pc]	L_{bol} [$L_{\odot}$]	Beamsize	Outflow type
B335	19345+0727	250	2.7	8 1″×5.0″ (2000 AU×1200 AU)	Edge-on
L483	18148-0440	250	8.0	5.6″×3.8″ (1400 AU×940 AU)	Edge-on
B1	03301+3057	350	2.8	6.4″×4.4″ (2300 AU×1500 AU)	Pole-on?, SiO
L723	19157+1907	300	6.5	7.1″×4.1″ (2100 AU×1200 AU)	Quadrupolar-looking
L1527	04368+2557	140	2.1	7.7″×4.3″ (1080 AU×600 AU)	Edge-on
TMC1A	04365+2535	140	2.4	8.1″×4.9″ (1100 AU×690 AU)	Compact ($r \lesssim 1'$)
VLA1623	—	160	≤1.5	8.7″×4.1″ (1400 AU×650 AU)	Edge-on

2. Results

We have detected both blueshifted and redshifted CO emission in B335, L483, L723, L1527, TMC1A and VLA1623, while only blueshifted one in B1. The physical parameters of the observed CO outflows are summarized in Table 2.

2.1 "Edge-On" CO Outflows: B335 and L483

Single-dish maps of the CO outflows in B335 (Hirano et al. 1988; Moriarty-Schieven & Snell 1989) and L483 (Parker, Padman, & Scott 1991) show that these outflows have biconical lobes centered at the IRAS sources and that their blueshifted lobes are accompanied by redshifted components and their redshifted lobes by blueshifted ones. These results suggest that both the B335 and L483 outflows are oriented almost perpendicular to the line of sight. Such " edge-on" configuration of the outflows provides us an opportunity to study the detail structure of the outflows in the close vicinity of their origin.

In Figure 1 we present the aperture synthesis maps of CO (J=1–0) and the 2.6 mm continuum emission of B335. The blueshifted and redshifted CO emission show distinct biconical structure. Two prominent peaks are separated by 10″ (= 2400 AU) in the blueshifted component and 15″ (= 3600 AU) in the redshifted one. The north-south extent of the outflow at its center is less than 8″ (= 2000 AU), indicating that the outflow is focused within ~ 1000 AU of its origin. The 2.6 mm continuum peak whose flux is ~ 80 mJy is located within 4" of the outflow center. In the vicinity of the origin, the flow velocity shows a sudden increase away from the center (see Figure 3 of Hirano et al. 1992). A simple dynamical model of the expanding "snowplowing" shell can explain the velocity field of B335.

Table 2. Physical Parameters of the CO Outflows

Object	V_{LSR} [km s^{-1}]		Velocity Range [km s^{-1}]	Integrated Intensity [Jy km s^{-1}]	Mass[a] [$\times 10^{-3} M_\odot$]
B335	8.3	Blue	4.20–8.25	38.37	1.1[b]–6.5[c]
		Red	9.06–12.31	29.12	0.82[b]–4.9[c]
		Total		67.49	1.9[b]–11.4[c]
L483	5.0	Blue	-1.89–4.60	36.96	1.0[b]–
		Red	5.41–13.51	21.04	0.59[b]–
		Total		58.00	1.64[b]–
B1	6.3	Blue	-1.40–5.90	87.22	4.82[b]–
		Red	NG	—	—
L723	11.0	Blue	-1.80–10.36	80.32	3.03[b]–
		Red	11.17–23.32	114.6	4.31[b]–
		Total		194.9	7.34[b]–
L1527	6.2	Blue	3.52–4.33	30.4	0.27[b]–4.4[d]
		Red	6.76–8.38	41.7	0.37[b]–9.5[d]
		Total		72.1	0.64[b]–13.9[d]
TMC1A	6.4	Blue	-2.15–5.14	63.6	0.56[b]–1.1[d]
		Red	6.76–8.38	11.8	0.10[b]–1.7[d]
		Total		75.4	0.67[b]–2.8[d]
VLA1623	4.0	Blue	-2.89–3.60	102.3	1.33[b]–
		Red	5.22–11.70	48.6	0.63[b]–
		Total		150.9	1.96[b]–

[a]T_{ex} is assumed to be 12 K for B335, B1, TMC1A, and L1527, 10 K for L723 and L483, and 15 K for VLA1623.
[b]Obtained by assuming optically thin CO emission.
[c]Obtained by assuming an averaged optical depth of ~6, which is derived from the J=2–1 and J=1–0 data of CO and ^{13}CO (Hirano et al. 1991).
[d]Calculated by using the value of optical depth, which is derived from the single-dish CO and ^{13}CO data.

The CO emission from the L483 outflow is detected mainly in the western lobe. On the other hand, no significant emission is found in the eastern lobe. The blueshifted emission peaks at 12″ (3000 AU) west of the IRAS position and is much stronger than the redshifted emission. Unfortunately, the signal to noise of our data is not enough to discuss the structure of the L483 outflow. However, our data show that the surface brightness of the CO outflow is rather low and that the outflow of L483 has no strong and compact component at its center.

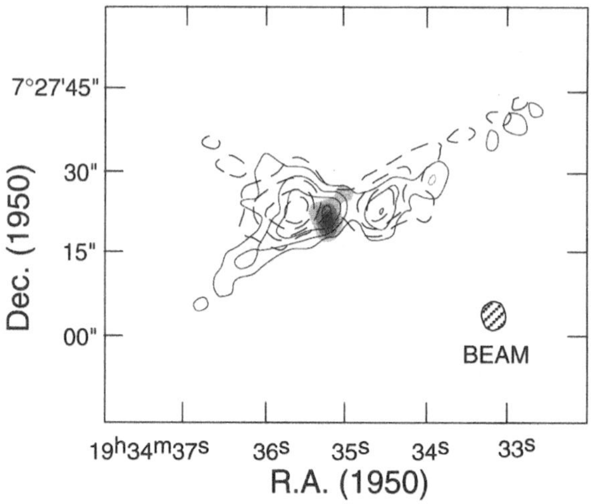

Fig. 1. Contour plots of the blueshifted emission (V_{LSR} = 7.84 km s^{-1}: solid), redshifted emission (V_{LSR} = 9.46 km s^{-1}: dashed), and the 2.6 mm continuum emission (gray scale) of B335.

2.2 CO Outflow in B1: A Pole-On View of a Shell-Like Outflow?

B1 contains a low-luminosity IRAS source 03301+3057, toward which high-velocity CO emission was detected (Nakayama 1988; Bachiller et al. 1990). Single-dish data show that the high-velocity emission is confined within ~40″ (0.07 pc at the distance of 350 pc) of the IRAS source and is dominated by the blueshifted material. The dynamical timescale of the outflow is estimated to be less than 10^4 yr (Nakayama 1988) or less than 10^3 yr (Bachiller et al. 1990), indicating that the outflow is in the early stage of its growth. In this source, the thermal emission lines of SiO were detected (Bachiller et al. 1990; Martin-Pintado, Bachiller, & Fuente 1992; Yamamoto et al. 1992). Since the SiO lines trace the shocked gas, this means the existence of some violent activities comparable to those of massive star formation.

The CO emission is detected in the velocity range from -1 to 5.5 km s^{-1}, which is blueshifted to the ambient cloud velocity of 6.3 km s^{-1}. No significant emission is seen at velocity larger than 6.3 km s^{-1}. This indicates that the outflow in B1 is asymmetric on scale of a few thousand AU and is dominated by blueshifted material. Figure 2 shows the total integrated intensity map of CO (J=1–0), superposed on the gray scale map of SiO (J=2–1: Yamamoto et al.

1992). The blueshifted gas shows distinct ring-like structure centered at IRAS 03301+3057. The size of the CO ring is 25″ (= 8700 AU)×14″ (= 5000 AU). The central hole of CO(J=1–0) emission coincides well with the IRAS position; this suggests that the observed structure is a nearly pole-on view of the molecular shell swept up by the wind from IRAS 03301+3057. The CO emission ring lies between two clumps of SiO; the distribution of the CO emission is anti-correlate with that of the SiO. The $H^{13}CO^+$ (Yamamoto et al. 1992) and NH_3 (Bachiller et al. 1990) results suggest that the dense gas exists at the southeast of the IRAS source. The morphological relation among the CO outflow, SiO clumps, and dense gas implies that the outflow interacts with the dense gas and causes the strong shock which produce the SiO.

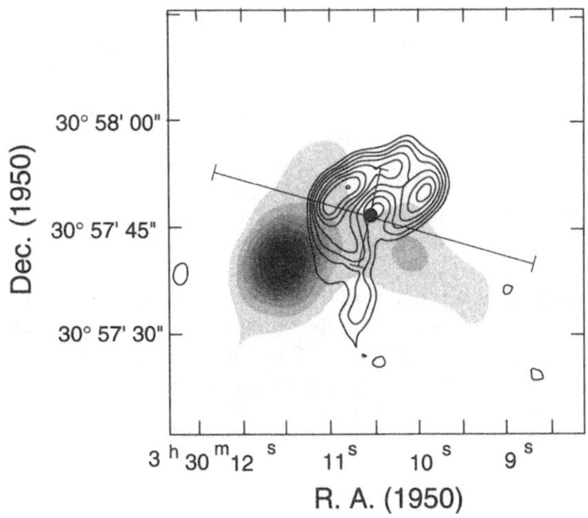

Fig. 2. A map of CO blueshifted emission integrated over the velocity interval -1–5.5 km s^{-1}, superposed on the gray scale map of SiO (Yamamoto et al. 1992). The cross mark indicates the IRAS position.

2.3 L723

The outflow in L723 is well known because of its unusual morphology. The high-resolution single-dish maps (e.g. Hayashi, Hasegawa & Kaifu 1991) show that the L723 outflow has two pairs of lobes; one pair lies along the axis north-south and another pair along an axis east-west. Such CO multilobe structure has been interpreted as (1) two outflows with two independent exciting sources (Anglada et al. 1991), (2) two outflows with a single exciting sources, (3) a precessing non-steady outflow (Moriarty-Schieven & Snell 1989), or as (4) a pair of limb-brightened cavity walls with a very wide opening angle (Avery, Hayashi, & White 1990).

In Figure 3, we present our interferometer maps of the L723 outflow. Figure 3 shows that the CO outflow in L723 has the point symmetric structure centered on the northeastern radio continuum source (L723-B). No CO emission is associated with the southwestern radio source (L723-A). Our results strongly suggest that only L723-B is driving the outflow and that the other source, L723-A, does not contribute to the structure of the outflow. The axis of the outflow lies almost north-south in the low velocity component ($\Delta v = 0.4$–4.5 km s^{-1}), however, it changes the direction to northeast-southwest in the intermediate velocity component ($\Delta v = 4.5$–8.5 km s^{-1}) and to east-west in the high velocity component ($\Delta v = 8.5$–12.6 km s^{-1}). This suggests that the quadrupolar-looking structure of the L723 outflow is settled within 3000 AU of its origin.

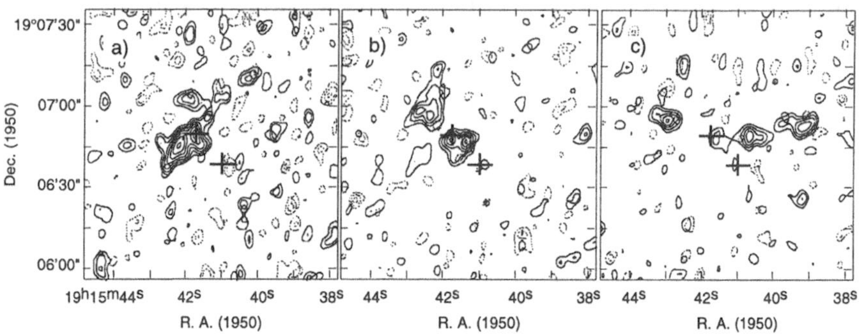

Fig. 3. Contour maps of the integrated intensity in the redshifted (solid lines) and blueshifted (dotted lines) CO emission in L723. The plus signs denote the positions of the 3.6 cm continuum sources (Anglada et al. 1991). a) The low-velocity components: $\Delta V = \pm 0.4$–4.5 km s^{-1}. b) The intermediate-velocity components: $\Delta V = \pm 4.5$–8.5 km s^{-1}. c) The high-velocity components: $\Delta V = \pm 8.5$–12.6 km s^{-1}.

2.4 Two Compact Outflows in Taurus; L1527 and TMC1A

These two outflows are found by the high-resolution surveys of Moriarty-Schieven et al. (1993) and Hirano et al. (1995). Both outflows are spatially compact and have been missed by the previous outflow surveys with moderate angular resolution.

The aperture synthesis CO observations of the L1527 and TMC1A outflows were made by Tamura et al. (1994, 1995). The interferometer maps show that the outflow in L1527 has distinct biconical lobes lying along east-west direction. Both eastern and western lobes contain the blueshifted and redshifted components, suggesting that the L1527 outflow is oriented almost perpendicular to the line of sight. At the common apex of two lobes, there is a mm-continuum source thought to be a protostar driving the outflow. The structure of the L1527 outflow is quite similar to that of B335.

The outflow in TMC1A is dominated by the blueshifted emission; the total flux of the blueshifted emission is ~5 times larger than that of the redshifted one (see Table 2). The blueshifted lobe and the redshifted lobe are overlapping near the center. Thus, this outflow is likely to be inclined ~45° to the line of sight.

The blueshifted lobe shows a conical shape whose apex almost coincides with the position of the mm-continuum source. The blueshifted lobe agrees well with the infrared nebula seen in the K-band image of Tamura et al. (1994), suggesting that there is a cavity carved out by the outflow.

2.5 Peculiar Outflow in VLA1623

VLA1623 is located in the main core of the ρ Ophuchi molecular cloud. Single-dish maps of André et al. (1990) show that the high-velocity gas is highly collimated and is confined in narrow lobes whose opening angles are $\sim 10°$. Since the blueshifted and redshifted emission is overlapping in the northwestern lobe, this outflow is likely to be highly inclined with respect to the line of sight. The central source of this outflow is detected only at wavelengths longer than sub-mm (André, Ward-Thompson, & Barsony 1993). Such spectral energy distribution strongly suggests that the central source, VLA1623, is in a very early stage of protostellar evolution.

The interferometer maps of CO and 2.6 mm continuum emissions are shown in Figure 4. We have detected strong blueshifted emission at $V_{LSR} = 2.4$ km s^{-1}, which is 1.6 km s^{-1} offset from the ambient cloud velocity of 4.0 km s^{-1}. The maximum CO intensity at this velocity is 5.5 Jy beam^{-1}. This corresponds to a brightness temperature of 14.3 K. The outflow shows a ridge-like structure that runs in a northwest to southeast direction. The width of the ridge is 1100–1500 AU. In this outflow, the flux of the blueshifted emission is twice as large as that of the redshifted emission.

The most notable feature of the VLA1623 outflow is that the blueshifted emission is strongly peaking at the position of the central source. This outflow is the only example that shows centrally peaked CO distribution. Such centrally peaked distribution is explained if the driving source is surrounded by a large amount of dense gas that strongly interacts with the stellar wind. The spectral energy distribution of the central source also supports this idea. Another characteristic of this outflow is that the outflow lobes consist of a chain of knot-like clumps. Each clump has size of 5–8″ (= 800–1300 AU) and mass of 10^{-5}–10^{-4} $M_{\odot}$. One possible interpretation is that these clumps are made by means of Kelvin–Helmholtz instability between the stellar wind and the dense ambient material. The small-scale structure of the outflow implies that the VLA1623 outflow is much younger than the B335 or L1527 outflows whose structure is characterized by biconical lobes without centrally peaked distribution.

3. Summary

We have observed seven molecular outflows associated with low-mass star forming regions by using the NMA. The interferometer maps show that the morphological characteristics of the outflows such as biconical shape in B335 or quadrupolar-looking structure in L723 are established in the close vicinity, at

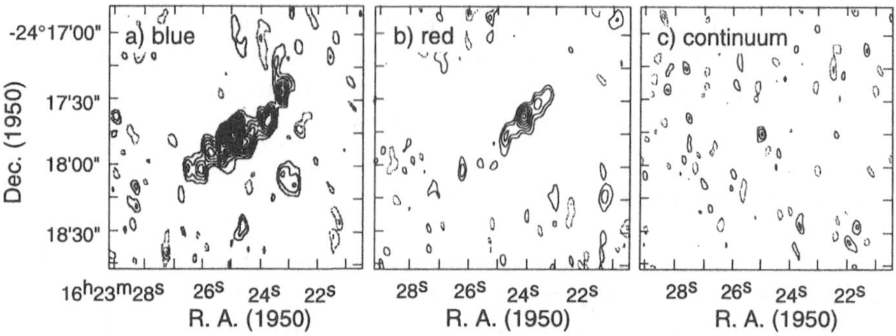

Fig. 4. Contour maps of the integrated CO emission in (a) the blueshifted component (V_{LSR} = -3.14–3.36 km s^{-1}), and (b) the redshifted component (V_{LSR} = 4.96–11.46 km s^{-1}). (c) Contour plot of the 2.6 mm continuum emission.

least a few thousand AU, of their origin. The CO emission generally peaks at 1000–3000 AU offset from the central source and shows no centrally peaked distribution. The only exception is the outflow in VLA1623, which shows strong blueshifted CO emission at the position of the central source. Such centrally peaked CO distribution suggests that the outflow in VLA1623 is in an earlier stage of evolution than other outflows and that the central source is still surrounded by a large amount of dense gas.

The observations presented here were carried out under collaborative works with M. Tamura, N. Ohashi, M. Hayashi, T. Kasuga, H. Mikami, S. Saito, S. Yamamoto, and S.S. Hayashi.

References

André, P., Martin-Pintado, J., Despois, D., & Montmerle, T. 1990, A&A, 236, 180
André, P., Ward-Thompson, D., & Barsony, M. 1993, ApJ, 406, 22
Anglada, G., Estalella, R., Rodgiguez, L.F., Torrelles, J.M., Lopez, R., & Canto, J. 1991, ApJ, 376, 615
Avery, L.W., Hayashi, S.S., & White, G.J. 1990, ApJ, 357, 524
Bachiller, R., Menten, K.M., & del Rio-Alvarez, S. 1990, A&A, 236, 461
Hayashi, S.S., Hasegawa, T., & Kaifu, N. 1991, ApJ, 377, 492
Hirano, N., Kameya, O., Nakayama, M., & Takakubo, K. 1988, ApJ, 327, L69
Hirano, N., Kameya, O., Kasuga, T., Hasegawa, T., Hayashi, S.S., & Umemoto, T. 1991, Proc. of the 7th Manchester Astronomical Conference on "Molecular Clouds", p115
Hirano, N., Kameya, O., Kasuga, T., & Umemoto, T. 1992, ApJ, 390, L85
Hirano, N., Hasegawa, T., Hayashi, M., Tamura, M., & Ohashi, N. 1995, Proc. of "Circumstellar Matter 1994", in press
Martin-Pintado, J., Bachiller, R., & Fuente, A. 1992, A&A, 254, 315
Moriarty-Schieven, G.H., & Snell, R.L. 1989, ApJ, 338, 952

Moriarty-Schieven, G.H., Wannier, P.G., Tamura, M., & Keene, J. 1992, ApJ, 400, 260
Nakayama, M. 1988, M.S. Thesis, Tohoku University
Parker, N.D., Padman, R., & Scott, p. 1991, MNRAS, 252, 442
Tamura, M., Ohashi, N., Moriarty-Schieven, G.H., Hayashi, M., & Hirano, N. 1994, Proc. of IAU Colloquium 140 "Astronomy with Millimeter and Submillimeter Wave Interferometry", p.268
Tamura, M., et al. 1995, in preparation
Terebey, S., Vogel, S.N., & Myers, P.C. 1989, ApJ, 340, 472
Yamamoto, S., Mikami, H., Saito, S., Kaifu, N., Ohishi, M., & Kawaguchi, K. 1992, PASJ, 44, 459

Evolution of Outflow Activity Around Low Mass Embedded Young Stellar Objects

S. Bontemps[1], P. André[1], S. Terebey[2], and S. Cabrit[3]

[1] CEA, Service d'Astrophysique, Centre d'Etudes de Saclay, F-91191 Gif-sur-Yvette Cedex, France
[2] I.P.A.C., IPAC 100-22, Caltech, Pasadena, CA 91125, USA
[3] Observatoire de Grenoble, B.P. 53 X, F-38041 Grenoble cedex, France

1. Introduction

In recent years, searches for molecular outflows driven by young stars have shown that most of the low-mass embedded young stellar objects (YSOs) excite a molecular outflow (e.g. Heyer et al. 1987, Myers et al. 1988, Terebey et al. 1989, Parker et al. 1991). It is now largely accepted that perhaps all young stars experience at least one outflow phase during their embedded protostellar phase. Nevertheless, it is not yet clear when the outflow phenomenom starts, what the duration of the outflow phase is, and what the basic agents for outflow mechanism are.

The embedded phase of the early evolution of the low-mass stars can be divided into, at least, two empirical stages. The first stage corresponds to the youngest observed YSOs, the "submillimeter" protostars or Class 0 objects recently introduced by André, Ward-Thompson, & Barsony (1993). These Class 0 YSOs are still surrounded by a dense circumstellar envelope which is more massive than the central star. The second stage comprises the Class I infrared sources of Lada (1987). They are interpreted as more evolved protostars, surrounded by residual circumstellar envelopes of substellar mass (André & Montmerle 1994 – hereafter AM). Millimeter continuum maps of optically thin dust emission from such envelopes can be used to make quantitative estimates of the involved masses, and are therefore a useful, practical tool to track the circumstellar evolution of embedded YSOs (e.g., AM).

In order to study outflow activity as a function of YSO evolutionary state in a systematic and quantitative way, we have conducted a homogeneous $J = 2-1$ ^{12}CO line survey of the close environment of a broad sample of nearby, low-luminosity YSOs still sufficiently embedded in their own circumstellar envelope (i.e. "self-embedded" YSOs) to show molecular outflows. The present CO observations benefit from the results of a coordinated millimeter continuum study of the same YSOs (e.g., AM; André, Cabrit et al., in prep.) which provide information on the total mass and spatial distribution of the circumstellar material. Here, we summarize the results of our CO survey, show that the outflow momentum flux is well correlated with the envelope mass of the driving source, and discuss the origin of this correlation.

2. Results

Our sample consists of 32 nearby ($d < 450$ pc), low-luminosity ($L_{\rm bol} < 50$ $L_\odot$) embedded YSOs, comprising 24 Class I and 8 Class 0 sources. In a related, complementary study, we obtained 1.3 mm continuum photometry of these sources using the IRAM 30 m telescope equipped with the MPIfR bolometer (AM; André, Cabrit et al., in prep.). These sources are mainly located in two of the nearest, low-mass star-forming complexes ρ Ophiuchi and Taurus (19 YSOs). Objects belonging to more distant star-forming regions were not considered in order to minimize confusion and distance-dependent effects.

The observations were carried out using three telescopes during six sessions: three sessions with the NRAO 12 m antenna at Kitt Peak in 1991 January and April, and 1992 January, one session with the CSO 10 m telescope at Mauna Kea in 1992 June, and a last one at Pico Veleta with the IRAM 30 m antenna in 1993 June. Our observing strategy was to systematically make small (3×3 or 5×5) ^{12}CO(2 − 1) maps of the close environment of the targeted YSOs. The spatial step of the rasters was typically 30″, except for the IRAM observations for which we used a 10″ step. We made maps of integrated high velocity emission in blue and red wings in order to obtain an outflow status for each source. And we measured the momentum fluxes of the detected flows. We obtained a high outflow detection rate of ∼ 90 % (29/32 objects), and in particular detected 8 new outflows (listed in Table 1). This very high detection rate, combined with the fact that the 3 non-outflow sources are weak 1.3 mm continuum emitters and very low luminosity YSOs, strongly suggests that virtually all low-mass self-embedded YSOs have a molecular outflow activity.

Table 1. The 8 new outflow sources

Star Forming Region	Driving Source	Evol. Class	$L_{\rm bol}$ [$L_\odot$]	$M_{\rm env}$ [$M_\odot$]	$F_{\rm CO}$ [10^{-6} $M_\odot$ yr^{-1} km s^{-1}]
Taurus	04166+2706	I	0.4	0.17	0.31
ρ Ophiuchi	16191-1936	I	1.8	0.12	0.04
ρ Ophiuchi	WL12	I	4.1	0.04	0.6
ρ Ophiuchi	EL29	I	41.0	0.09	1.5
ρ Ophiuchi	IRS43	I	10.1	0.06	1.5
ρ Ophiuchi	IRS51	I	1.3	0.03	0.55
L1152	20353+6742	I	3.3	0.006	0.47
L1082A	20520+6003	I	3.0	0.012	2.5

3. Discussion

3.1 F_{CO} is Correlated with M_{env}

In order to study in a quantitative way this widespread outflow activity, we plot, in Figure. 1, the CO outflow momentum flux F_{CO} derived from our observations against the circumstellar envelope mass M_{env} of the driving source. M_{env} was derived from the integrated 1.3 mm flux $S_{1.3mm}^{int}$ measured by André and collaborators (see previous section).

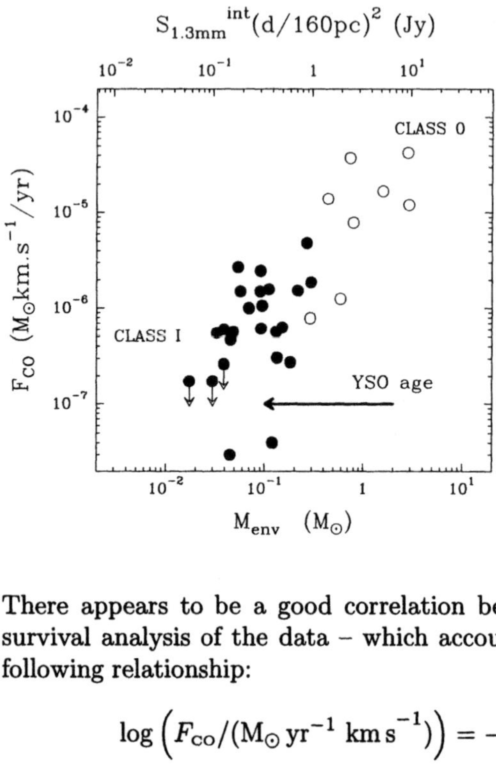

Fig. 1. log – log plot of outflow momentum flux versus circumstellar envelope mass of the driving source. Class 0 protostars are plotted as open circles and Class I YSOs as filled circles. A suggested arrow of time is indicated.

There appears to be a good correlation between F_{CO} and M_{env}. A statistical survival analysis of the data – which accounts for upper limits – provides the following relationship:

$$\log\left(F_{CO}/(M_\odot\,yr^{-1}\,km\,s^{-1})\right) = -5.1 + 1.05\log\left(M_{env}/M_\odot\right) \qquad (1)$$

with a probability of no correlation lower than 0.01 %.

3.2 What Does This Correlation Mean?

As mentioned in the introduction, the circumstellar envelope mass decreases as the central YSO evolves from the Class 0 to the Class I evolutionary stage. In a sample of objects spanning a small range of luminosities, the circumstellar mass may be used as an indicator of YSO age. While Class 0 protostars (open symbols) are on the right-hand side of Figure. 1, more evolved Class I YSOs (filled symbols) lie on the left-hand side. Thus, following our evolutionary description, the energetics of CO outflows decrease as YSOs evolve from the right to the left of Figure. 1.

However, we must check that the correlation seen in Figure. 1 does not result from a luminosity effect before concluding in favor of an outflow activity evolution. It is well-known that the driving source luminosity correlates with outflow energetics (e.g., Lada 1985, Levreault 1988, Cabrit & Bertout 1992). In addition, the millimeter continuum flux $S_{1.3mm}^{int}$, which is used to measure M_{env}, does also depend on YSO luminosity. Indeed, Reipurth et al. (1993) found a good correlation between 1.3 mm continuum flux and bolometric luminosity for a sample of Herbig-Haro exciting sources.

In order to visualize the luminosity effect on outflow activity in our sample, we plot F_{CO} versus bolometric source luminosity L_{bol} in Figure. 2. The slope obtained by Cabrit & Bertout (1992) on a sample of sources spanning 5 decades in luminosity is also shown.

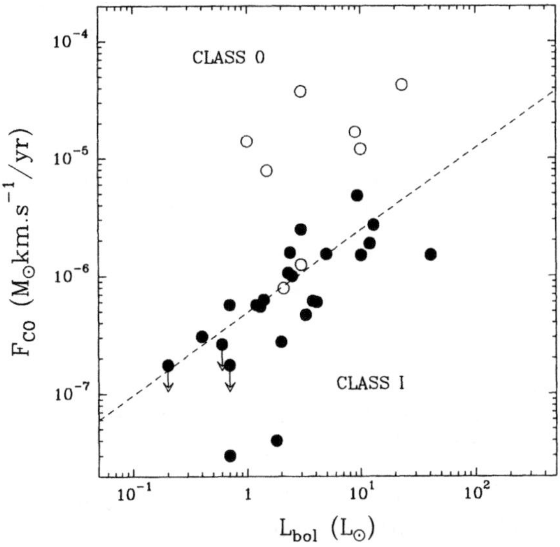

Fig. 2. log – log plot showing the influence of source luminosity on F_{CO}. The linear slope obtained by Cabrit & Bertout (1992) (of power index $\alpha \sim 0.7$) is shown as a dashed line. The symbols are the same as in Figure. 1.

As expected, F_{CO} is well correlated with L_{bol} in our sample. On the other hand, three converging pieces of evidence indicate that M_{env} is not governed by L_{bol} among our sources. First, we do not find a good correlation between M_{env} and L_{bol} (probability of no correlation of ~ 10 %). Second, by looking carefully at Figs. 1-2, we see that whereas Class 0 YSOs fall nicely on the correlation of Figure. 1, they lie systematically above the luminosity correlation in Figure. 2. Class 0 sources which are the youngest protostars are thus more efficient CO outflow drivers than Class I YSOs (at least in our sample). Third, we have tried to remove the dependences of F_{CO} and M_{env} on luminosity by dividing these two quantities by the luminosity laws found by Cabrit & Bertout (1992) and Reipurth et al. (1993) ($F_{CO} \propto {L_{bol}}^{0.7}$ and $M_{env} \propto {L_{bol}}^{0.6}$, respectively). The resulting plot (Figure. 3) shows that the correlation between outflow energetics and envelope mass remains after removing luminosity effects.

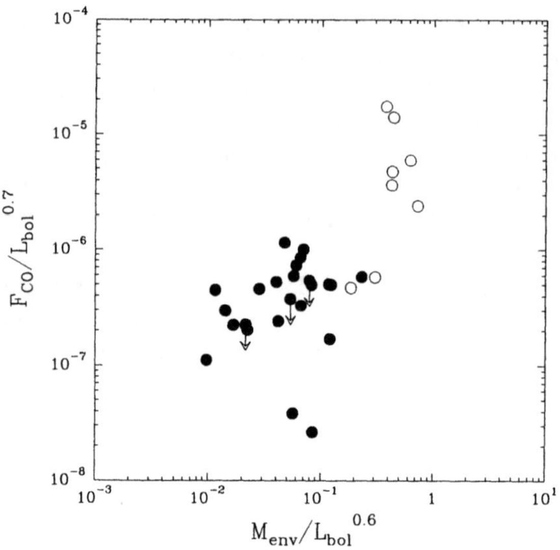

Fig. 3. log – log plot of $F_{CO}/L_{bol}^{0.7}$ versus $M_{env}/L_{bol}^{0.6}$. The indexes 0.7 and 0.6 come from correlations obtained by Cabrit & Bertout (1992) and Reipurth et al. (1993), respectively.

Therefore we conclude that the correlation seen in Figure. 1 most likely arises from a progressive decrease of outflow activity during the self-embedded protostellar phase of YSOs.

4. Conclusions

We have carried out a survey of outflow activity during the self-embedded protostellar phase of YSOs and obtained the following results:

1. Potentially *all* low-mass YSOs have an outflow activity when they are still embedded within a significant circumstellar envelope.
2. Outflow momentum flux and circumstellar envelope mass are well correlated with each other among self-embedded YSOs. We interpret this correlation as a progressive *decline of outflow activity* in the course of protostellar evolution.
3. Our successful use of M_{env} to trace the evolution of outflow activity illustrates its interest as a practical *age ordering* indicator for embedded YSOs.

References

André, P., Montmerle, T. 1994, ApJ 420,837 (AM)
André, P., Ward-Thompson, D., Barsony, M. 1993, ApJ 406,122
Cabrit, S., Bertout, C. 1992, A&A 261,274
Heyer, M.H., Snell, R.L, Goldsmith, P.F., Myers, P.C. 1987, ApJ 321,370
Lada, C.J. 1985, ARA&A 23,267

Lada, C.J. 1987, in IAU Symposium 115, "Star Forming Regions", Ed. M. Peimbert and J. Jugaku, p.1
Levreault, R.M. 1988, ApJ 330,897
Myers, P.C., Heyer, M.H., Snell, R.L, Goldsmith, P.F. 1988, ApJ 324,907
Parker, N.D., Padman, R., Scott, P.F. 1991, MNRAS 252,442
Reipurth, B., Chini, R., Krügel, E., Kreysa, E., Sievers, A. 1993, A&A 273,221
Terebey, S., Vogel, S.N., Myers, P.C. 1989, ApJ 340,472

Forbidden Line Emission and Correlations with the Infra-Red Excess in Herbig Ae/Be Stars

M.T.P. Corcoran and T.P. Ray

Dublin Institute for Advanced Studies, School of Cosmic Physics,
5 Merrion Square, Dublin 2, Ireland.

Summary. In a sample of 22 Herbig Ae/Be stars, covering a range of equivalent widths in Hα from 10 to 125 Å, we find a correlation between the [OI]λ6300 and Hα line luminosities and the excess mid-IR luminosity ($L_{\rm MIR}$) between 0.6μm and 10.2μm over three orders of magnitude. This matches the trend seen in T-Tauri stars (Cabrit et al. (1990)) and extends the range of these correlations to over five orders in magnitude in $L_{\rm MIR}$ and over a mass range of 0.4 to 25 $M_\odot$.

We argue that such correlations observed in both Herbig Ae/Be and T-Tauri stars arise from a proportionality between the mass accretion and mass-loss processes occurring in and about these young stars. Statistical tests indicate that the [OI] line luminosities correlate better with $L_{\rm MIR}$ than with L_*, the underlying stellar luminosity. However the Hα line luminosity correlates equally well with $L_{\rm MIR}$ or L_*. This suggests that, at least in part, the Hα line arises from a different source than the [OI] line. The smooth rise in [OI] line luminosity with $L_{\rm MIR}$, from T-Tauri stars to the much more luminous Herbig Ae/Be stars, implies that the *same physical process is responsible of the forbidden line emission in all these stars.* This is in conflict with recent models (Böhm and Catala (1994)) which propose that the forbidden line luminosity in Herbig Ae/Be stars is not disk related as in the T-Tauri stars.

1. Introduction

Finkenzeller and Mundt (1984) have catalogued and classified over fifty Herbig Ae/Be stars (HAeBeSs) according to the profile of their Hα line. The categories separated the stars into single peaked, double peaked, P-Cygni and complex emission profiles. Roughly 20% of the sample showed P-Cygni profiles implying the presence of a strong stellar wind. The broad wings of the single and double peaked profiles also indicated a stellar wind in a majority of the other stars in their sample. Many other indicators of mass loss have been observed in HAeBeSs; HH objects and jets (for examples see Ray et al. (1990), Corcoran, Ray and Bastien (1995), Mundt and Ray (1994)), molecular outflows such as PV Cep and LkHα 234 (Staude and Elsässer (1993)) and occasionally blue-shifted forbidden lines (Corcoran and Ray (1994)) The mechanism or mechanisms whereby these outflow phenomena arise are still uncertain but circumstellar accretion disks may play a role. Recent papers in the field have invoked the presence of circumstellar disks around HAeBeSs with a view towards explaining the observed infra-red excess and the shape of the spectral energy distributions (SEDs) of these stars. There is however, considerable controversy as to whether disks exist around HAeBeSs (see for example Hillenbrand et al. (1992), Berrilli et al. (1992), or Di Francesco et al. (1994) for opposing views).

In their study of forbidden line emission in T-Tauri stars (TTSs), Cabrit et al. (1990) proposed accretion driven mass loss as the explanation for the observed correlations between the luminosities of the [OI]λ6300 and Hα lines and the infra-red excesses measured for 36 TTSs. The fact that the luminosities of these lines were not found to correlate well with the stellar photospheric luminosities is taken to indicate that the forbidden lines are more likely due to a disk corona/wind. Cabrit et al. suggested that accretion occurring via the disk gives rise to the mass-loss processes (i. e. the stellar winds) that produce the forbidden line and Hα line emission. Here we extend their study of forbidden lines to include the more luminous Herbig Ae/Be stars.

2. Observations and Results

From our spectroscopic data-set of over 50 Herbig Ae/Be stars, we have selected a subset of 22 HAeBeS with good photometric data. Of these 21 show both [OI]λ6300 emission and an infra-red excess. One other star (LkHα 215) shows not [OI]λ6300 and is included in the data-set as a censored data point in the statistical analysis. The complete data-set consists of long slit spectra (R=10000–30000) taken during observing runs at La Palma on the Isaac Newton Telescope(INT) with the IDS spectrometer in September 1991 and July 1993 and an observing run at La Silla on the ESO 2.2m Telescope (EFOSC-II instrument.) Data reduction was carried out using standard IRAF routines. Table 1 shows the data-set of the 22 HAeBeSs. The broad band photometry values come from Hillenbrand et al. (1992) with the exception of the stars PV Cep which comes from Hamann and Persson (1992).

The line luminosities of Hα and [OI] are calculated from the observed equivalent width and the extinction corrected R_c band magnitudes, assuming isotropic emission and the distances given in Thé, de Winter and Pérez (1993). The typical error bars shown in Figs. 1 and 2 derive from errors in A_v, distance estimates and spectral type uncertainty. The figures show the dereddened [OI] and Hα emission line luminosities against the mid infra-red excess luminosity, L_{MIR}, as defined by Cabrit et al. (1990): $L_{MIR} = L_{total} - L_*$ where L_* is the photospheric luminosity of a standard star of matching spectral type over the range 0.6μm to 10.2 μm and L_{total} is the total dereddened luminosity of the observed star over the range. The wavelength range is deliberately chosen to allow a direct comparison with mid-IR excess luminosity calculated for 36 T-Tauri stars by Cabrit et al. (1990). Excess emission at wavelengths 0.6μm to 10.2μm will come predominantly from the inner disk regions and will therefore be more sensitive to any variations in the disk accretion rates and disk activity in general. It must be noted however that while this range may be the optimum for measuring disk activity in T Tauri stars, this may not be the case for Herbig Ae/Be stars due to various features. For example dust emission has been observed at 3.29 μm (Brooke Tokunaga and Strom, 1993) which might complicate the correct calculation of a mid-IR excess over the 0.6–10.2 μm range. Hillenbrand et al. (1992) have also observed a "dip" in the SED of many HAeBeSs in the 1–2μm range

Table 1. Stellar and line luminosities of Herbig Ae/Be Stars. $L_{\rm total}$ and $L_{\rm MIR}$ are measured over a wavelength range of 0.6μm–10.2μm. All luminosities are in units of log $L_\odot$.

Star	SpType	log L_*	log $L_{\rm total}$	log $L_{\rm MIR}$	log $L_{\rm H\alpha}$	log $L_{\rm [OI]}$
HD 259431	B2	3.80	2.93	2.92	0.35	-1.62
HK Ori	A5	1.09	3.26	1.10	-1.30	-2.85
KK Oph	A6	0.76	0.99	0.66	-1.77	-3.26
LkHα 25	A0	0.96	1.91	1.84	-1.55	-3.09
LkHα 198	A5	0.75	1.82	1.78	-1.52	-3.04
LkHα 215	B5	2.95	2.41	2.38	-0.43	< -5.00
LkHα 233	A5	1.45	1.61	1.55	-1.31	-2.89
LkHα 234	B5	3.24	2.21	2.15	0.01	-2.21
MWC 297	O9	4.98	3.37	4.09	1.71	0.46
MWC 1080	B0	4.59	3.49	3.48	0.98	-1.70
NX Pup	A0	1.65	1.83	1.75	-0.97	-2.70
PV Cep	A5	0.95	1.78	1.74	-1.57	-2.33
R Mon	B0	3.55	2.66	2.56	-0.23	-1.57
V376 Cas	F0	0.45	1.77	1.75	-2.12	-3.09
V380 Ori	B9	1.93	1.88	1.80	-0.96	-1.92
BD+61° 154	B8	2.52	2.34	2.31	-0.43	-2.77
VV Ser	B9	1.80	1.87	1.56	-1.09	-3.92
AB Aur	A0	1.76	1.69	1.56	-1.09	-3.27
HD 97048	B9	1.84	1.65	1.50	-0.84	-3.28
HD 150193	A2	1.43	1.39	1.09	-1.77	-4.09
HD 163296	A0	1.79	1.68	1.54	-1.18	-3.78
HD 200775	B2	4.25	3.35	3.34	0.35	-1.68

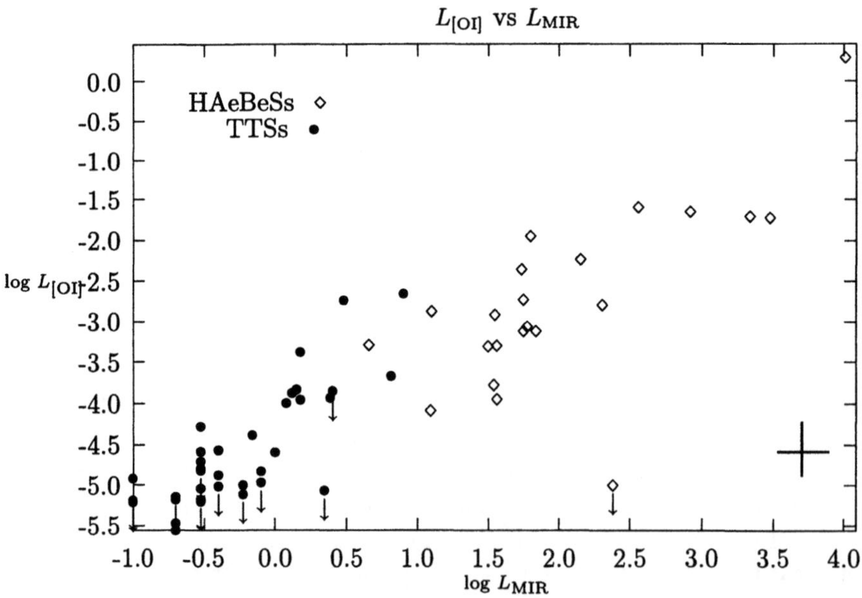

Fig. 1. A logscale plot of the forbidden [OI]λ 6300 emission line luminosity against the excess mid-IR luminosity for the combined sample of Herbig Ae/Be and T-Tauri stars. The TTSs data comes from Cabrit et al. (1990.)

and model it using a circumstellar disk with a central hole. As shown in Figs. 1 and 2 the HAeBeSs fill a region in the line luminosity vs mid-IR excess logscale plots that extends the trend seen in the TTSs described by Cabrit et al. (1990).

3. Discussion

Table 2. The results of the statistical tests. The results for the TTSs are from Cabrit et al. (1990). All probabilities are those of *no* correlation existing between the compared variables.

Correlation	Correlation Coefficient	Cox's Probability	Kendall's Probability
$L_{[OI]}$ vs. L_{MIR}	0.97	0.0005	0.0008
$L_{H\alpha}$ vs. L_{MIR}	0.99	0.0006	< 0.0001
$L_{[OI]}$ vs. L_*	0.53	0.0074	0.0204
$L_{H\alpha}$ vs. L_*	0.60	0.0015	0.0002
L_{MIR} vs. L_*	0.51	0.0029	0.2041
$L_{[OI]}$ vs. $L_{H\alpha}$	0.72	0.0008	0.0015
TTS $L_{[OI]}$ vs. L_{MIR}	0.83	< 0.0001	...
TTS $L_{H\alpha}$ vs. L_{MIR}	0.74	< 0.0001	...
TTS $L_{[OI]}$ vs. L_*	0.64	0.0100	...
TTS $L_{H\alpha}$ vs. L_*	0.48	...	...
TTS $L_{[OI]}$ vs. $L_{H\alpha}$	0.78	< 0.0001	...

We have carried out statistical tests on the HAeBeSs data to determine the degree of correlation between the line luminosities and L_{MIR} and L_*. Linear regression analysis was carried out on the HAeBeSs data set. Two "survival analysis" tests were also applied (the Cox's Proportional Hazard and the Generalized Kendall's Tau) to the data. The software for both these tests comes from the ASURV statistics package provided by Isobe, LaValley and Feigelson (1986 and 1992.)

The results of the statistical tests in Table 2 indicate that while the luminosity of the Hα emission is equally correlated with both the photospheric and the excess IR luminosity, the [OI] emission is more strongly correlated with the excess IR luminosity. This suggests that while one physical process, probably a disk driven wind, gives rise to the forbidden lines, the Hα lines, in part at least, come from the star. For example, the Hα luminosity may form from a combination of accretion-driven wind and a simple radiatively driven stellar wind, particularly in the earlier stars with high photospheric luminosities.

The low probability of an L_{MIR}–L_* correlation indicates that the excess mid-IR luminosity is likely not due solely to disk reprocessing of incident starlight. This suggests that there is dust heating independent of the central star, be it from viscous heating in the disk or from energy released by an accretion shock at the star or boundary layer.

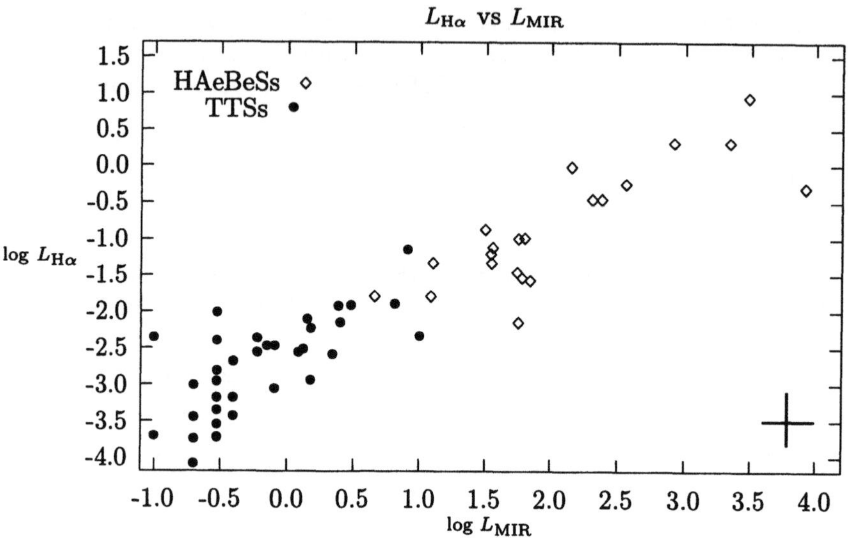

Fig. 2. A logscale plot of the Hα emission line luminosity against the excess mid-IR luminosity for the combined sample of Herbig Ae/Be and T-Tauri stars. The TTSs data comes from Cabrit et al. (1990.)

4. Conclusions

The line luminosities of Hα and [OI]λ6300 correlate with the mid-IR excess over nearly 5 orders of magnitude for a sample of 36 T-Tauri stars and 21 Herbig Ae/Be stars. The smooth progression of the [OI]λ6300 luminosity with L_{MIR} implies that the same physical process is responsible for the forbidden line emission in both the low mass TTSs and the more massive HAeBeSs. Statistical tests on the HAeBeSs data set show that, while the Hα line correlation with the excess luminosity could also be explained by a simple dependence on photospheric luminosity, the forbidden [OI]λ6300 line luminosity shows a dependence on the mid-IR excess luminosity. The proportionality observed suggests that whatever drives the IR excess also drives, perhaps indirectly, the source of the [OI] emission. We believe that accretion processes in the disk give rise to outflow phenomena which are responsible in turn for exciting the [OI]λ6300 emission.

References

Berrilli, F., Corciulo, G., Ingrosso, G., Lorenzetti, D., Nisini, B., and Strafella, F., 1992 ApJ, **398**, 254.
Böhm, T. and Catala, C. 1994, A&A, **290**, 167.
Brooke, T. Y., Tokunaga, A. T., and Strom, S. E., 1993, AJ, **106**, 656.
Cabrit, S., Edwards, S., Strom, S. E. and Strom, K. M., 1990, ApJ **354**, 687.
Corcoran, D., Ray, T. P., Bastien, P., 1995, A&A, **293**, 550.

Corcoran, M. and Ray, T. 1994 *Proceedings of the First Conference on the Nature and Evolutionary Status of Herbig Ae/Be Stars*, P. S. Thé, Pérez, M. R., and E. P. J., van den Heuvel, Astronomical Society of the Pacific Conference Series, p 151.
Di Francesco, J., Evans, N. J., Harvey, P. M., Mundy, L. G., and Butner, H. M., 1994, ApJ, **432**, 710.
Finkenzeller, U. and Mundt, R., 1984 A&ASS **55**, 109.
Hamann, F., and Persson, S.E. 1992, ApJ, **394**, 628.
Hillenbrand, L.A., Strom, S.E., Vrba, F.J. and Keene, J. 1992, ApJ, **397**, 613.
Isobe, T., LaValley, M. and Feigelson, E. D. 1986, ApJ, **306**, 490.
Isobe, T., LaValley, M. and Feigelson, E. D. 1992, ASURV software supplied with STSDAS/IRAF.
Mundt, R., and Ray, T. P., 1994, *Proceedings of the First Conference on the Nature and Evolutionary Status of Herbig Ae/Be Stars*, P. S. Thé, Pérez, M. R., and E. P. J., van den Heuvel, Astronomical Society of the Pacific Conference Series, p 237.
Ray, T.P., Poeztel, Solf, J., and Mundt, R., 1990, ApJ Lett, **357**. L45
Kwan, J. and Tademaru, E. 1988, ApJ, **322**, L41.
Staude, H.J., Elsässer, H. 1993, Ann. Rev A&A, **5**, 165.
Thé, P.S., de Winter, D., Pérez, M.R. 1993, A&ASS, **104**, 315.

Theory of Outflows

Arieh Königl

Department of Astronomy & Astrophysics, The University of Chicago, Chicago, IL 60637, U.S.A.

Summary. Following a brief review of the current observational picture of outflows from low- and intermediate-mass young stellar objects, various theoretical model ingredients are discussed, focusing on a scenario in which the outflows correspond to centrifugally driven winds from magnetized accretion disks. Among the topics considered are the wind kinematics and dynamics, the outflow thermal and ionization structure, and nonsteady ejection. Particular attention is paid to some of the outstanding questions in both theory and observations.

1. Observational Highlights

Many of the recent observational developments on the properties of outflows from young stellar objects (YSOs) have been reviewed in these Proceedings by R. Mundt. Before turning to the theoretical interpretation, it is, however, useful to list the basic elements of the picture that has emerged so far as well as the main results of recent studies and the modeling issues that they have raised. In this summary, I first focus on low-luminosity YSOs – their outflow components, the relationship between accretion and outflow, and the evidence for nonsteady mass ejection, and then turn briefly to higher-luminosity objects.

1.1 Outflow Components

Bipolar outflows in low-luminosity ($L_{\rm bol} \lesssim 10^3\, L_\odot$) YSOs have originally been discovered through their associated molecular flows (Snell et al. 1980). It is now recognized that the latter are driven by higher-velocity outflows that emanate from the vicinity of the central object and carry momentum discharges that are typically much larger than the force $L_{\rm bol}/c$ that can be imparted by the YSO radiation field under optically thin conditions (e.g., Lada 1985). The mode of coupling between the driving outflows and the ambient molecular gas has been the subject of several recent studies (e.g., Chernin et al. 1994), but inasmuch as this contribution is mainly concerned with the origin of the outflows, this topic will not be elaborated on here.

It has also been recognized that the YSO outflows consist of both a *neutral* and an *ionized* component. The neutral gas is detected in either atomic or molecular form (e.g., Edwards et al. 1993), and, when observed directly, is often inferred to carry enough momentum to drive the bipolar molecular lobes. The ionized component is detected through its thermal radio and high-velocity optical emission signatures, and has originally been inferred to carry relatively little momentum. However, it was subsequently realized that, if most of the jet material contributes to the observed low-excitation line emission (after being excited by weak shocks or turbulent entrainment), and if the emitting gas is only partially ionized, then this dichotomy disappears and the gas in the optical/radio jets could have sufficient momentum to drive the outer lobes (Raga 1991; Hartigan et al. 1994). The ionized components are generally highly collimated, but, in many of the studied cases, the collimation appears to be relatively poor near the origin (opening half-angle $\sim 10° - 15°$) and to be established only on scales $\lesssim 10^2$ AU (Mundt et al. 1991).

One of the main diagnostics of outflows from low-luminosity YSOs (specifically, T Tauri stars) has been their forbidden optical line emission ([O I], [S II], [N II]). It is detected on scales of $\sim 10 - 100$ AU, and one of the chief puzzles associated with its interpretation has been the nature of the heating mechanism that keeps the emitting gas at a temperature of $\sim 10^4$ K so far away from the central object. Important advances in the study of this gas have been recently made with the help of high-resolution, ground-based long-slit spectroscopy (e.g., Solf & Böhm 1993; Hirth et al. 1994; Böhm & Solf 1994) as well as through *HST* imaging (e.g., Kepner et al. 1993). These observations have revealed the existence of two distinct components: a low-velocity ($\sim 5 - 40$ km s^{-1}) component (LVC) that is relatively narrow ($\Delta V_{\rm FWHM} \approx 20 - 40$ km s^{-1}), although one can occasionally identify a red wing that extends to $\gtrsim 10^2$ km s^{-1}, and a high-velocity ($\gtrsim 100$ km s^{-1}) component (HVC) that is relatively broad ($\Delta V_{\rm FWHM} \approx 50 - 100$ km s^{-1}), although it appears that the width decreases with distance from the origin (possibly signifying collimation). The [O I] LVC emission peaks at $\sim 15 - 20$ AU from the YSO and extends to ~ 50 AU, whereas the [S II] LVC emission usually peaks further out, and one finds evidence for acceleration between the [O I] and [S II] emission regions. The LVC is characterized by high density but relatively low excitation. In contrast, the HVC emission extends to $\gtrsim 10^{16}$ cm (an order of magnitude larger than the LVC emission region), and it appears to continue smoothly into the large-scale optical jet. Furthermore, it is characterized by high excitation, which is likely shock-induced (as in the outer jet).

There are at present apparently conflicting inferences about the relative magnitude of the electron densities in the LVC and the HVC. On scales $\lesssim 100$ AU, the average densities as deduced from the [S II] λ6716 / [S II] λ6731 line ratios are $n_{\rm e,HVC} \gtrsim 10^4$ cm$^{-3} \approx 10\, n_{\rm e,LVC}$ (e.g., Mundt et al. 1987). In contrast, on smaller scales where the electron densities are $\sim 10^6 - 10^8$ cm^{-3}, the absence of HVC [O I] λ5577 emission suggests that $n_{\rm e,LVC}$ exceeds $n_{\rm e,HVC}$ (Edwards et al. 1989; Hamann 1994). One possible explanation for the apparent absence of this line is that the HVC is not yet strongly excited near the base of the outflow: this situation could arise if, for example, the shocks that give rise to this emission only develop further out along the jet (see Section 1.3). Recent [Fe II] infrared line measurements (Hamann et al. 1994) have indicated that both the LVC and the HVC emission regions may have similar sizes and densities in certain cases, which suggests that these components do not necessarily occupy different regions of the outflow.

1.2 The Accretion – Outflow Connection

There is strong evidence for an intimate link between accretion and outflow signatures in YSOs (e.g., Edwards et al. 1993, and poster in this conference). For example, among T Tauri stars, forbidden line emission (arising in the outflow) and optical veiling (produced in the course of accretion onto the YSO) are found only in objects that exhibit a near-infrared color excess (a signpost to distributed circumstellar mass, most likely a disk). This immediately suggests that the outflows are powered by accretion. The measured high velocities indicate an origin in the region near the stellar surface, where the liberated gravitational potential energy is high, and this conclusion is strengthened by the apparent correlation between the measured radial velocities in the [O I] λ6300 line and the presence of P Cygni profiles in the observed Hα and Na D lines, which is discussed in Mundt's contribution. It is as yet unclear, however, whether the outflows originate right at the stellar surface or nearby in the accretion disk. The detection of high-velocity emission from *neutral* species (such as Na and CO) argues, particularly in the case of high-surface-temperature sources (e.g., Mitchell et al. 1989), against the possibility that the gas started out on the stellar surface itself. The presence of [O I] LVC emission on scales of $\sim 0''.1$ in objects like DG Tau (Solf & Böhm 1993) suggests that at least the LVC originates in the disk. There is also evidence that the high-velocity gas expelled during an FU Orionis outburst comes out of the disk surface rather than the star (Calvet et al. 1993).

1.3 Nonsteady Ejection

Another key property of the outflows that has come to light in recent years is their apparent variability. A nonsteady outflow is likely to lead to shock formation along the jet, which may be manifested as emission knots and bow structures. Various groups have studied the evolution of such flow disturbances both analytically and numerically (e.g., Raga & Kofman 1992; Hartigan & Raymond 1993; Gouveia Dal Pino & Benz 1994). Schematically, if the jet velocity $V_{\rm j}$ varies by an amount $\Delta V_{\rm j}$ on a time scale $t_{\rm var}$, then a pair of shocks (an internal "working surface") will first form at a distance X from the origin after a time $t_{\rm on}$, with X and $t_{\rm on}$ related by $X \approx V_{\rm j} t_{\rm on}$. Subsequent shocks will have a separation $\Delta X \approx V_{\rm j} t_{\rm var}$, with $\Delta X / X \approx \Delta V_{\rm j} / V_{\rm j}$. Based on this picture, Hirth et al. (1994) have inferred $t_{\rm on} \lesssim 10$ yr for CW Tau, which is consistent with the direct measurement of velocity variability in this source, whereas Reipurth et al. (1992) have deduced a time scale of decades for $t_{\rm var}$ in HH 111. In the case of HH 46/47, Raga et al. (1990) inferred a time scale between major outbursts (leading to the formation of bow shocks along the jet) of $\gtrsim 10^3$ yr, which is compatible with the estimated typical time between FU Ori outbursts in YSOs (Bell & Lin 1994). These values are consistent with the suggestion (Herbig 1989) that there may be a progression of variability time scales, with $t_{\rm var}$ increasing with source luminosity on going from flash variables to EX Ori-type stars to FU Ori stars. In fact, it is also conceivable that individual star/disk systems may experience a variety of nonsteady outflow episodes, characterized by different time scales and kinetic luminosities, with the dominant mode of eruption possibly depending on the evolutionary stage of the star.

1.4 Herbig Ae/Be Stars

Rapid progress is also being made in the study of higher-luminosity YSOs, particularly the intermediate-mass Herbig Ae/Be stars. Several investigations have revealed a similarity between Herbig Ae/Be stars and classical T Tauri stars in showing evidence both for outflows (P Cyg profiles, thermal radio continuum, forbidden line emission) and for circumstellar dust (infrared and millimeter excesses). However, as the UV emission is not a useful accretion diagnostic in these comparatively high-surface-temperature stars, the presence of circumstellar disks is at present still controversial (e.g., Hillenbrand et al. 1992; Berrilli et al. 1992; Natta et al. 1993; Di Francesco et al. 1994; Prusti et al. 1994). Much of the uncertainty centers on the interpretation of the infrared observations, particularly on whether the mid- and far-infrared emission is dominated by the contribution of an extended envelope. Another problem concerns the interpretation of the $\sim 3\,\mu$m peak in the near-infrared spectral energy distribution, which has been attributed to dust evaporation in the disk or envelope. Hartmann et al. (1993) have pointed out that either one of these interpretations would lead to accretion rates approaching $10^{-5}\, M_{\odot}\, \mathrm{yr}^{-1}$,[1] resulting in bolometric luminosities well above those that are observed, and suggested that the near-infrared "bumps" might originate instead in transient heating of small grains by UV photons.

Forbidden line emission, which serves as an outflow diagnostic in T Tauri stars, has also been detected in Herbig Ae/Be stars (e.g., Hamann 1994; Böhm & Catala 1994). One puzzling aspect of these observations is that, in many cases, only a broad ($\Delta V \approx 30-100$ km s^{-1}), symmetric LVC component is detected in [O I], even in sources (such as LkHα 234; Ray et al. 1990) where a high-velocity, well-collimated jet is seen further out.

2. Model Ingredients

A viable outflow model for YSOs must satisfy the following observational constraints: it should be directly tied to the accretion process and produce winds with kinetic powers that are correlated with (but remain less than) the bolometric luminosities; it should account for the large momentum discharges inferred from measurements of the associated molecular bipolar flows; and it should be able to reproduce the radiative properties of the observed jets and winds. The fact that the requisite momentum discharges in low-luminosity objects can

[1] A similar conclusion might apply even if the dust is associated with an *outflowing* disk wind (see Section 2.2 below), for which the outflow rate would typically be lower than the inflow rate associated with the *infalling* envelope considered by Hartmann et al., since the accretion rate through the disk is generally higher than the outflow rate in disk-driven wind models (see Section 2.1).

greatly exceed $L_{\rm bol}/c$ rules out radiatively driven winds, whereas thermally driven winds are generally excluded by the expected short radiative cooling times (which, in turn, are consistent with the apparent absence of significant quantities of coronal gas near YSOs). The most promising models involve magnetohydrodynamically driven winds, which readily satisfy the momentum and energy discharge constraints. In this section, I summarize the salient features of these models and attempt to relate them to some of the observational highlights described in the previous section.

2.1 Centrifugally Driven Outflows

The most popular models for magnetically driven outflows in YSOs assume that they originate as centrifugally driven winds from circumstellar disks (see Pudritz et al. [1991] and Königl & Ruden [1993] for reviews). In this picture, the disks are threaded by open magnetic field lines and material is expelled centrifugally from their surfaces like beads on inclined, rotating wires. The outflows are powered by the liberated gravitational potential energy of the accreted gas and could in principle transport most of the angular momentum of the inflowing matter, which is removed by magnetic torques in the disk. This scenario could account in a natural way for the intimate connection between disks and jets: the outflows are the means by which the gas in the rotating disk disposes of its angular momentum and potential energy in order to reach the stellar surface.

Blandford & Payne (1982) were the first to consider centrifugally driven winds in a disk threaded by a large-scale, ordered magnetic field. As they pointed out, the basic outflow mechanism can be understood with the help of a mechanical analogy to a bead on a rotating, rigid wire: for a Keplerian rotation law, the bead (representing the matter) will be flung out if the wire (representing the magnetic field line in the ideal-MHD limit) is inclined at an angle $> 30°$ to the disk normal. Using a self-similar (in the spherical radial coordinate), cold wind model, they constructed explicit solutions of flows that are magnetically dominated above the disk surface and become super-Alfvénic and magnetically collimated further out.[2] Furthermore, they showed that such winds are an efficient means of extracting angular momentum from the disk because any given field line (rotating with the angular velocity of the material at the radius ϖ_0 where it leaves the disk) enforces corotation out to the (cylindrical) Alfvén radius $\varpi_{\rm A}$, thereby increasing the lever arm of the torque that acts back on the disk. Since $\varpi_{\rm A}$ could be $\gg \varpi_0$, the mass outflow rate that is required for the removal of all the liberated angular momentum might be only a small fraction of the mass accretion rate: $\dot{M}_{\rm wind}/\dot{M}_{\rm acc} \approx (\varpi_0/\varpi_{\rm A})^2$. The wind could also tap a significant fraction of the gravitational potential energy given up by the accreted

[2] In the centrifugally driven wind scenario, the toroidal magnetic field component is not important in driving the outflow near the disk surface, although it plays a central role in collimating the wind by hoop stresses beyond the Alfvén point. It is in principle possible for magnetized outflows to be driven from the start by the gradient in B_ϕ (e.g., Lovelace et al. 1991), but such winds might be hard to maintain in a steady state (e.g., Cao & Spruit 1994).

matter and carry it out in the form of a Poynting flux. Given the comparatively low mass outflow rate, the terminal (poloidal) speed of the wind would be correspondingly large, $V_{p\infty} \approx (\varpi_A/\varpi_0)V_{K0}$ (where V_{K0} is the Keplerian speed at ϖ_0).

In the self-similar model of Blandford & Payne (1982), the scaling of the speed V, magnetic field amplitude B, and gas density ρ with the spherical radial coordinate R is determined from the relations $B/\sqrt{\rho} \propto V \propto V_K \propto R^{-1/2}$ and from the assumption that the mass outflow rate from any given decade of radius, $\dot{M}_{\rm wind}(\varpi_0) \propto \rho_0 V_0 \varpi_0^2$, is constant, which imply $\rho_0 \propto \varpi_0^{-3/2}$ and $B_0 \propto \varpi_0^{-5/4}$. Contopoulos & Lovelace (1994) generalized this class of self-similar solutions by considering winds with a density scaling $\rho \propto R^{-\alpha}$, for which $B \propto R^{-(\alpha+1)/2}$ and $\dot{M}_{\rm wind}(\varpi_0) \propto \varpi_0^{(3/2-\alpha)}$ (with $\alpha = 3/2$ corresponding to the Blandford & Payne solution). A favored "minimum energy" solution, corresponding to a constant axial current $I = 2\pi c r B_\phi$ across the jet, is given by $\alpha = 1$. These self-similar solutions are characterized by constant values of the (normalized) specific angular momentum (including kinetic and magnetic contributions) and mass-to-flux ratio throughout the flow, and all the flow lines cross the disk at the same angle to its surface.

Pelletier & Pudritz (1992) further generalized the centrifugally driven wind solutions to a class of non–self-similar outflows. In particular, they considered outflows in which the quantity $\frac{V_{K0}^2 \varpi_A^2}{V_A^2 \varpi_0^2}|\nabla\varpi_A|_A^2$ (where the subscript A denotes the Alfvén point) is everywhere a constant, and derived solutions in which the poloidal magnetic field component at the disk surface scales as ϖ_0^{-x} (with x expected to lie between 1 and 2). The asymptotic outflow velocity along a flow line that intersects the disk at ϖ_0 scales as $V_{p\infty}(\varpi_0) \propto \varpi_0^{2(1-x)}$, whereas $\dot{M}_{\rm wind}(\varpi_0) \propto \varpi_0^{(4x-5)}$. The Blandford & Payne solution is recovered by setting $x = 5/4$: for $x < 5/4$, the angle between the flow lines and the disk surface increases with ϖ_0, whereas for $x > 5/4$ the flow lines converge toward the axis and the wind has a finite lateral extent. The $I = const.$ solution is given in this case by $x = 3/2$, which corresponds to $V_{p\infty} \propto \varpi_0^{-1}$ and $\dot{M}_{\rm wind}(\varpi_0) \propto \varpi_0$. Note that this differs from the self-similar $I = const.$ solution discussed above, which is characterized by $V_{p\infty} \propto \varpi_0^{-1/2}$ and $\dot{M}_{\rm wind}(\varpi_0) \propto \varpi_0^{1/2}$. Although the simple Blandford & Payne model captures the essential features of the more general wind solutions, the latter are important for investigating the properties of real sources and more work is needed on their development (see also the contribution by Sauty).

The origin of the magnetic field that threads the disk is still an open question. One likely possibility is that it corresponds to interstellar field lines that are frozen into the parent molecular cloud and that have been advected inward either during the collapse of the molecular cloud core (cf. Fiedler & Mouschovias 1993; Galli & Shu 1993) or subsequently by accretion through the disk (e.g., Lubow et al. 1994; see also poster by Agapitou & Papaloizou in this conference). This possibility is supported by the close alignment between the ambient field and the outflow direction that has been inferred in many sources (e.g., Strom et al. 1986).

Another possibility is that the magnetic field in the disk has been amplified by a local dynamo process (e.g., Stepinski et al. 1993). Alternatively, the outflow may be driven by stellar magnetic field lines (e.g., Arons 1987; Shu et al. 1994 and these Proceedings; see also the contribution by Fendt & Camenzind). It is worth noting that, even in the latter class of models, the mass driven out in the wind is most often postulated to come from the disk rather than from the star.

Another fundamental question is how exactly the magnetic field mediates the accretion and outflow processes. Specifically, how is the angular momentum transferred from the disk to the wind and how does part of the inflow turn into an outflow? Wardle & Königl (1993) addressed these issues in the context of weakly ionized disks and constructed steady-state disk/wind models in which the radial advection and azimuthal shearing of the field lines are counteracted by ambipolar diffusion. Analogous models have been constructed also for disks in which Ohmic diffusivity is the main field dissipation mechanism (Königl 1994, 1995; Ferreira & Pelletier 1995). Much more work remains, however, to be done on this subject.

2.2 Wind Kinematics and Dynamics

Centrifugally driven winds are characterized by rapid acceleration near the disk surface, which gives rise to a density distribution that is highly stratified in the vertical direction. The velocity distribution is also stratified: for example, in a self-similar Keplerian disk, V scales as $R^{-1/2}$ and increases with decreasing polar angle θ (see Figure 1). The outflow typically carries enough momentum to uplift dust from the disk through grain-neutral collisions. The radiation field of the YSO will evaporate the dust in the inner regions of the wind, but, beyond the sublimation radius $\varpi_{\rm sub} \approx 0.1\,(L_{\rm bol}/L_{\odot})^{1/2}$ AU, the dust will survive and will exhibit the same density distribution as the gas in which it is embedded.

The strong density stratification that is generic to these winds has important observational consequences. For example, if part of the forbidden line emission in YSOs originates in such outflows (see Section 2.3), then the measured forbidden line ratios (such as L_{6731}/L_{6717} of [S II], which is used as a density diagnostic) will be influenced by the detailed density distribution in the source (Safier 1992). In particular, since the critical densities of the above lines are related by $n_{\rm c}^{6731} \approx 3.4 n_{\rm c}^{6717}$, the emission region of the 6731 Å line would typically be smaller that that of the 6717 Å line, implying that the luminosity ratio L_{6731}/L_{6717} would generally yield an *under*estimate of the mean electron density in the emission region.

The stratified dust distribution in the outer regions of disk-driven winds would lead to angle-dependent obscuration that could strongly influence the statistics of optically visible YSOs (see Königl & Kartje [1994] for a discussion of the analogous effect that such winds might have on the classification of active galactic nuclei). Figure 2 shows the expected angular dependence of the visual extinction for two representative centrifugally driven wind models as well as for a rotational infall model. In the case of a freely falling envelope, the main contribution to the obscuring column density comes from regions where the

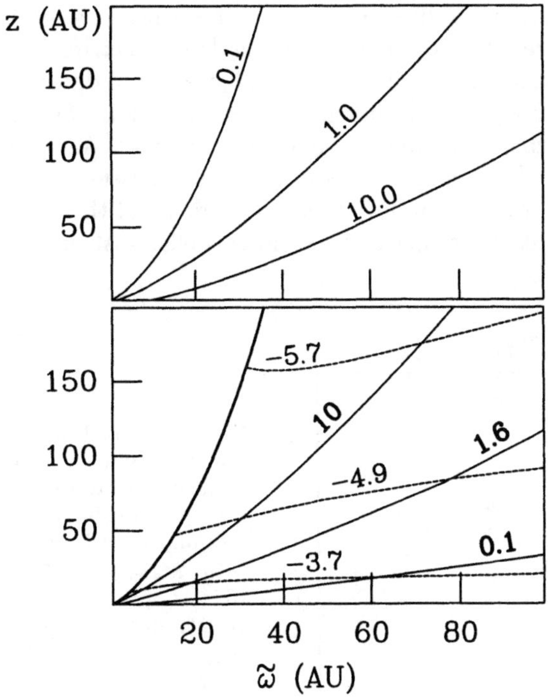

Fig. 1. Properties of a self-similar ($\alpha = 3/2$), centrifugally driven wind model from a Keplerian accretion disk (model C of Safier 1993a). In this solution, the central mass is $0.5\,M_\odot$ and the mass outflow rate between 0.1 and 10 AU is $\sim 1 \times 10^{-7}\,M_\odot\,\mathrm{yr}^{-1}$. (*top*) Meridional projections of the flow lines (labeled by the value, in AU, of the radius where they intersect the disk). (*bottom*) Contours of constant $V_{p\infty}/V_{K1}$ (*solid* lines), where $V_{p\infty}$ is the asymptotic poloidal speed and V_{K1} is the Keplerian speed at 1 AU, and $log(\rho/\rho_1)$ (*dashed* lines), where ρ_1 is the density at the base of the wind at 1 AU. The heavy line on the left indicates the flow line that originates at 0.1 AU.

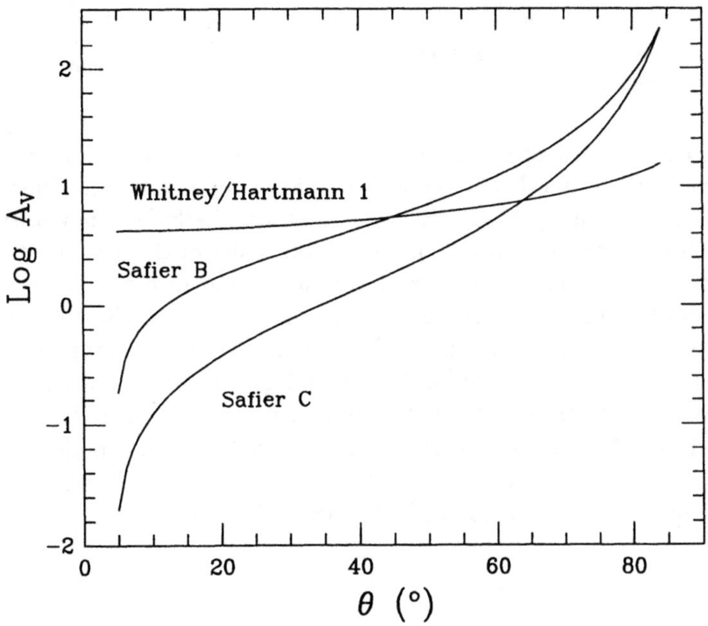

Fig. 2. Variation of the visual extinction to the YSO with polar angle θ for two centrifugally driven wind models (models B and C of Safier 1993a) and for a rotating, freely falling envelope model (model 1 of Whitney & Hartmann 1993), with $\dot{M}_{acc} = 2 \times 10^{-6}\,M_\odot\,\mathrm{yr}^{-1}$ in each case.

particle trajectories are not strongly modified by rotation, resulting in a weak dependence of A_V on θ – in contrast to the behavior predicted for a centrifugally driven outflow. As the different angular distributions would give rise to different cumulative distributions of A_V for properly chosen YSO samples, this offers a potential observational test of the dust infall vs. outflow scenarios.[3] Another potential test could be provided by measurements of scattered emission lines: Staude & Neckel (poster in this conference) have reported evidence of this type for outflowing dust that most likely originates in a circumstellar disk in the luminous YSO IRAS 08159-3543.

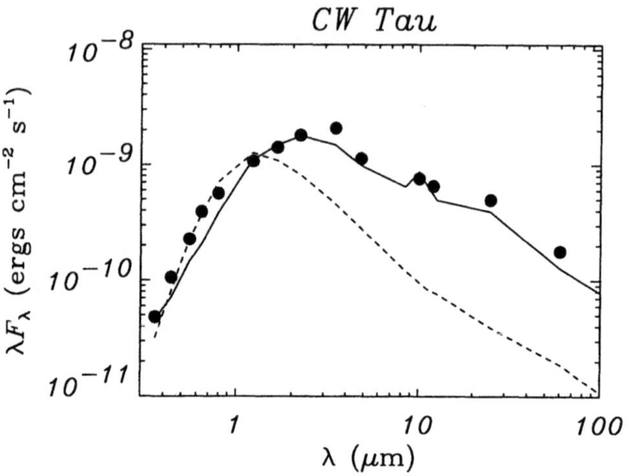

Fig. 3. Model fit to the spectrum of CW Tau, calculated assuming an optically thick disk and a dusty wind with $\dot{M}_{\rm wind} = 5 \times 10^{-8}\, M_\odot\, {\rm yr}^{-1}$ whose density decreases with height as $z^{-3/2}$. The central object has a photospheric temperature of 4000 K and a "boundary layer" temperature of 10^4 K.

The stratified dust distribution is also expected to account naturally for the centrosymmetric + parallel equatorial polarization pattern (e.g., Bastien 1989) that characterizes many YSOs (Elsässer & Staude 1978; Whitney & Hartmann 1993). Furthermore, the scattering and reprocessing of the YSO continuum radiation by the dust in the wind and the disk could give rise to the observed infrared spectra. In particular, one can readily show (e.g., Natta 1993) that a wind with the $\alpha = 1$ density power-law index of the "minimum energy" self-similar wind solution (see Section 2.1) would give rise to a flat λF_λ near- and mid-infrared spectrum akin to that exhibited by actively accreting YSOs (e.g., Rucinski 1985). Detailed calculations have verified that this model can also reproduce the 9.7 μm Si features as well as the 3 μm "bumps" observed in some sources (Figure 3; see also Königl & Kartje 1994).

[3] These scenarios are not necessarily mutually exclusive: one can imagine a disk supplied near its outer boundary by inflowing gas and driving a dusty wind closer to the central source.

2.3 Thermal and Ionization Structure

The magnetic field that accelerates the mostly neutral gas in centrifugally driven winds from weakly ionized disks is coupled to the bulk of the matter through ion-neutral collisions. As was discussed in detail by Safier (1993a), the accompanying collisional dissipation could be the dominant heating mechanism in winds from low-luminosity YSOs, and, through a robust feedback process, it could in fact regulate the thermal structure of these outflows.

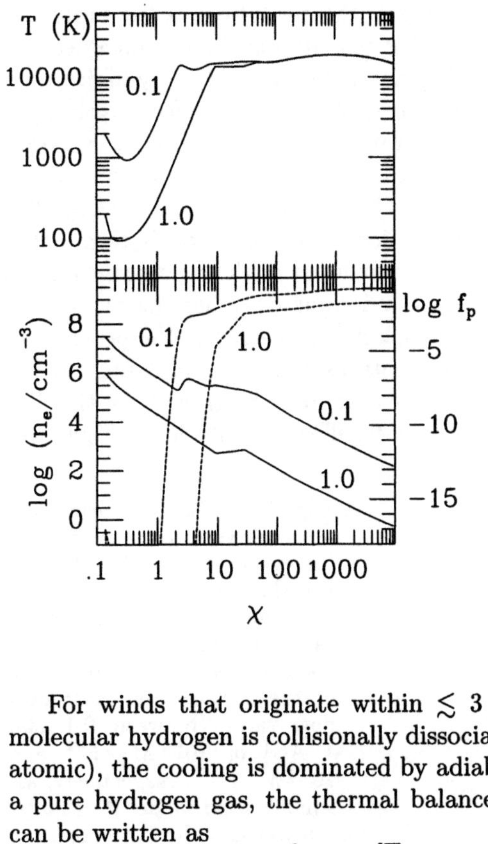

Fig. 4. The distribution of the temperature (*top*) and the degree of ionization (*bottom*) as a function of $\chi \equiv z/\varpi_0$ along two flow lines (labeled by the distance, in AU, of the footpoint from the central source) for the wind model of Figure 1. The *solid* lines in the bottom panel corrspond to the electron density n_e, whereas the *dashed* lines describe the ionization fraction f_p. The effects of ambipolar diffusion heating, adiabatic and radiative cooling, as well as of collisional and radiative ionization have been included (from Safier 1993a).

For winds that originate within $\lesssim$ 3 AU from a solar-mass YSO (where molecular hydrogen is collisionally dissociated and the outflow is predominantly atomic), the cooling is dominated by adiabatic expansion losses. Specializing to a pure hydrogen gas, the thermal balance equation in this region of the wind can be written as

$$\frac{3}{2} n_H k_B \frac{dT}{dt} = \Gamma + k_B T \frac{dn_H}{dt} ,$$

where T is the temperature, k_B is Boltzmann's constant, d/dt is the convective time derivative along a flow line, and Γ is the heating rate per unit volume. The heating rate due to ion-neutral collisions scales as

$$\Gamma \propto \frac{|(\nabla \times \mathbf{B}) \times \mathbf{B}|^2}{n_H^2 f_p (1 - f_p)} ,$$

where $f_p \equiv n_{H^+}/n_H$ is the ionization fraction. Since the large-scale structure of the wind is not affected by the thermal pressure, the distribution of the density and magnetic field in the wind is effectively fixed, and the only unknown quantities in the above relations are T and f_p. They can be determined with the help of the ionization balance equation

$$\frac{df_p}{dt} = \{I(T)f_p[1-f_p] \;-\; R(T)f_p^2\}\; n_H \,,$$

where $I(T) \propto T^{1/2}\exp(-1.6\times10^5/T)$ and $R(T) \propto T^{-1/2}$ are the ground state collisional ionization and recombination coefficients, respectively.

Figure 4 exhibits the behavior of T and f_p along two flow lines of a representative wind model. The striking property of these solutions is the rapid rise of the temperature to a value of $\sim 10^4$ K that is maintained over three decades in distance irrespective of the initial radius, and the correspondingly high values attained by f_p. This behavior is a manifestation of a feedback effect involving T and f_p: a low degree of ionization leads to an increase in the temperature (since the heating rate scales as $\sim 1/f_p$), which in turn increases the collisional ionization rate (through the exponential term in $I[T]$) and hence the value of f_p. The coupling between the temperature and the degree of ionization is terminated when the drop in n_H brought about by the expansion of the flow quenches the growth of f_p. Adiabatic cooling then comes to dominate the heating and causes the temperature to fall. Although the feedback link is severed in the presence of a strong photoionizing continuum from the central source, this mechanism could possibly still operate in high-luminosity objects at small heights above the disk surface, where the central radiation field may be effectively shielded (Figure 5).

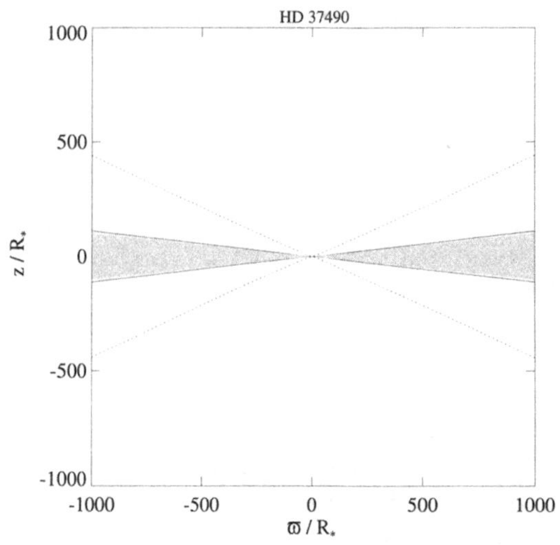

Fig. 5. Strömgren surfaces in a disk-driven wind near a high-luminosity YSO, calculated using the direct stellar ionizing flux but neglecting the diffuse radiation field. The *dashed* and *solid* lines represent the shape of the surfaces for mass outflow rates (between 0.1 and 1 AU) of $10^{-6}\, M_\odot\, \mathrm{yr}^{-1}$ and $10^{-7}\, M_\odot\, \mathrm{yr}^{-1}$, respectively, assuming wind model B of Safier (1993a) and the parameters of the Herbig Ae/Be star HD 37490 ($L_{\mathrm{bol}} = 2\times10^4\, L_\odot$, $T_{\mathrm{eff}} = 20,400$ K, $R_* = 11.4\, R_\odot$). The shaded region represents an accretion disk of height $h(\varpi_0) = 0.1\varpi_0$.

The predicted temperatures and electron densities in winds that are subject to this "ambipolar diffusion" heating are consistent with the values inferred from the forbidden line emission in T Tauri stars on scales of a few tens of AU. In fact (Figure 6), this model can account quantitatively for the observed line

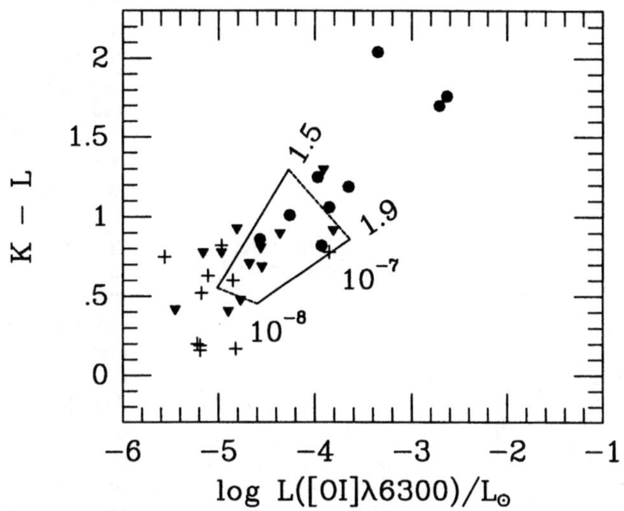

Fig. 6. Observed and predicted values of the K–L index and the dereddend [O I] λ6300 luminosity. The K–L data are from S. Beckwith (1991, private communication) and the [O I] luminosities from Cabrit et al. (1990). The computed colors and luminosities correspond to wind solutions with a fixed angle (30°) between the flow lines at the base of the wind and the surface of the disk. The different solutions are connected by lines of constant $\dot{M}_{\rm wind}$ (*dashed*, labeled below the respective lines in units of $M_\odot\ {\rm yr}^{-1}$) and *log*($\dot{M}_{\rm acc}/\dot{M}_{\rm wind}$) (*solid*, labeled above the respective lines).

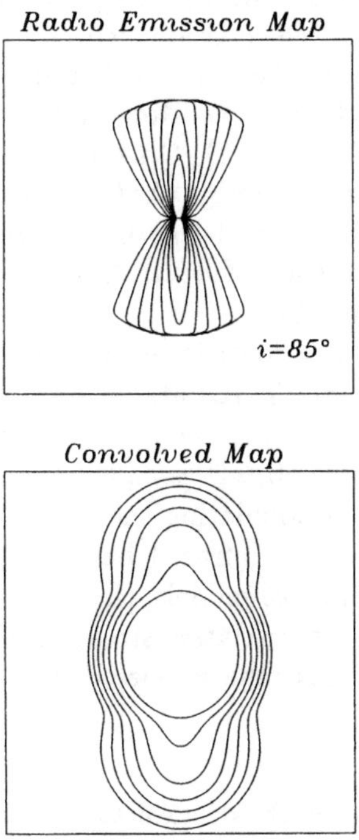

Fig. 7. Thermal radio emission from an ambipolar diffusion-heated outflow, calculated for wind model B of Safier (1993a) assuming a mass outflow rate (between 0.1 and 10 AU) of $10^{-7}\ M_\odot\ {\rm yr}^{-1}$. The *top* panel shows the 5 GHz specific intensity contours for an outflow in the Taurus molecular cloud that is observed at an angle $i = 85°$ to the symmetry axis. The innermost contour corresponds to 10^{-7} ergs cm^{-2} s^{-1} Hz^{-1} ster^{-1}, and each successive contour represents a factor of 10 decrease in intensity. The *bottom* panel shows the result of a convolution of this map with a 0."35-FWHM beam (from Martin 1995).

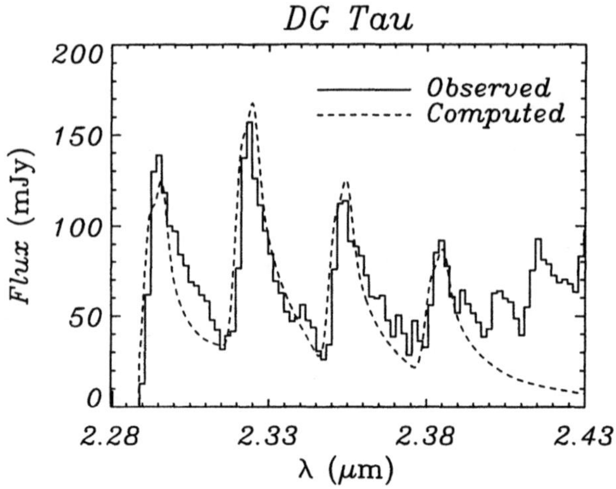

Fig. 8. Wind-model fit to the first overtone band emission of CO in DG Tau (from Safier et al. 1995; data from Chandler et al. 1993). Pumping by the stellar UV and IR radiation field has been taken into account, with optical depths calculated in the Sobolev approximation. The bulk of the emission comes from flow lines that originate at distances $0.05 \leq \varpi_0 \leq 0.5$ AU from the star and carry a mass discharge of $\sim 7.5 \times 10^{-7}\, M_\odot\, \mathrm{yr}^{-1}$. For wavelengths longer than $\sim 2.38\,\mu$m, the observed spectrum may contain contributions from H Pfund and H_2 Q-branch emission.

luminosities as well as for the correlation between these luminosities and the measured infrared excesses (which, in this picture, are produced by reprocessing of the stellar radiation in the dusty outer regions of the wind). Disk-driven winds could thus contribute to the observed forbidden line emission, especially to the low-velocity component (see Safier 1993b). However, the simple self-similar models considered so far have difficulty reproducing the observed line profiles, particularly the single, low-V peak of the LVC (see Section 1.1). On the other hand, the absence of an HVC in the [O I] emission from Herbig Ae/Be stars (Section 1.4) may have a simple explanation in the context of this model (S. Martin 1994, private communication), namely, the absence of neutral O near the symmetry axis of high-luminosity YSOs[4] (see Figure 5).

The ambipolar diffusion wind-heating mechanism may contribute to several other emission signatures of YSOs, including the thermal radio radiation (Martin 1995; see Figure 7) and the Ca II triplet line emission, which has been inferred to exhibit a puzzling temperature inversion pattern in several objects (e.g., Hamann & Persson 1989). It may also play an important role in populating the low vibrational levels of CO and in producing the observed CO infrared bandhead emission (Safier et al. 1995; see Figure 8). Note that, in these models, the population of the higher vibrational levels, and, accordingly, the relative strength of the upper rovibrational bands, depends on the magnitude of the stellar UV flux.

[4] Recall that the ionization potential of neutral oxygen is nearly identical to that of hydrogen.

2.4 Nonsteady Outflows

The magnetized disk/wind models may provide a physical basis for some of the attributes of the nonsteady ejection episodes (notably FU Ori outbursts) observed in low-mass YSOs. One promising model for these events invokes a disk thermal ionization instability of the type previously applied to the outbursts of dwarf novae (e.g., Bell & Lin 1994). In this picture, accretion in the innermost disk proceeds in a nonsteady fashion, with a "gate" at $\lesssim 0.25$ AU opening every $\sim 10^3$ yr or so after the accumulated column density has become large enough to trigger the instability. During the "high" phase of the instability (which is identified with an FU Ori outburst) the gas is hot ($T \gtrsim 10^4$ K) and almost completely ionized, and mass rains in at a rate $\gtrsim 10^{-5}\, M_\odot\, \mathrm{yr}^{-1}$, whereas during the "low" phase the temperature and degree of ionization decline sharply, and the accretion rate drops to $\lesssim 10^{-7}\, M_\odot\, \mathrm{yr}^{-1}$.[5] In the context of the magnetized disk model, the presence of the "gate" is explained by the decline in the coupling between the magnetic field and the gas as the density and column density in the disk become large (see Königl 1994, 1995). Once the temperature in the inner disk increases to $\gtrsim 10^4$ K and the gas becomes ionized, good coupling is restored and the field can transfer angular momentum from the accreting gas to a centrifugally driven wind. This could account both for the marked increase in $\dot{M}_{\mathrm{acc}}$ and for the strong disk outflow that evidently accompanies it (e.g., Calvet et al. 1993).

3. Desiderata

In conclusion, I present a partial "wish list," based on the preceding discussion, of possible steps that might be taken to clarify the observational picture and sharpen the focus for some of us struggling theorists...

- Separate the effects of stratification and geometry on density-sensitive line ratios by comparing, to the extent that good data can be taken, several different line ratios (see Safier 1992).
- Try to determine the density and excitation distribution of the HVC forbidden line-emitting gas as a function of the distance from the source, in an attempt to clarify whether it is purely shock excited and whether it is present very close (on scales where the electron density exceeds $\sim 10^6\ \mathrm{cm}^{-3}$) to the origin.
- Ditto for the LVC gas: does the low-velocity component have similar properties in T Tauri stars and in Herbig Ae/Be stars?
- Attempt to carry out additional searches for scattered lines of the type reported in this conference by Staude & Neckel. An even more powerful probe

[5] A similar mechanism could in principle operate also in disks around Herbig Ae/Be stars, where it might account for the apparent discrepancy found in several cases between the inferred mass accretion rate on large scales and the upper limit (deduced from the measured bolometric luminosity) on $\dot{M}_{\mathrm{acc}}$ near the stellar surface (Bell 1994).

of the different outflow components could be achieved through spectropolarimetry in the optical (e.g., Hα), near-IR (e.g., Brγ; see poster by Murdoch et al. in this conference), and mid-IR (e.g., Si 9.7 μm; see Aitken et al. 1993). Theoretical modeling of the polarization signatures of magnetized disk-driven winds is already under way (e.g., Kartje 1995).

- Carry out correlated variability studies to test various outflow scenarios and model predictions. For example, do flux variations at the source accompany the appearance and separation of HVC forbidden line-emission knots (cf. Hirth et al. 1994)? Is the UV flux correlated with the amplitude of the upper (4-2, 5-3, 6-4, etc.) CO infrared bandheads in objects where both can be measured (cf. Safier et al. 1995)?

Acknowledgement. I am grateful to John Kartje, Steve Martin, and Pedro Safier for valuable discussions and for assistance with the preparation of the figures. I also thank the organizers of this conference for encouraging and facilitating my participation. This work has been supported in part by NASA grants NAGW-2379 and NAG 5-2766.

References

Aitken, D. K., Wright, C. M., Smith, C. H., & Roche, P. F. 1993, MNRAS, 262, 456
Arons, J. 1987, in The Origin and Evolution of Neutron Stars, ed. D. J. Helfand & J.-H. Huang (Dordrecht: Reidel), 207
Bastien, P. 1989, in Polarized Radiation of Circumstellar Origin, ed. G. V. Coyne et al. (Vatican City: Vatican Press), 541
Bell, K. R. 1994, in The Nature and Evolutionary Status of the Herbig Ae/Be Stars, ed. P. S. Thé, M. P. Pérez, & E. P. J. van den Heuvel (ASP Conf. Ser. 62), 215
Bell, K. R., & Lin, D. N. C. 1994, ApJ, 427, 987
Berrilli, F., Corciulo, G., Ingrosso, G., Lorenzetti, D., Nisini, B., & Strafella, F. 1992, ApJ, 398, 254
Blandford, R. D., & Payne, D. G. 1982, MNRAS, 199, 883
Böhm, K.-H., & Solf, J. 1994, ApJ, 430, 277
Böhm, T., & Catala, C. 1994, A & A, 290, 167
Cabrit, S., Edwards, S., Strom, S. E., & Strom, K. M. 1990, ApJ, 354, 687
Cao, X., & Spruit, H. C. 1994, A & A, 287, 80
Calvet, N., Hartmann, L., & Kenyon, S. J. 1993, ApJ, 402, 623
Chandler, C. J., Carlstrom, J. E., Scoville, N. Z., Dent, W. R. F., & Geballe, T. R. 1993, ApJ, 412, L71
Chernin, L., Masson, C., Gouveia Dal Pino, E. M., & Benz, W. 1994, ApJ, 426, 204
Contopoulos, J., & Lovelace, R. V. E. 1994, ApJ, 429, 139
Di Francesco, J., Evans, N. J. II, Harvey, P. M., Mundy, L. G., & Butner, H. M. 1994, ApJ, 432, 710
Edwards, S., Cabrit, S., Ghandour, L. O., & Strom, S. E. 1989, in Low Mass Star Formation and Pre-Main Sequence Objects, ed. B. Reipurth (Garching bei München: ESO), 385
Edwards, S., Ray, T., & Mundt, R. 1993, in Protostars and Planets III, ed. E. H. Levy & J. I. Lunine (Tucson: Univ. of Arizona Press), 567
Elsässer, H., & Staude, H. J. 1978, A & A, 70, L3
Ferreira, J., & Pelletier, G. 1995, A & A, **295**, 807
Fiedler, R. A., & Mouschovias, T. Ch. 1993, ApJ, 415, 680

Galli, D., & Shu, F. H. 1993, ApJ, 417, 243
Gouveia Dal Pino, E. M., & Benz, W. 1994, ApJ, 435, 261
Hamann, F. 1994, ApJS, 93, 485
Hamann, F., & Persson, S. E. 1989, ApJ, 339, 1078
Hamann, F., Simon, M., Carr, J. S., & Prato, L. 1994, ApJ, 436, 292
Hartigan, P., & Raymond, J. 1993, ApJ, 409, 705
Hartigan, P., Morse, J., & Raymond, J. 1994, ApJ, 436, 125
Hartmann, L., Kenyon, S. J., & Calvet, N. 1993, ApJ, 407, 219
Herbig, G. H. 1989, in Low Mass Star Formation and Pre-Main Sequence Objects, ed. B. Reipurth (Garching bei München: ESO), 233
Hillenbrand, L. A., Strom, S. E., Vrba, F. J., & Keene, J. 1992, ApJ, 397, 613
Hirth, G. A., Mundt, R., & Solf, J. 1994, A & A, 285, 929
Kartje, J. F. 1995, ApJ, 452, 565
Kepner, J., Hartigan, P., Yang, C., & Strom, S. 1993, ApJ, 415, L119
Königl, A. 1994, in Theory of Accretion Disks – 2, ed. W. J. Duschl, J. Frank, F. Meyer, E. Meyer-Hofmeister, & W. M. Tscharnuter (Dordrecht: Kluwer), 53
Königl, A. 1995, in Research Trends in Plasma Astrophysics, ed. R. Kulsrud (New York: AIP Press), in press
Königl, A., & Kartje, J. F. 1994, ApJ, 434, 446
Königl, A., & Ruden, S. P. 1993, in Protostars and Planets III, ed. E. H. Levy & J. I. Lunine (Tucson: University of Arizona Press), 641
Lada, C. J. 1985, ARAA, 23, 267
Lovelace, R. V. E., Berk, H. L., & Contopoulos, J. 1991, ApJ, 379, 696
Lubow, S. H., Papaloizou, J. C. B., & Pringle, J. 1994, MNRAS, 267, 235
Martin, S. C. 1995, in preparation
Mitchell, G. F., Curry, C., Maillard, J.-P., & Allen, M. 1989, ApJ, 341, 1020
Mundt, R., Brugel, E. W., & Bührke, T. 1987, ApJ, 319, 275
Mundt, R., Ray, T. P., & Raga, A. C. 1991, A & A, 252, 740
Natta, A. 1993, ApJ, 412, 761
Natta, A., Palla, F., Butner, H. M., Evans, N. J., II, & Harvey, P. M. 1993, ApJ, 406, 674
Pelletier, G., & Pudritz, R. E. 1992, ApJ, 394, 117
Prusti, T., Natta, A., & Palla, F. 1994, A & A, **292**, 593
Pudritz, R. E., Pelletier, G., & Gomez de Castro, A. I. 1991, in The Physics of Star Formation and Early Stellar Evolution, ed. C. J. Lada & N. D. Kylafis (Dordrecht: Kluwer), 539
Raga, A. C. 1991, AJ, 101, 1472
Raga, A. C., Cantó, J., Binette, L., & Calvet, N. 1990, ApJ, 364, 601
Raga, A. C., & Kofman, L. 1992, ApJ, 386, 222
Ray, T. P., Poetzel, R., Solf, J., & Mundt, R. 1990, ApJ, 357, L45
Reipurth, B., Raga, A. C., & Heathcote, S. 1992, ApJ, 392, 145
Rucinski, S. M. 1985, AJ, 90, 2321
Safier, P. N. 1992, ApJ, 392, 492
Safier, P. N. 1993a, ApJ, 408, 115
Safier, P. N. 1993b, ApJ, 408, 148
Safier, P. N., Martin, S. C., & Königl, A. 1995, in preparation
Shu, F. H., Najita, J., Ostriker, E., Wilkin, F., Ruden, S., & Lizano, S. 1994, ApJ, 429, 781
Snell, R. L., Loren, R. B., & Plambeck, R. L. 1980, ApJ, 239, L17
Solf, J., & Böhm, K.-H. 1993, ApJ, 410, L31
Stepinski, T. F., Reyes-Ruiz, M., & Vanhala H. A. T. 1993, Icarus, 106, 77
Strom, K. M., Strom, S. E., Wolff, S. C., Morgan, J., & Wenz, M. 1986, ApJS, 62, 39
Wardle, M., & Königl, A. 1993, ApJ, 410, 218
Whitney, B. A., & Hartmann, L. 1993, ApJ, 402, 605

Formation and Structure of Magnetized Protostellar Jets

Christian Fendt[1] and Max Camenzind[2]

[1] Lund Observatory, Box 43, S-22100 Lund, Sweden
[2] Landessternwarte, Königstuhl, D-69117 Heidelberg, Germany

Summary. Protostellar jets most probably originate in the closest environment of a fully convective young stellar object which presumably carries a magnetosphere built up by a strong stellar dynamo and is surrounded by an accretion disk. Interaction between the magnetic field and the accretion disk leads to the formation of a gap between the stellar surface and the disk. The observed jet opening angles may be less than 5°.

We present numerical solutions treating the relativistic 2D force–balance of the magnetic field, described by the Grad–Schlüter–Shafranov (GSS) equation, and the steady motion of a cold plasma flow along the calculated magnetic flux surfaces in the collimation domain of the jet. Our model takes into account the topology of the star–disk–jet scenario mentioned above. The resulting flows have a finite asymptotic jet radius. From the observed rotational periods of T Tauri stars it follows that the derived light cylinder of a stellar magnetosphere is of the order of the observed jet radii. This fact requires a relativistic treatment although the jet velocities are clearly non-relativistic.

The resulting magnetic field structure allows simultaneously for wind outflow towards an asymptotically cylindrical jet and for mass accretion towards the central star along dipolar field lines. The outflow is initially poorly collimated near the source with an opening angle of 65° and then rapidly collimates within a distance of 0.3 jet radii along the jet axis.

The 2D velocity structure of the flow along the flux surfaces strongly depends on the magnetization σ of the plasma flow. For the asymptotic poloidal jet velocity u_∞ we find a power law $u_\infty = A c \sigma^{1/3}$. Since the factor $A \geq 1$, this implies that the acceleration along the collimated flux surfaces is more efficient than in a purely conical magnetic structure. The asymptotic fast-magnetosonic Mach-number of the flow turned out to be independent of the magnetization and generally is of the order of 2.5. This has far–reaching consequences for the interpretation of the knot–spacing in protostellar jets.

1. Introduction

Today there is observational evidence that young stellar objects may consist of three main components: the central protostellar object (a T Tauri star (TTS) or an IR protostar), a surrounding accretion disk and, occasionally, but always in connection with the appearance of a disk signature, bipolar outflows in the form of highly collimated high velocity jets or weakly collimated low velocity molecular flows and winds (Appenzeller and Mundt 1989, Camenzind 1990, Montmerle et al. 1993, see also the reviews by Königl in this proceedings). In addition, there is strong evidence that magnetic fields play an important or even dominant role in the evolution and interaction of these components. Therefore, in a physical model both accretion and wind/jet phenomena has to be considered in a self-consistent way.

In particular, VLA and VLBI radio observations suggest that these stars have a dipolar kG magnetic field (André et al. 1991). The magnetic structure of the stellar surface is also plausible from measurements of rotational periods of TTSs: cold and hot spots or flares on the surface lead to photometric variations (Gahm et al. 1993, Bouvier et al. 1993). Some observations directly point to a magnetic character of protostellar jets. As suggested by Mundt et al. (1990), a magnetic current carrying jet would be able to explain the observed jet bending due to the cloud field in the HL Tauri region. Bow shock observations give estimates of the pre-shock jets magnetic field strength of the order of $10\,\mu$G (Morse et al. 1992, 1993).

The first theoretical treatment of the scenario of magnetized disk winds was given by fundamental papers of Blandford & Payne (1982) and Pudritz & Norman (1983). However, for solving the GSS equation they assumed self-similarity (BP) or another special scaling (PN) which both imply e.g. an infinite jet radius. Sakurai investigated the GSS equation for a stellar wind (1985) as well as for a disk wind topology (1987) and showed that field lines of an initial monopole type geometry will bend towards the rotational axis.

Camenzind (1986, 1987) developed a fully relativistic description of hydromagnetic flows in magnetospheres, essentially applicable to any field topology. These results, originally derived for active galactic nuclei, were extended to a self-consistent model of protostellar systems (Camenzind 1990; Fendt 1994; Fendt et al. 1994, hereafter FCA; Paatz & Camenzind 1994; Camenzind et al. 1994), considering the accretion process onto the star as well as the ejection of plasma in magnetized winds or jets. The stellar magnetosphere, developed by the highly convective protostar, strongly influences the entire star–disk–jet system. In this framework the relatively low rotational periods of TTSs can be explained by a magnetic link between the magnetic star, the accretion disk and the escaping wind, decelerating the star by removing angular momentum and transferring it to the disk and the jet. A similar model for the star braking was published by Königl (1991).

Since the work by Heyvaerts and Norman (1989) for the Newtonian case and Chiueh et al. (1991) for the relativistic case we know that magnetized winds generally will collimate along the symmetry axis. Appl & Camenzind (1993a,b) presented the first non-linear analytical model of a stationary relativistic magnetized jet in the asymptotic regime.

In this talk we presented a numerical solution of the 2D force-balance asymptotically collimated to a cylinder with finite radius. The jet magnetosphere originates in a dipolar stellar magnetic field. Along the calculated flux surfaces the wind equation is solved, providing the dynamical properties of the plasma flow.

In our basic assumptions, we follow the model presented in FCA. The numerical method of finite elements integrating the GSS equation as a boundary value problem, in principle allows us an application to *any* field topology. Problems were involved with the circumstance that this boundary value problem is in fact a free boundary problem since the location and the shape of the outer jet boundary is not known *a priori*.

To avoid any misunderstanding from the beginning: here our aim is *not* to consider how the collimation process works (in particular, this is done in the papers on the asymptotic solutions cited above), but to look whether 2D solutions under "protostellar boundary conditions" do exist, and how those collimated jet magnetospheres and the accompanying parameters of the plasma motion look like. In fact, it turned out that the investigation of the basic problem given by the unknown shape of the outer jet boundary is very important in order to calculate global solutions of the GSS. We consider this as a remarkable point of our results since a *global* jet solution *together* with a *finite* asymptotic jet radius were not yet obtained elsewhere.

However, to escape further misjudgement, we note that our calculations are not yet self-consistent. The GSS equation is solved using a force-free "ansatz", while the wind equation, computed along the resulting field lines, is not force-free by definition. But the presented results are the very first step towards a self-consistent solution which will be achieved by an iterative process, considering the interrelation of the plasma inertia with the magnetic field structure. This will be subject of our future work.

2. MHD Description

In our approach the magnetohydrodynamic (MHD) description of the collimating jet magnetosphere is based on the assumptions of axisymmetry, stationarity and a high plasma conductivity (ideal MHD). We apply cylindrical coordinates (R, ϕ, Z) or (x, ϕ, z), normalised to the LC radius (see Section 3), respectively.

In this framework the magnetic flux function $\Psi(R, Z)$ describes the structure of the magnetosphere,

$$\Psi = \frac{1}{2\pi} \int \mathbf{B}_{\mathrm{P}} \cdot d\mathbf{A}, \qquad R\mathbf{B}_{\mathrm{P}} = \nabla\Psi \wedge \mathbf{e}_{\phi}. \tag{1}$$

It has to be calculated from the GSS equation,

$$R\nabla \cdot \left\{ \frac{1}{R^2} \nabla\Psi \right\} = -\frac{4\pi}{c} j_{\phi}, \tag{2}$$

which expresses the force-balance perpendicular to the magnetic flux surfaces. The GSS equation is equivalent to Ampère's law, rewritten for the magnetic flux function Ψ. Projection of the spatial part of the relativistic equation of motion perpendicular to the magnetic flux surfaces provides the toroidal current density j_{ϕ}.

Applying a force-free "ansatz" in the equation of motion the source current density j_{ϕ} can be expressed in terms of the poloidal current $I(\Psi) = (2/c)\, RB_{\phi}$. It remains a free function and depends on the applied model. Normalising with

$\Psi \to \Psi \Psi_{\max}$, $I \to I I_{\max}$, Eq. (2) can be evaluated to the pulsar-equation (Michel 1973),

$$x\nabla \cdot \left\{ \frac{D}{x^2} \nabla \Psi \right\} = -\frac{1}{x} g_I \, I \, \partial_\Psi I \,, \qquad (3)$$

where $D = 1 - x^2$. In this force-free approach the LC, where $D = 0$, provides a singular surface of the partial differential equation. At this surface a special boundary condition, the regularity condition, $\partial_R \Psi = -0.5 \, g_I I \, \partial_\Psi I$, has to be satisfied. g_I is the coupling constant for the source current of the force-free GSS equation (3),

$$g_{\rm I} = \frac{4 I_{\max}^2 R_{\rm L}^2}{c^2 \Psi_{\rm D}^2} = 4 \left(\frac{I_{\max}}{10^{15} \rm A} \right)^2 \left(\frac{R_{\rm L}}{10^{15} \rm cm} \right)^2 \left(\frac{\Psi_{\max}}{10^{25} \rm G} \right)^{-2} . \qquad (4)$$

The projection of the equation of motion parallel to the field, i.e. the wind equation, gives the dynamical properties of the plasma flow: energy, velocity, density or Mach-number. In the cold wind limit (neglecting the gas pressure) the wind equation can be written as a polynomial in the poloidal plasma velocity $u_{\rm P} = \gamma v_{\rm P}/c$ of degree of 4,

$$\sum_{m=0}^{4} A_m(x; E, L, \Phi, \sigma) \, u_{\rm P}^m = 0 \,, \qquad (5)$$

(Camenzind 1987), where the coefficients A_m depend explicitly on the plasma magnetization,

$$\sigma = \frac{\Psi_{\max}^2}{2 \dot{M}_{\rm jet} c R_{\rm L}^2} = 5\,10^{-7} \dot{M}_{\rm jet,-10}^{-1} R_{\rm L,15}^{-2} \Psi_{\rm max,25}^2 \,, \qquad (6)$$

and the flux tube function, $\Phi = B_{\rm P} R^2$. The constants of motion, the total plasma energy $E(\Psi)$ and the total angular momentum $L(\Psi)$ are fixed by the requirement of a *critical* wind motion, i.e. a wind solution which passes smoothly the critical points where the flow velocity equals the speed of the MHD waves.

In a self-consistent approach, both the GSS equation (2) and the wind equation (5) have to be solved simultaneously. But in general, the current density j_ϕ has a very complicated structure and also depends on plasma parameters, which are *a priori* unknown. The solution of the wind equation however, requires the knowledge of the magnetic field structure.

Due to these difficulties, it has not yet been possible to solve the problem in general without assuming further simplifications. Other authors use self-similarity (e.g. Blandford & Payne 1982), other special scaling (e.g. Pelletier & Pudritz 1992, Sauty & Tsinganos 1994; Sauty, this proceedings) or monopole type topologies (Michel 1973; Sakurai 1985, 1987) for the magnetic field distribution.

We applied a force-free "ansatz", i.e. we neglect the back-reaction of the plasma inertial forces on the magnetic field. Together with the above mentioned model this enables us to take into account several astronomical boundary conditions and therefore allows for a good approximation of the magnetic field structure underlying the wind motion. The future goal however, is to consider the inertial forces in the GSS equation in an iterative way in order to finally arrive at the self-consistent solution.

3. The Star–Disk–Jet Model

In the following the main characteristics of the model underlying our numerical calculations are reviewed. For additional information we refer to previous publications of our work (Camenzind 1990; Fendt 1994; FCA, and references therein).

Basically, the model follows the observational constraints resulting in a magnetized star–disk–jet scenario (see Figure 1). The central protostellar object is surrounded by a stellar dipolar magnetic field and an accretion disk. The strong stellar field of the order of about 1000 Gauss is believed to be generated by a dynamo in the fully convective protostar. There also exists some observational evidence for the derived field strength as well as for its topology (André et al. 1991; Montmerle et al. 1993; Bouvier et al. 1993). Interaction of the accretion disk with the stellar magnetic field produces a gap (Gosh & Lamb 1978; Camenzind 1990). The inner disk radius is determined by pressure equilibrium between the stellar magnetic field and the accretion flow of the disk and is typically of the order of a few stellar radii (Camenzind 1990).

A rather strong model constraint is given by the observed rotational periods of TTSs. These are of the order of days (Bouvier et al. 1993) and thus, the light cylinder of the stellar magnetosphere (LC, where the rotational velocity of the field lines equals the speed of light) is located at $R_{\rm L} \simeq 5000 R_*$ which is of the order of the observed jet radii of about 10^{15} cm (Mundt et al 1987, 1990). Therefore, a special relativistic treatment of the star-jet magnetosphere is required. The rapid rotation (compared e.g. to the sun) generates electric fields which cannot be neglected.

This might be surprising since the observed jet velocities of about 400 km s^{-1} (Mundt et al. 1987, 1990; Reipurth 1989) are clearly non-relativistic. We argue that the application of relativity could never be wrong since it is the more general theory and it also allows for a smooth transition towards the Newtonian case.

The LC, on the other hand, has no direct influence on the dynamics of the moving plasma since the "decoupling" of the plasma from the field lines due to inertial forces already happens at the Alfvén surface. Because we finally assume a force-free "ansatz", in the GSS equation the Alfvén surface becomes identical to the LC.

A second argument supporting a relativistic treatment is the fact that the plasma loading rate from the disk or the stellar surface into the magnetosphere may not necessarily be uniform for different flux surfaces. The value for a certain flux surface may substantially differ from the observed mass loss rate as a mean

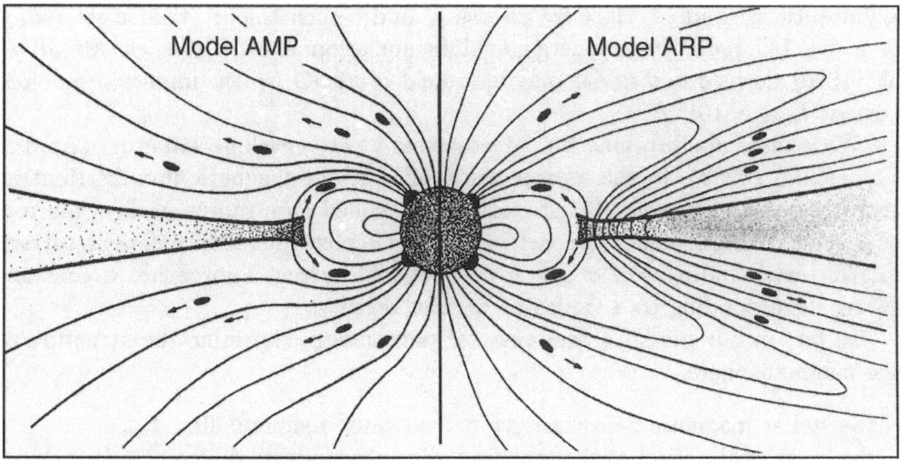

Fig. 1. Model of the innermost region of a protostellar jet source. The asymptotical topology of the field (here indicated as poloidal field lines or axisymmetric flux surfaces) is determined by the assumption of the resistivity in the disk (see text). The sketch shows on the l.h.s. the AMP topology and on the r.h.s. the ARP topology. Arrows indicate the direction of the mass flow (filled bubbles). Mass infall along the dipolar field lines leads to an accretion ring on the stellar surface at high stellar latitudes. The lines indicate magnetic flux surfaces or field lines, respectively. Picture taken from Fendt (1994)

across the jet. Low mass loss rates (equivalent to a high plasma magnetization) imply high plasma velocities which may be in the relativistic regime.

Heyvaerts and Norman (1989) have shown that jets carrying a poloidal current collimate asymptotically to a cylindrical geometry. Indeed, highly collimated outflows with asymptotic opening angles of a few degrees are observed (Mundt et al. 1987). In our model we assume a cylindrical shape of the asymptotic jet.

The intrinsic topology of the jet magnetosphere is determined by the assumption of the resistivity in the disk. For a disk with high resistivity, the protostellar magnetic field will penetrate the disk. This will lead to a *reversed field pinch* in asymptotic regime (model ARP). For a disk with low resistivity, the stellar dipole is closed only in the gap between star and disk. All field lines reaching the asymptotic regime originate from the stellar surface leading to an *asymptotic monotonous* flux distribution (model AMP). However, since we did not find collimated jet solutions for model topology ARP (Fendt 1994, FCA), we concentrate in the following on the topology AMP.

The asymptotic jet is determined by the distribution of the integral poloidal current as well as the strength of the current. In the calculations for the AMP topology the poloidal current distribution $I(\Psi)$ is chosen from the non-linear analytical 1D solution in the asymptotic domain given by Appl & Camenzind (1993b), $I(\Psi) \sim (1 - e^{-b\Psi})$. Then, the strength of the current, $g_{\rm I}$, and the shape of the profile ($b \ll 1$ diffuse pinch, $b \gg 1$ sharp pinch) control the magnetic structure and the kinematics of the jet. In particular, they determine the asymptotic jet radius. Here we choose g_I and b such that it yield a jet radius of a few LC radii. From energy equilibrium arguments in the jet, Mundt et al. (1990) derived a toroidal magnetic field of 60 μG, which implies a poloidal current $I_{max} \simeq 10^{12}$ A.

With these assumptions the jet possesses a core-envelope structure (Appl & Camenzind 1993b). In the asymptotic profiles of the magnetic flux distribution and the poloidal current distribution the so called core radius a plays the role of a typical length scale. For radii $x << a$ magnetic flux and poloidal current increase with radius. For $x >> a$ the poloidal current approaches a constant value, corresponding to a vanishing current density.

So far, in our model 4 fundamental parameters determine the structure of the magnetosphere,

- the stellar magnetic field strength or maximum magnetic flux Ψ_{max}
- the rotational period of the star providing the position of the LC, R_L
- the asymptotic jet radius R_{jet}
- the parameter b of the current distribution across the jet. It provides, together with the jet radius the maximum poloidal current in the jet, I_{max}

The magnetization σ is the fundamental parameter for the wind motion. In our cold wind limit it remains as a free parameter.

4. Numerical Approach

The GSS equation is solved by means of the method of finite elements. In this approach the GSS equation is integrated over the 2D star-jet domain as a boundary value problem (Camenzind 1987; Fendt 1994; FCA). In principle, this allows us for a calculation of *any* field topology, i.e. we are not restricted to e.g. self-similar or monopole type solutions of the magnetic field. Furthermore, our solver allows to integrate the GSS equation either on domains separated by the singular surface or on the entire jet domain, alternatively. A grid of 64x64 curvilinear elements resolves the star–gap–disk–jet system with a scaling of about

$$100\, R_* = R_L \leq 0.2 R_{jet}. \tag{7}$$

Note, that this numerical stellar radius is larger than the physical value by a factor of about 10–50. This implies some limitations in the calculation of the wind equation, especially we are limited by $\sigma \geq 10^{-5}$.

The critical wind solution was achieved using a bracketing procedure. It numerically investigates the topology of the wind solution in the neighbourhood of the fast magnetosonic point and varies the plasma energy parameter until the flow smoothly passes the fast magnetosonic point.

Further details concerning the finite element solver, including several test solutions, could be found in a previous paper (FCA).

5. Results and Discussion

5.1 Structure of the Magnetosphere

A fundamental problem for the calculation of relativistic force-free magnetospheres is introduced by the singularity of the GSS equation. As known from the treatment of pulsar magnetospheres, global solutions extending beyond the LC are found to show kinks at the LC (see Michel 1991).

Obviously, it is not possible to solve the wind equation along such a kinky flux surface and it became a major prospect of our work to find global solutions of the GSS equation, which are well behaved at the LC.

Indeed, in the case of model topology AMP, we found solutions without kinks at the LC. For a detailed description of the search and finding of well behaved solutions we refer to our previous papers (Fendt 1994, FCA). As an important result, we mention that the adjustment towards the well behaved solution finally follows from the adjustment of the shape and the position of the jet boundary $R_{\rm jet}(Z)$ together with the choice of a certain poloidal current distribution. Or from an alternative point of view, these parameters are defined by the well behaved solution uniquely.

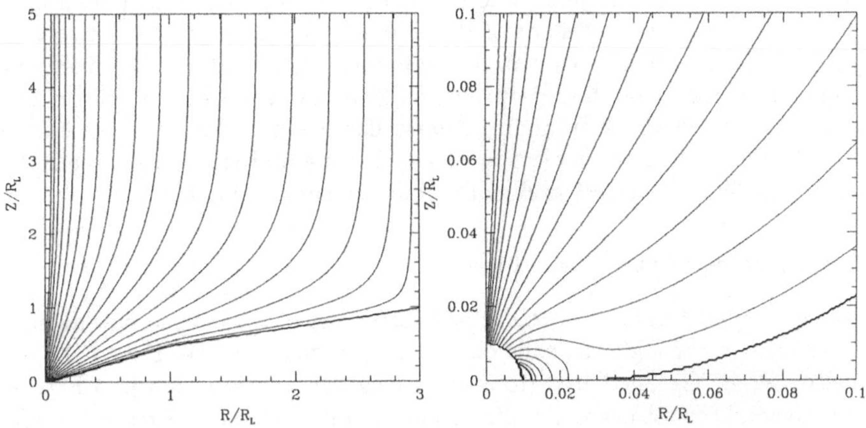

Fig. 2. Magnetic flux distribution $\Psi(x,z)$. Model topology AMP. Contour levels: $10^{-n^2}, n = 0, 1.8, 0.1$. Jet parameters: $R_{\rm jet} = 3.0\,R_{\rm L}$, poloidal current $\simeq 10^{11}$ A, magnetic flux $\Psi = 10^{25}\,{\rm G\,cm^2}$, ($\equiv B_\star = 1000\,{\rm G}$). *Left frame:* Global jet magnetosphere; *right frame:* Subset from the global solution

Figure 2 shows a typical example of a our solutions. The flux surfaces calculated for the *global* collimated jet magnetosphere extend from the stellar surface to the asymptotic jet. The parameters of the solution are $x_{\rm jet} = 3$, $b = 1.0$, $\Psi_{\rm max} = 10^{25}$. The field configuration simultaneously allows for wind outflow and mass accretion towards the central star. This is consistent with the observed link between the existence of outflows and the appearance of an accretion signature in the spectrum of these sources.

The initial jet opening angle is 65°. The outflow rapidly collimates within a distance of 0.3 jet radii along the jet axis towards a cylindrical jet of finite radius. Again we like to mention, that we do *not* prescribe the shape of the outer jet boundary: the cylindrical shape down to a distance of about one LC along the jet axis only results from the requirement of a well behaved solution. In fact, we also considered e.g. re-collimated solutions (Fendt 1994; FCA). The only constraint is that of a *asymptotically* cylindrical topology, which is introduced by the appropriate current distribution.

For the presented solution, the maximum poloidal net current flowing in the jet is about

$$I_{\max} = 7\,10^{10}\,\mathrm{A}\left(\frac{\Psi_{\max}}{10^{25}\mathrm{G}}\right)\left(\frac{10^{15}\mathrm{cm}}{R_{\mathrm{L}}}\right) \tag{8}$$

Equivalently, the toroidal magnetic field at the asymptotic jet radius is of the order of $B_\phi = 5\mu\mathrm{G}$. This is in agreement with the observations (Morse et al. 1992, 1993).

Results of other authors (Sakurai 1985, 1987) show that even for solutions which take into account the plasma inertia, a *fast* self-collimated magnetic field structure could hardly be obtained. Thus, from our fast collimated jet magnetosphere we conclude that the plasma inertia does not strongly influence the structure of the magnetosphere and the presented solutions do well describe the magnetic field in a protostellar jet. We believe that fast collimated jet solutions only could be obtained with a fixed asymptotic jet boundary. The fixed outer jet boundary then may be interpreted as given by a pressure equilibrium between the jet and the surrounding interstellar medium. In this sense the fast collimation is due to the molecular cloud pressure and therefore it seems that protostellar jets are not self-collimated, but pressure-collimated.

5.2 Dynamics of the Wind Flow

In the cold wind scenario the 2D velocity structure of the plasma flow along the flux surfaces Ψ strongly depends on the magnetization σ. We solved the wind equation along the calculated (force-free) magnetic surfaces shown in Figure 2. Here the wind solutions are investigated in two different ways. First, we compare the plasma flow along a single flux surface for different magnetizations. Then we compare the flow parameters on different flux surfaces, but keeping the magnetization fixed. Together, this provides an information on the 2D structure of the plasma flow. Note that for each point shown in the figures the wind equation was solved separately. Each point represents a critical wind solution (a solution which passes the (fast) magnetosonic point smoothly) to the specified flux surface and magnetization.

In Figure 3 we show how the dynamical properties of the plasma (flowing along the surface $\Psi = 0.813$, near the jet boundary) depend on the magnetization σ. Interestingly from the observational point of view, the asymptotic jet velocity u_M and the asymptotic particle density N follow a power law,

$$u_M = \gamma(v_{\mathrm{jet}}/c) = 1.653\,\sigma^{1/3}\,, \qquad N = 10^{-9}\sigma^{-4/3}\,, \tag{9}$$

while the fast magnetosonic Mach-number remains constant for a low magnetization, $M_F \simeq 2.5$. Note that in the case of a monopole topology of the magnetic field the asymptotic velocity was analytically found to be $u_M = 1.0\,\sigma^{1/3}$ (Michel 1973). In comparison with our results we see that the collimated structure is accelerating the plasma more efficiently. This is due to the curvature of the field lines which provides an additional term in the Lorentz force.

In the subset displaying the radial position of the critical points, we see that for lower magnetizations the location of the (fast) magnetosonic point $R_{\rm FM}$ remains nearly constant, while the Alfvén point of the plasma motion, $R_{\rm A}$, continuously moves from the LC towards the stellar surface. In particular, due to the latter fact we were not able to solve the wind equation for a typical jet magnetization of $\sigma \simeq 10^{-8}$ (s. Eq. 4), since in this case the Alfvén point is located within the *numerical* stellar surface.

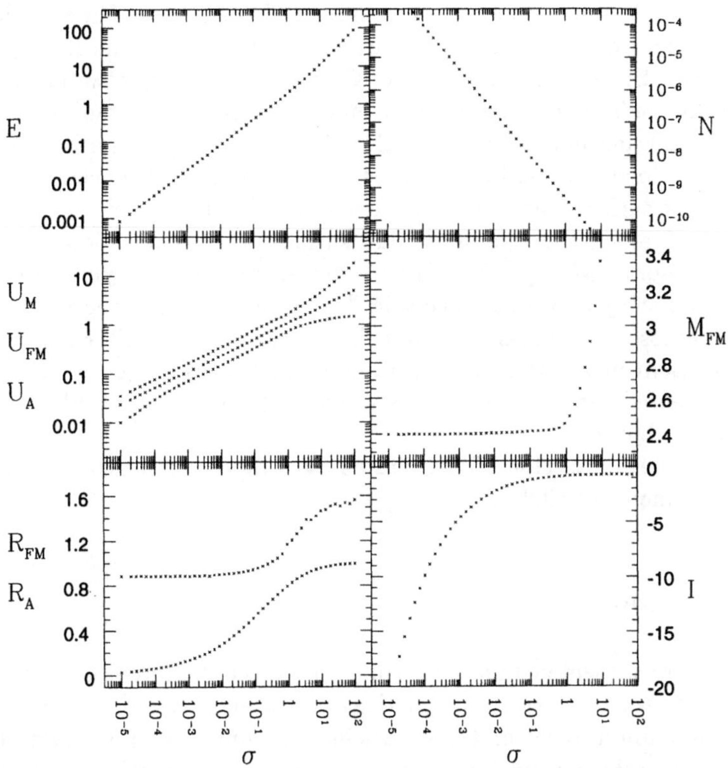

Fig. 3. Dependency of the dynamical parameters of the plasma on the magnetization σ along the flux surface $\Psi = 0.813$. Total plasma energy $(E(\Psi)/m_p c^2 - 1)$, poloidal plasma velocities at the asymptotic radius, u_M/c, at the fast magnetosonic point, u_{FM}/c, at the Alfvén point u_A/c. Position of the fast magnetosonic point, R_{FM}/R_L and the Alfvén point R_A/R_L. Particle density N, fast Mach number M_F and the poloidal current $I \sim RB_\phi$

But with the power laws derived above we are able to extrapolate towards a lower magnetization. Taking $\sigma = 10^{-8}...10^{-9}$, we easily calculate a typical protostellar jet velocity of 500 km s^{-1} and a particle density of 50 cm^{-3}. Again, this result is in good agreement with the observations.

The fast magnetosonic Mach number M_F is important when considering the stability of the asymptotic jet. Stability analysis of magnetized jets (Appl & Camenzind 1992) show that the growth rates of stable or unstable modes are determined by M_F. Moreover, the fast magnetosonic Mach number is directly connected with the spacing of shocks or knots observed in the jet. The ratio (knot distance) / (knot diameter) is then just given by the value of M_F. In the case of the "most beautiful symmetric jet to date" observed by Zinnecker et al. (this proceedings), the ratio is of about 3, which is similar to our calculations.

In Figure 4 we show the distribution of the dynamical parameters of the plasma across the asymptotic jet. As mentioned above, here we have calculated the critical wind solutions along a certain set of flux surfaces. The plot shows the asymptotic parameters of each flux surface Ψ, i.e. the dynamical properties of the plasma at the asymptotic radius of Ψ. A constant magnetization across the jet generally leads to an increasing asymptotic velocity with radius (not shown here). A decreasing magnetization with radius, $\sigma \sim \Psi^{-2}$, results in a decreasing asymptotic velocity across the jet. Then the plasma flow is concentrated within the outer flux surfaces, as shown in the density distribution N.

Note that along the innermost flux surfaces the plasma flow remains *sub* fast magnetosonic, $M_F \leq 1$. The fast magnetosonic surface asymptotically does *not* approach the jet axis but a certain flux surface dividing the jet in sub and super magnetosonic regimes. The same is indicated for the Alfvén surface. This important result means that the asymptotic structure of the jet dynamics across the jet is oriented parallel to the jet axis. It would have to be considered in terms of the jet's stability.

For a further discussion of the plasma dynamics we refer to a future paper (Fendt & Camenzind 1995).

6. Conclusion

In this talk we presented numerical solutions of the 2D force balance for collimated jets driven by strongly magnetized protostellar systems. The model underlying the numerical calculations takes into account several observational constraints, such as the observed disk-jet connection, a strong stellar magnetic field, stellar rotational periods of the order of days (which in turn introduces a relativistic treatment of the jet magnetosphere) and a high degree of jet collimation.

The structure of the magnetosphere was calculated from the GSS equation using a force-free "ansatz". We found solutions with appropriate "protostellar" boundary conditions. Besides the achievement of a *global* solution of the problem together with a finite asymptotic jet radius, the main characteristics of the resulting field configuration are the possibility of both accretion along dipolar

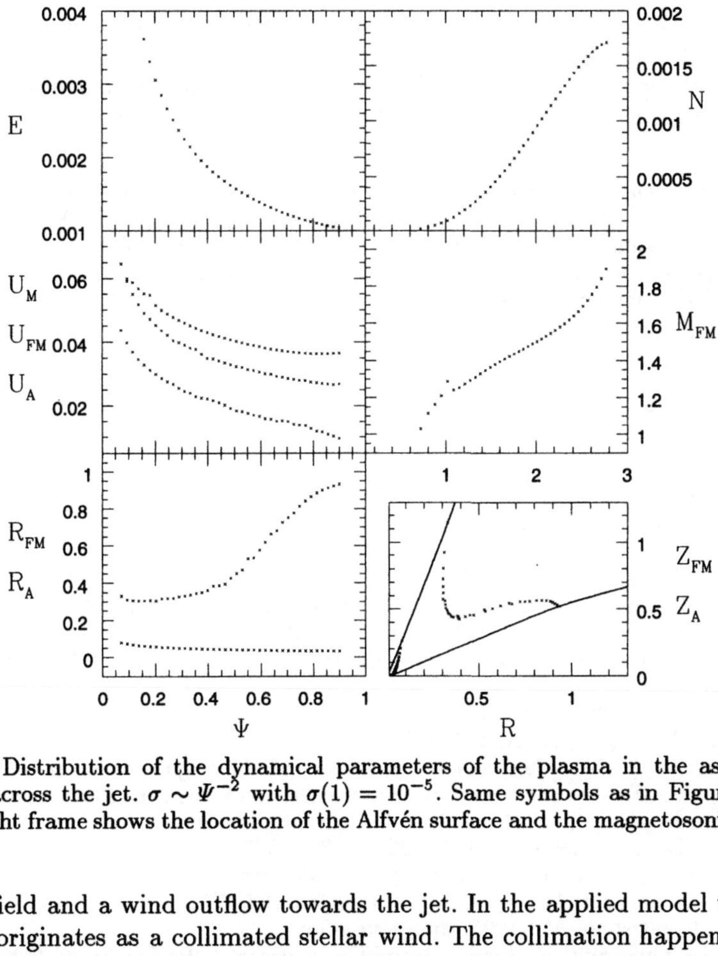

Fig. 4. Distribution of the dynamical parameters of the plasma in the asymptotic regime across the jet. $\sigma \sim \Psi^{-2}$ with $\sigma(1) = 10^{-5}$. Same symbols as in Figure 3. The lower right frame shows the location of the Alfvén surface and the magnetosonic surface

stellar field and a wind outflow towards the jet. In the applied model topology the jet originates as a collimated stellar wind. The collimation happens within 0.3 jet radii distance along the jet axis. We found arguments that the fast and strong collimation observed for protostellar jets could be explained by a pressure collimation by the ambient medium.

Along the calculated (force-free) magnetic flux surfaces the dynamics of the plasma flow was investigated. The comparison of critical solutions of the wind equation for different magnetic surfaces and for different plasma magnetizations gave information about the 2D distribution of the parameters of the plasma flow. We found that the plasma is accelerated more efficiently in a collimated (and therefore bended) magnetosphere. The terminal jet velocity and density follow a power law in terms of the plasma magnetization, while the magnetosonic Mach-number remains constant. In principle, the resulting equation enables us to calculate these parameters for different magnetizations, or, equivalently, for different mass loss rates along each magnetic flux surface.

The number values derived for the dynamical parameters are in nice agreement with the observations. However, we mention that our aim was not yet to fit the observations but to go the other way round and investigate whether something about the intrinsic jet driving mechanism could be derived from the observation of the asymptotic jet. In that sense a velocity profile observed across the asymptotic jet would also provide information about the magnetization and the mass loading at the foot points of the jet's magnetic field.

Acknowledgement. Our special thanks go to S. Appl for valuable comments. We are also grateful to R. Mundt and G. Hirth for fruitful discussions. C.F. likes to thank D. Dravins for reading the manuscript. This work was partly supported by the Deutsche Forschungsgemeinschaft (DFG). C.F. thanks the Swedish Natural Science Research Council (NFR) for financial support.

References

Appenzeller, I., Mundt, R. (1989), A&AR, 1, 291
Appl, S., Camenzind, M. (1992), A&A, 256, 354
Appl, S., Camenzind, M. (1993a), A&A, 270, 71
Appl, S., Camenzind, M. (1993b), A&A, 274, 699
André, P., Phillips, R.B., Lestrade, J.-F., Klein, K.L. (1991), ApJ, 376, 630
Blandford, R.D., Payne, D.G. (1982), MNRAS, 199, 883
Bouvier, J., Cabrit, S., Fernandez, M., Martin, E.L., Matthews, J.M. (1993), A&A, 272, 176
Camenzind, M. (1986), A&A, 162, 32
Camenzind, M. (1987), A&A, 184, 341
Camenzind, M. (1990), Magnetized disk-winds and the origin of bipolar outflows, in: Klare, G. (ed) Rev. Mod. Astron., 3, Springer, Heidelberg, p. 234
Camenzind, M., Fendt, C., Paatz, G. (1994), in preparation
Chiueh, T., Li, Z., Begelmann, M.C. (1991), ApJ, 377, 462
Fendt, C. (1994), PhD thesis, University of Heidelberg
Fendt, C., Camenzind, M., Appl, S. (1994), A&A, 300, 791
Fendt, C., Camenzind, M. (1995), in preparation
Gahm, G.F., Gullbring, E., Fischerström, C., Lindroos, K.P., Lodén, K. (1993), A&A Suppl. Series, 100, 371
Gosh, P., Lamb, F.K. (1978), ApJ, 223, L83
Heyvaerts, J., Norman, C.A. (1989), ApJ, 347, 1055
Königl, A. (1991), ApJ, 370, L39
Michel, F.C. (1973), ApJ, 180, 207
Michel, F.C. (1991), Theory of neutron star magnetospheres, The University of Chicago Press, Chicago
Montmerle, T., Feigelson, E.D., Bouvier, J., André, P. (1993), Magnetic fields, activity and circumstellar material around young stellar objects, in: Levy, E.H., Lunine, J.I. (eds) Protostars and Planets III, The University of Arizona Press, p. 689
Morse, J.A., Hartigan, P., Cecil, G., Heathcote, S., Raymond, J.C. (1992), ApJ, 399, 231
Morse, J.A., Heathcote, S., Cecil, G., Hartigan, P., Raymond, J.C. (1993), ApJ, 410, 764
Mundt, R., Brugel, E.W., Bührke, T. (1987), ApJ, 319, 275
Mundt, R., Ray, T.P., Bührke, T., Raga, A.C., Solf, J. (1990), A&A, 232, 37
Paatz, G., Camenzind, M. (1994), submitted to A&A
Pelletier, G., Pudritz, R. (1992), ApJ, 394, 117
Pudritz, R., Norman, C.A. (1983), ApJ, 274, 677

Reipurth, B. (1989), A&A, 220, 249
Sakurai, N.I. (1985), A&A, 152, 121
Sakurai, N.I. (1987), Publ. Astr. Soc. Japan, 39, 821
Sauty, C., Tsinganos, K. (1994), A&A, 287, 893

On the Magnetic Collimation of Jets from Young Stellar Objects

C. Sauty[1], K. Tsinganos[2], and E. Trussoni[3]

[1] Observatoire de Paris, DAEC, 92195 Meudon, France
[2] Dept. of Physics, University of Crete, 71409, Heraklion, Greece
[3] Osservatorio di Torino, Pino Torinese, Italy

Summary. Self-similar assumptions are commonly used to model stationary axisymmetric jets from keplerian discs. Conversely, we propose in this paper a self-similar model where the outflow is expelled from the star itself and its close environment. A comparison of these two self-similar approaches is given. We apply our analysis to describe the central optical part of jets from young stars. A simple scheme is drawn to link the parameters with the observables. Thus the jet morphology can be reproduced and a scenario is given for the transition of the flow from jet to a solar-type wind as the star evolves towards the main sequence. Finally, we briefly outline the properties of critical points under self-similarity through the example of the present model.

1. Introduction

Outflows are a widespread phenomenon through the various scales of the universe from the solar wind to the jets of AGNs, but they are best studied in closeby star forming regions. Observational as well as theoretical evidence of jets associated with young stellar objects surrounded by accretion discs have been accumulated in the past decades and reviews on this topic may be found in the present proceedings by R. Mundt, A. Königl, F. Shu etc... Despite the various efforts to model such outflows and their connection to the source, only a few solutions describing the initial formation of the wind up to the point where it is collimated as a jet are available. This means there are few solutions that can follow the flow from its origin where it is subsonic and subalfvénic up to the point where it has become super-fastmagnetosonic. None of them can explain consistently the morphological as well as the dynamical structure of all the components of the jets at once.

Although various hydrodynamical models have been well developed (Cantó 1980, Königl 1982) and may well explain some of the time dependent features (Raga & Cantó 1989), the observed ambient magnetic field and theoretical results suggest that magnetic collimation may be at work even with weak magnetic field strength. Hartmann & MacGregor (1982) first proposed a magnetic model based on the theory of equatorial winds (Weber & Davis 1967) with collimation in addition through a magnetic pressure gradient. Although a lot of insight has been gained through dimensional analysis (Pudritz & Norman 1983), solving consistently the stationary problem of a magnetized axisymmetric wind remains an open question. Numerical simulations by Sakurai (1985, 1987, 1990) have been successfull in producing a collimated wind but are restricted to a few sets of boundary conditions. Time dependent simulations (Shibata & Uchida 1987,

Washimi & Shibata 1993) are not satisfying because they cannot converge to physical solutions after some time of integration. Promising numerical results have been obtained by Shu et al. (1994a,b) where the source of the jet is a cold plasma emitted from a small zone between the accretion disc and the magnetosphere. However these simulations still cannot go beyond the alfvénic transition where collimation is actually taking place.

On the other hand, the asymptotical analysis by Heyvaerts & Norman (1989) predicts that all magnetized polytropic winds should collimate into jets, at least partially. However, it does not predict neither the distance from the source where the collimation takes place nor the final jet radius. Consequently, some winds may be loosely collimated like the solar wind whilst winds from young star show a rapid collimation. Parallely, different analytical approaches have tried to model the dynamical as well as the geometrical structure of the jets. The first way to reduce the difficult numerical problem is to integrate over the jet cross section all quantities either using the thin jet approximation (Koupelis 1990) or the thin hollow jet approximation (Lovelace et al. 1991). The first approach does not lead to transonic jets while the second is not consistent everywhere in the jet. The most promising class of analytical solutions so far is composed of the so-called self-similar models. We shall summarize in the following section how self-similar assumptions lead to a set of ordinary differential equations though keeping the non linearity of the system. Essentially two types of self-similar models exist that can follow the flow from its source to infinity including the gravitationnal potential of the star. The first is composed of the r-self-similar model where the wind source is the keplerian disc. r is the spherical radius in spherical coordinates (r, θ, ϕ). Initiated by Bardeen & Berger (1978) for hydrodynamical galactic winds, the magnetized version by Blandford & Payne (1982) remains the most famous. This model has been explored in the context of YSOs by several authors (Pudritz & Norman 1986, Wardle & Königl 1989, Safier 1993a,b, etc...). A recent generalization to more complex distributions of the electrical current has been given both by Contopoulos & Lovelace (1994) and Rosso & Pelletier (1994). A second kind of self-similar models developed by Tsinganos et al. exists which consider θ-self-similar no polytropic flows ejected from the central source, the star and possibly the surrounding internal part of the disc. Such models also include magnetic field and rotation, considering either fixed (Tsinganos & Trussoni 1991, Trussoni & Tsinganos 1993) or free (Sauty & Tsinganos 1994, Sauty 1993) geometry of the flux tubes. They are successful in proposing solutions for modelling the inner part of jets where disc wind model cannot be valid.

In the following we review some of the results obtained in the frame of this θ-self-similar model where the geometrical shape of the flux tubes is a free parameter resulting from the dynamical equilibrium. Thus, we are interested in modelling the optical part of jets assuming this is the innermost part that can be detected. The first result of the modelization is that oscillations in the jet cross section appear naturally. Therefore it can be used in the context of jets from young stars, in particular T Tauris, to ajust the parameters. This method also allows to follow the magnetic collimation process and to give a criterion whether it will be efficient or not to form a jet. Then a brief comment on the critical points in self-similar flows is in order to emphasize the underlying difficulty of solving the non-linear problem. It is also interesting to compare these results with those obtained by disc wind models. It finally suggests a global picture for the various components of the jet. Most of all, this model gives a scenario for the evolution of the jet towards an ordinary wind as the star evolves towards the main sequence and predicts quantitatively the disappearance of the jet during the evolution.

2. Stationary Axisymmetric MHD Flows

From a theoretical point of view, jets may be considered as plasma flowing out of the gravitational potential of a rotating central star surrounded by an accretion disc. To first approximation, flows are stationary and axisymmetric structures made of inviscid fluid with an infinite conductivity (ideal MHD). The governing equations are mass and magnetic flux conservation, the frozen flux condition of ideal MHD, momentum equilibrium and energy conservation,

$$\nabla \cdot \rho \mathbf{V} = 0 \ , \quad \nabla \cdot \mathbf{B} = 0 \ , \quad \nabla \times (\mathbf{V} \times \mathbf{B}) = 0 \, , \tag{1}$$

$$\rho(\mathbf{V} \cdot \nabla)\mathbf{V} = -\nabla P + \frac{(\nabla \times \mathbf{B}) \times \mathbf{B}}{4\pi} + \frac{\rho \mathcal{G} \mathcal{M}}{r^2}\mathbf{r} \, , \tag{2}$$

$$q = \rho \mathbf{V} \cdot \left[\frac{\Gamma}{\Gamma - 1}\nabla\left(\frac{P}{\rho}\right) - \frac{\nabla P}{\rho}\right] . \tag{3}$$

Thus, dynamical equilibrium appears as a set of three equations in the three spatial directions. In the following we may use either spherical (r, θ, ϕ) or cylindrical (ϖ, ϕ, z) coordinates. As it is well known (see Tsinganos 1982 and also in Königl's review in the present proceedings), axisymmetry introduces free integrals that are constant along each fieldline and streamline: A the magnetic flux, $\Psi_A(A)$ the mass flux to magnetic flux ratio, $L(A)$ the total angular momentum and $\Omega(A)$ the angular velocity of the line footpoints. Fieldlines and streamlines are roped on the same surfaces. The seconds rotate with respect to the firsts at a constant speed Ω (isorotation law). These surfaces of constant magnetic flux carry a constant total angular momentum extracted from the wind source. Thus the velocity and the magnetic fields have two components, one in the poloidal plane and the second in the azimuthal direction:

$$\mathbf{B} = \nabla \times \left(\frac{A}{\varpi}\mathbf{e}_\varphi\right) - \frac{L\Psi_A}{\varpi}\frac{1 - \varpi^2 \dfrac{\Omega}{L}}{1 - M_a^2}\mathbf{e}_\varphi \tag{4}$$

$$\mathbf{V} = \frac{\Psi_A}{4\pi\rho}\nabla \times \left(\frac{A}{\varpi}\mathbf{e}_\varphi\right) + \frac{L}{\varpi}\frac{\varpi^2 \dfrac{\Omega}{L} - M_a^2}{1 - M_a^2}\mathbf{e}_\varphi = \frac{\Psi_A}{4\pi\rho}\mathbf{B} + \varpi\Omega\mathbf{e}_\varphi \tag{5}$$

where

$$M_a^2 = 4\pi\rho\frac{\mathbf{V}_p^2}{\mathbf{B}_p^2} = \frac{\rho_a}{\rho} = \frac{\Psi_A^2}{4\pi\rho}\,. \tag{6}$$

is the poloidal Alfvén Mach number. Note that the expressions of the azimuthal components have a singularity when the poloidal Alfvén Mach number is one. To avoid this singularity it is demanded that the ratio $L/\Omega = \varpi_a^2$ is equal to the square of the cylindrical radius of the jet at this particular point along the fieldline. ϖ_a is the so-called magnetic lever arm, a constant along each field line equal to the jet radius at the Alfvén transition. Hence the azimuthal components are entirely determined from the density and the free integrals.

The global system is reduced to two equations in the poloidal (meridional) plane. One of the two, the component along each fieldline, is usually integrated to give the Bernoulli $E(A)$ constant which expresses energy conservation,

$$E \equiv \frac{1}{2}V_\mathrm{p}^2 + \frac{1}{2}V_\varphi^2 + h - \frac{\mathcal{GM}}{r} - \frac{\Omega}{\Psi_A} r\sin\theta B_\varphi - \int_{s_o}^{s} \frac{q(s,A)}{\rho(s,A)V_\mathrm{p}(s,A)}\mathrm{d}s\,, \tag{7}$$

where s is the path length along a fieldline A, h the enthalpy of the gas and q the local heating rate. However, this formulation is really relevant only when a polytropic relation is assumed. In such a case, the implicit integral can be reduced and put into the form of some enthalpic term. When no polytropic relation is assumed, one has to go back to the initial system of coupled partial differentiel equations.

3. Self-similarity

Yet, this set of partial differential equations (PDE) remains of mixed type changing from hyperbolic to elliptic several times. At this stage, we may avoid this numerical difficulty and work almost analytically by reducing the PDEs to ordinary differential equations (ODE), using self-similarity. Self-similarity consists of the following assumptions: **first** separation of variables in some key physical quantities, **second** scaling of all functions with one of the two variables and **third** care that this scaling is such as to remove all dependence from the scaling variable. In $x_\parallel$-self-similar model for jets, where coordinates are $(x_\perp, x_\parallel, \varphi)$, it is assumed that the magnetic flux A has a separable form,

$$A \propto f(x_\perp)g(x_\parallel)\,, \tag{8}$$

where $g(x_\parallel)$ is a known function of $x_\parallel$. Moreover the poloidal Alfvén Mach number does not depend on the second coordinate $x_\parallel$. Instead M_a is chosen to be a function of $x_\perp$ only, $M_a(x_\perp)$. Thus surfaces of constant Alfvén Mach number are also surfaces of self-similarity.

For a wind emitted from a keplerian disc, the spherically symmetric gravitational field and the lack of any intrinsic scale of the keplerian law naturally leads to a scaling with the spherical radius. Thus disc wind models are using r-self-similarity. The initial hydrodynamical model of Bardeen & Berger (1978) has been rapidly extended to MHD flows by Blandford & Payne (1982). Lovelace & Contopoulos (1992), Contopoulos & Lovelace (1994), and Rosso & Pelletier (1994) have independently generalized Blandford & Payne's model, by using a parametrized distribution of the magnetic flux on the disc. They use the following assumptions for the Alfvén Mach number M_a, the magnetic flux A, the mass to magnetic flux ratio Ψ_A (or ρ_a the Alfvén density), the current distribution $L\Psi_A$, the magnetic lever arm ϖ_a and the pressure,

$$M_a^2 = M_a^2(\theta)\,, \tag{9}$$

$$A \propto f(\theta) r^x\,, \tag{10}$$

$$\Psi_A^2 = 4\pi\rho_a \propto A^{2-3/x}\,, \tag{11}$$

$$L\Psi_A \propto A^{1-1/x}\,, \tag{12}$$

$$\frac{L}{\Omega} = \varpi_a^2 \propto A^{2/x}\,, \tag{13}$$

$$P \propto A^{2-2\gamma+(3\gamma-4)/x}\rho^\gamma \propto A^{2-4/x}\frac{1}{M(\theta)^{2\gamma}}\,, \tag{14}$$

where x is a constant index. Blandford and Payne (1982) restrict themselves to the special case $x = 3/4$ and assume a cold plasma ($P = 0$), mostly because of Bardeen & Berger's hypotheses (1978). Rosso & Pelletier (1994) assume a cold plasma and $\gamma = 1 - 1/(2x - 3)$ which corresponds to a constant entropy through the disc: $S(A) \propto A^{2-2\gamma+(3\gamma-4)/x}$=cst. On the other hand, Contopoulos & Lovelace (1994) remain more general and do not consider a cold plasma. Nevertheless, they choose to start beyond the slow magnetosonic point in order to simplify the mathematical treatment. Surfaces of constant Alfvén Mach number are cones and there are two kinds of distribution of the electrical current in the jet ($L\Psi_A$). If $x < 1$, the disc wind carries a return current going into the disc and balancing a current line along the singular axis. The total current carried by the jet is zero in this case. If $x > 1$, the disc wind carries a net current outflowing from the disc. No current exists along the axis although it remains singular because the Keplerian law breaks down there.

Conversely, θ-self-similar models are by construction adapted to describe the internal part of jets close to the axis, *i.e.* the part flowing out from the central star and its surrounding. They can be seen as resulting from a formal expansion of the equations with the magnetic flux or the colatitude: as the stellar source is spherical as well as the gravitational field, everythings scales naturally with the colatitude θ. Six assumptions are necessary to construct solutions as previously,

$$M_a^2 = M_a^2(r)\,, \tag{15}$$

$$A = A_\star f(r)\sin^2\theta\,, \tag{16}$$

$$\Psi_A^2 = 4\pi\rho_a = 4\pi\rho_\star(1 + \delta\frac{A}{A_\star})\,, \tag{17}$$

$$L\Psi_A = r_\star B_\star \lambda \frac{A}{A_\star}\,, \tag{18}$$

$$\frac{L}{\Omega} = \varpi_a^2 = r_\star^2 \frac{A}{A_\star}, \tag{19}$$

$$P = \frac{1}{2}\rho_\star V_\star^2 \Pi(r)(1 + K(r)\frac{A}{A_\star}). \tag{20}$$

$r_\star$, $B_\star$, $V_\star$ and $\rho_\star$ are values taken along the pole at the Alfvén point while $A_\star$ is simply $r_\star^2 B_\star/2$. In this case, surfaces of constant Alfvén Mach number are spheres. A net poloidal current is carried away by the flow. The bigger is the cross section, the bigger is the enclosed current. Moreover, and contrary to disc wind models, neither the current density nor the velocity distribution are singular along the axis. For cylindrically collimated jets, asymptotically the model is quite close to the ϖ-self-similar model of Chan & Henriksen (1980). In particular Alfvén surfaces are flatter and flatter, almost planar as the distance from the source increases. The main difference resides in the form of the pressure: no polytropic relation is used which would lead to drop some important dynamical terms in the equations. Instead, either the geometry is fixed by giving $f(R)$ (Tsinganos & Trussoni 1991, Trussoni & Tsinganos 1993) or a simple relation is used for the variation of the pressure across the jet, K (Sauty & Tsinganos 1994, Sauty 1993).

In the results presented hereafter, we assume $K = \kappa$=cst following Sauty & Tsinganos (1994). Once self-similar hypotheses are set up, only unknown functions of $x_\perp$ remain while the nonlinearity of the equations is kept. In the present case, we need to determine how functions vary with $R = r/r_\star$ the dimensionless spherical radius. The dynamical equations split into three equations for the Alfvén Mach number $M_a(R)$, the pressure $\Pi(R)$ and the radius of the jet cross section,

$$G(R) = R/\sqrt{f(R)}. \tag{21}$$

The geometry of the flux tubes and thus the whole jet structure emerges consistently by solving these equations of motion.

We may also introduce a crucial conserved quantity by writing the difference of energy density across the field lines. As no polytropic assumption is made the Bernoulli integral, Eq. 7, remains implicit. However, because of our assumptions (Eqs. 15-20, $K = \kappa$ =cst.), the difference of energy density between the pole and any fieldline is a constant along each fieldline, as it is the case in polytropic models. Moreover, we may introduce in our case, a real constant ϵ by normalizing this term with $L\Omega$ (note that this ϵ corresponds to $\epsilon/(2\lambda^2)$ in Sauty & Tsinganos, 1994),

$$\epsilon = \frac{\rho(r,A)E(A) - \rho(r,\text{pole})E(\text{pole})}{\rho(r,A)L(A)\Omega(A)} \tag{22}$$

This parameter measures the excess of the volumetric energy along a given field line, as compared to that energy along the pole, normalized to the corresponding volumetric magnetic rotator energy, $\rho\Omega L$. The energy of the magnetic rotator $L\Omega$ is the energy necessary to bring the fieldlines in rotation in the comoving frame of the fluid (Belcher & MacGregor 1976).

4. Cylindrical Collimated Jets and Oscillations

In the limit of purely hydrodynamical θ-self-similar flows with free geometry (Tsinganos & Sauty 1992a), jets are produced only through the surrounding medium forming a De Laval Nozzle. Constraints on the length of the nozzle and on the energetics of the flow are incompatible with the observations. On the other hand, using a θ-self-similar model with fixed geometry reproducing collimated flows, Trussoni & Tsinganos (1993) have shown that a MHD flow with no rotation could not collimate. The physical reason is mainly energetic because pushing the jet into a narrow channel without magnetic pinching forces demands a strong pressure gradient.

In conclusion, collimation seems to be a natural result of the coupling of rotation with the magnetic field. As a matter of fact, once we consider both magnetic field and rotation, strongly collimated jets are obtained even assuming a spherically symmetric pressure distribution, $\kappa = 0$ (Sauty & Tsinganos 1994). Note that spherically symmetric pressure implies that there is no external over-pressure that may help confining the jet. Two examples of such jets are presented in Figs 1a and 1b. Consistently with the results of Heyvaerts & Norman (1989), those jets carry a net poloidal current so they are collimated with a cylindrical shape.

In the asymptotic region where the jet is cylindrical, the bulk velocity is constant in strength and direction. Thus forces act only across the jet in the ϖ-direction. The dominant forces are the magnetic pinching force and the magnetic pressure gradient, both acting *towards* the axis and the centrifugal force of rotation acting *away* from the axis. The pressure gradient is acting *towards* this axis if there is an external overpressure ($\kappa > 0$) or *away* from it in the opposite case ($\kappa < 0$). But in the two cases of Figure 1 the transverse difference of pressure is negligible ($\kappa = 0$) and magnetic forces are exactly balanced by centrifugal forces.

However, before the asymptotic region where equilibrium is reached, the collimated jet undergoes a series of pinchings in its cross section. In fact, forces do not completely balance each other in this region for a cylindrical shape without oscillations. The net result is the bending of the streamlines and the formation of a sausage-like structure. Let us write the total momentum equation in the limit of large distances from the source,

$$\rho(\mathbf{V}_p.\nabla)\mathbf{V}_p - \frac{1}{4\pi}(\nabla\times\mathbf{B}_p)\times\mathbf{B}_p = \frac{B_\phi}{4\pi\varpi}\nabla(\varpi B_\phi) - \frac{\rho V_\phi^2}{\varpi}\nabla\varpi - \nabla P\,. \quad (23)$$

We see that the RHS is composed of the previous magnetic forces (pinching + magnetic pressure), the pressure gradient and the centrifugal force, while the LHS is the difference of the advective term and the Lorentz force associated with the poloidal flow and field. Using assumptions 15-20 with a constant κ, the previous equation becomes,

$$\frac{1}{R^2}\frac{\mathrm{d}^2\ln G^2}{\mathrm{d}\ln R^2} =$$

$$\frac{2\lambda^2}{1-M_a^2}\left[2\left(\frac{1-G^2}{1-M_a^2}\right)^2 - \frac{1}{M_a^2}\left(\frac{M_a^2-G^2}{1-M_a^2}\right)^2 + \frac{\kappa}{\lambda^2}\Pi G^2\right] . \qquad (24)$$

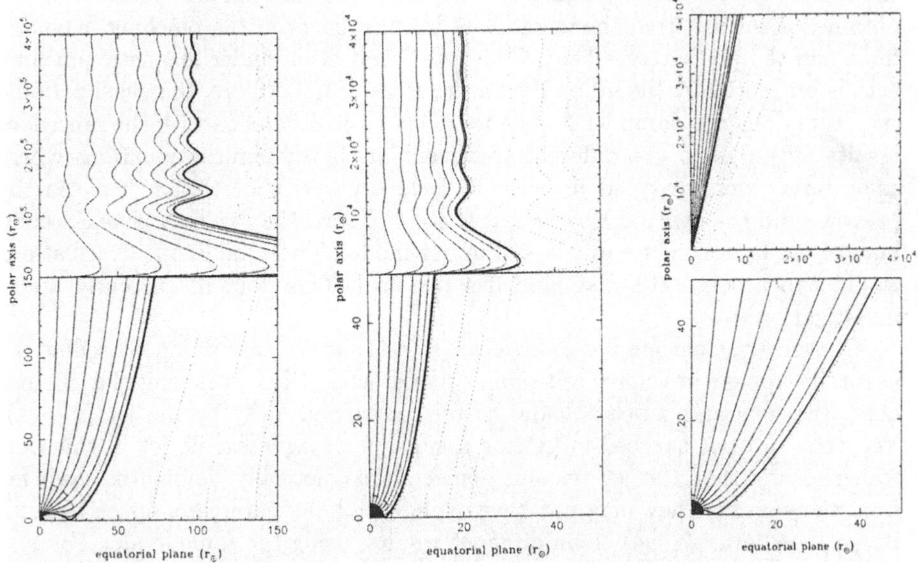

Fig. 1. In (a): streamlines of a cylindrically collimated jet represented in the first quarter of the poloidal plane. The central star possess a big dead zone. The jet has two components, one originating from the star (solid lines) and the second from the disc (dotted lines). The wavelength corresponds to the spacing of the knots in HH34. In (b): same as (a) for a collimated jet which parameters are deduced from typical observations for T Tauri optical jets following the procedure described in the text. In (c): same as in (b) once the central star has lost enough of its magnetic rotator energy. The jet has become an ordinary wind. In (a) and (b), the apparent kink in the lines is an artifact due to the change of vertical scale between the upper and the lower part of the plot (From Sauty & Tsinganos 1994).

The first term of the RHS is the pinching force, the second is the centrifugal restoring force, the third term is the pressure gradient across the lines. Note that the LHS of Eq. 23 decreases like $1/R^2$ which means that these oscillations are damped asymptotically, a result showing up in Figs. 1a,b. In these plots ($\kappa = 0$), the magnetic forces dominate over the centrifugal forces where the cross section of the jet is maximal, and the lines are bent towards the axis. The centrifugal force reacts as a restoring force where the cross section of the jet becomes minimal, thus opening the fieldlines away from the axis. If the transverse pressure gradient is not strictly zero, $\kappa \neq 0$, this conclusion is modified, as we see from Eq. (24). Extra external pressure ($\kappa > 0$) may help the pinching forces against centrifugal opening, whilst over internal pressure in the jet ($\kappa < 0$) plays the same role as centrifugal forces against magnetic pinching. However, we see that force balance across the jet remains qualitatively unchanged if the plasma in the jet and in the surrounding medium is asymptotically cold ($\Pi_\infty = 0$) which sounds reasonable.

Some different kind of oscillations were obtained by Chan & Henriksen (1980) in their ϖ-self-similar model. There, the jet is assumed to be already far from the source region and consequently Alfvén surfaces are planar. Oscillations are obtained resulting from the incomplete balance between the pinching magnetic force and the pressure gradient. Complete focalisation under magnetic pinching is thus prevented by the internal pressure alone. Furthermore, the quasi periodic structure does not damp with distance. This basic difference with the precedent results may due to the different treatment of the dynamical equations where some terms were dropped in order to keep a polytropic relation between the pressure and the density. However, it is not excluded that we may have a second kind of oscillations if we take $\kappa < 0$ in our model. Those undamped oscillations would appear once the first kind has vanished. This point is currently under investigation.

Conversely, Contopoulos & Lovelace (1994) showed that damped oscillations naturally appear in their r-self-similar model when there is no singular current along the axis and a net poloidal current is carried away by the jet ($x > 1$). Yet, they did not succeed in getting a super fastmagnetosonic jet in this case which means that the jet remains somehow magnetically dominated even far from the source. They may not have explored all the parameter space because Rosso & Pelletier (1994) claim different results, using the same model. For $x > 1$, their cylindrically collimated jet is super fast magnetosonic, exhibiting one recollimation point. However, their solution is only an approximate solution of the problem. In the case where $x < 1$, r-self-similar models are not able to obtain super fastmagnetosonic cylindrical jets. As in Blandford & Payne (1982), the jet recollimates along the axis. This can be understand because $x < 1$ means there is an infinite current density along the axis. Fieldlines are focalized along the axis and cannot decollimate further downstream because the extra pinching force, created by the singular current along the axis, is so large that it cannot be balanced neither by the centrifugal force nor by the pressure gradient.

As a result, it is tempting to relate the wavelength of the oscillations, Λ_{osc}, to the quasi-periodic spacing of the emission knots seen in various jets. In Figure 1a, the wavelength is close to the quasi-periodic spacing of the Herbig Haro objects in HH 34. However, at least in the context of jets from YSOs where proper motions of the knots are measured, a stationary model may provide an idea for the origin of the knot, while a full time-dependent study is needed in order to explain these motions. With this second point of view, the first point of recollimation may be considered as the point where the jet becomes visible (Pelletier & Pudritz, 1992) and another mechanism such as the variability of the source (Raga et al., 1993) is invoked to explain the regular spacing of the knots.

5. Modelling Optical Jets from T Tauris

In the framework of the present model, a complete solution is determined by knowing the parameters ϵ (or δ), κ, λ, $\nu = V_\star/V_{\rm esc}$ (where $V_{\rm esc}$ is the escape velocity at $r_\star$) and the asymptotic pressure Π_∞ (or the temperature T_∞). Furthermore, by specifying $r_\star$ and $\rho_\star$, all physical quantities can be expressed in dimensional form which can be compared with observables, such as the terminal speed, rotational rate, etc. $V_\star$ is also determined from the solution. Altogether then, we need seven parameters which can be fixed by observations in the case of an optical jet from a T Tauri star.

A representative T Tauri star has a mass of the order of one solar mass (Malbet 1993), a radius of the order of a few solar radii, while its rotational speed at the equator is generally in the interval of 10 – 20 km/s (Bouvier et al., 1986). Such a star usually has a well collimated jet which typically becomes visible at 100 AU (Pudritz 1991). This scale approximately determines the wavelength of the oscillations, *if* we assume that it corresponds to the location of the first oscillation/knot. Finally, the terminal speed is generally measured to be of several hundred km/s, although we may expect that the velocity along the jet axis is likely to be greater than the observed one (Pudritz 1991). Moreover, as proper motion of the knots is usually measured to be a few hundred kilometers, it is likely that we are still underestimating the terminal velocity of the gas itself flowing through the knots. The sixth parameter is given by the asymptotic density in the jet n_∞. It should be of the order of $10^4\,\mathrm{cm}^{-3}$ in the surrounding region but this means actually $n_\infty = 10^3\,\mathrm{cm}^{-3}$ within the jet, along its axis, if we assume $\kappa = 0$ as we do in the following. Last but not least, the parameter κ is given by the drop of the temperature across the jet. To first approximation we assume that the drop of the temperature is solely due to the increase of the density with the radius and we take $\kappa = 0$.

Table 1. Parameters used in Figure 1

parameters	initial guess	Figure 1b	Figure 1c	Figure 1a
n_∞ [cm^{-3}]	10^3	10^3	0	10^3
$\mathcal{M}$ [$M_\odot$]	1	1	1	1
r_o [$r_\odot$]	3	3	3	3
$V_{\varphi o}$ [km/s]	15	14.8	9.5	2.9
Λ_{osc} [AU]	100	44/58*	—	320/642*
V_∞ [km/s]	700	543	115	950
ϵ	0.0447	0.0447	-1.27	0.0501
λ	1.77	1.77	1	0.83
ν	22	22	22	4.5
$r_\star$ [$r_\odot$]	5.66	4.65	4.11	23.4
$V_\star$ [km/s]	15.7	13	13	28.4
$\rho_\star$ [10^{18}g.cm^{-3}]	—	1.7	1.7	0.96

* the second value corresponds to the distance of the first recollimation point from the source while the first value is the asymptotic value of the wavelength.

In view of the constraints on $\mathcal{M}$, r_o, $V_{\varphi o}$, Λ_{osc}, V_∞ and n_∞ (see Table 1, first column), we use some approximate analytical relations to fix the parameters.

Details of the calculations can be found in Sauty & Tsinganos (1994). First, from the definition of λ (Eqs. 18-19), we have,

$$\frac{V_{\varphi o}}{r_o} \approx \Omega \approx \lambda \frac{V_\star}{r_\star} . \tag{25}$$

Then making the assumption that the asymptotic Alfvén Mach number M_∞ is large, an asymptotic analysis of the dynamical equations shows that the parameter $1/\epsilon$ is also expected to be large. In this limit, a linear analysis of the oscillating modes gives an analytical form for the wavelength, together with a simplified form of the asymptotic velocity on the axis,

$$\Lambda_{\rm osc} \approx \frac{2\pi r_\star}{\lambda \epsilon^2} , \qquad V_\infty \approx \frac{2V_\star}{\epsilon} . \tag{26}$$

From these equations we get approximately the value of $V_\star$, ϵ and $\mu = \lambda r_o / r_\star$. Then, we use the fact that the parameter ϵ introduced in Eq. 22 is a constant together with the crude assumption that the fieldlines are radial up to the corotation radius $r_\star$. Thus, we get an equation for λ,

$$\frac{\mu^3}{\lambda^2} - 2\mu^3 (1-\epsilon) + 4\lambda - 2\sqrt{2\lambda^2+1} - 2\lambda\epsilon \approx 0 . \tag{27}$$

Solving these four equations, we find ϵ, λ, $r_\star$ and $V_\star$ (see first column of Table 1). A value of ν of the order of $\nu \approx 22$ corresponds to a typical one solar mass star.

The next step is to fix the parameters of the problem and calculate the solution numerically. We actually need three dimensionless parameters, namely ϵ, λ and ν to define a solution, plus three reference values, i.e. $\mathcal{M}$, r_o and n_∞ (the terminal density along the jet axis), to translate all dynamical and thermodynamical quantities into physical units. At this point, the equations of motion must be integrated, taking special care to cross consistently the various critical points of the system. We shall come back later briefly to this issue. This gives the values of $V_{\varphi o}$, $\Lambda_{\rm osc}$ and V_∞ as well as those of $r_\star$, $V_\star$ and $\rho_\star$ (see Table 1, column 2). We note that the wavelength we obtain is lower by a factor of two than our initial guess to fit the observations. However, the distance of the first recollimation point is farther than the asymptotic wavelength due to some non linear effect of the pressure there. This is the second value we give in the table. Thus, we finally get a rapid way of constructing a consistent solution from the initial data. In addition, this scheme may be the first step of an iterative process if one wants a better fitting with the observations from a specific object.

A two dimensional poloidal configuration of this final solution is plotted in Figure 1b. The collimated jet has three components: one from the star, an adjacent one from the surrounding disk, and a sandwiched small part ejected from the transition region. The thick line separates the stellar wind from the disc wind. The mass loss rate ejected from the star itself is of the order of the solar mass loss rate. Thus, most of the visible material is ejected from the surrounding disc. We obtain a typical mass loss rate for an optical jet (Mundt et al. 1990) of $10^{-9}M_{\odot}$/yr within an asymptotic jet radius of 19 AU (corresponding to $10^{-8}M_{\odot}$/yr within 42 AU). The density is of the order of 10^4 cm^{-3} and the temperature approximately 10^4 K at this distance from the axis. By introducing a slight positive pressure gradient accross the jet ($\kappa > 0$), we expect that the same mass loss rate and temperature will be found much closer to the axis. In other words, an increase of the pressure towards the edges of the jet would produce a thiner jet with the same other characteristics. More material would be ejected from the very internal part of the central source.

Various interpretations of the oscillation wavelength are possible as we already mentioned. For instance, the regular spacing of the HH objects in HH 33–34 (Reipurth 1991) which is of the order of 0.004 pc = 825 AU, may be used to determine the wavelength. In Figure 1a, we show the two dimensional configuration of the corresponding solution in the poloidal plane. Its parameters are given in the last column of Table 1. The resulting jet has a more extended magnetic lever arm and a much bigger dead zone around the star. For such an object, the velocity of the gas inside the jet should definitely be larger than the proper motions of the knots in the jet; such a study needs to be carried in the framework of a time dependent model.

6. From Jets to Winds

In the previous section, we have been interested mainly on producing strongly collimated jets. However, a second class of non collimated winds also exists depending on the parameters. The final morphology of the outflow actually depends on a single parameter has it is shown in Sauty & Tsinganos (1994). This parameter is ϵ (introduced in Eq. 22) which measures the excess of the total volumetric energy along a given field line as compared to that energy along the pole, normalized to the corresponding volumetric energy of the magnetic rotator, $\rho\Omega L$.

In the case of spherically symmetric pressure, we may easily express ϵ in terms of the conditions at the source boundary r_o where the cylindrical radius is $\varpi_o(A)$, the escape speed $V_{esc,o}$, the polar density ρ_o(pole) and the density at any other streamline $\rho_o(A)$. We find,

$$\frac{\Delta(\rho E)}{\rho\Omega L} = \epsilon = 1 - \left[\frac{E_{\mathrm{R}} + E_{\mathrm{G}}}{E_{\mathrm{MR}}}\right], \tag{28}$$

where E_{MR} is the magnetic rotator energy and E_{R}, E_{G} the rotational and corrected gravitational energies at the stellar base, respectively,

$$E_{\mathrm{MR}} = \Omega L\,, \qquad E_{\mathrm{R}} = \frac{V^2_{\varphi,o}}{2} = \frac{\varpi_o^2\Omega^2}{2}\,, \tag{29}$$

$$E_{\mathrm{G}} = \frac{V^2_{esc,o}}{2}.\frac{\Delta\rho}{\rho_o} = \frac{\mathcal{GM}}{r_o}\left[1 - \frac{\rho_o(\mathrm{pole})}{\rho_o(A)}\right] \tag{30}$$

Thus, in a fast magnetic rotator (FMR), $E_{\rm MR}$ is larger than the sum of $E_{\rm R}$, the rotational energy, and $E_{\rm G}$, the gravitational energy corrected by the density latitudinal anisotropy factor $1-\rho_o({\rm pole})/\rho_o(A)$, such that $\epsilon > 0$ and the flow is asymptotically collimated by the magnetic forces. Note that the ratio $E_{\rm R}/E_{\rm MR}$ equals the ratio of the cross section of the flow tube at the stellar surface to the cross section of the same tube at the Alfvén surface and therefore one expects that if this ratio is small compared to unity, the magnetic lever arm is much larger than the stellar radius and thus angular momentum is efficiently extracted from the central object. Examples of such FMR's are shown in Figure 1a,b, where ϵ is positive (see Table 1, columns 2 and 4).

On the other hand, if the magnetic rotator energy $E_{\rm MR}$ is smaller than the sum of the rotational energy $E_{\rm R}$ and the corrected gravitational energy $E_{\rm G}$ such that $\epsilon < 0$, the flow is asymptotically radial. Thus, we do not expect slow magnetic rotators (SMR) to be collimated. Note that the latitudinal anisotropy of the density ($\Delta\rho > 0$) favours radial asymptotics ($\epsilon < 0$) together with the rotational energy $E_{\rm R}$. We present an example of such non collimated solutions in Figure 1c. Actually, we just keeped the same values of $\Delta\rho/\rho = \delta$ (Eq. 17) and ν as in Figure 1b and lowered λ (see Table 1 column 3). Reducing the magnetic rotator strength, through λ^2, by about a factor of two, the jet has undergone a sudden transition from a purely collimated configuration with cylindrical shape (Figure 1b) towards an asymptotically radial shape presumably similar to the solar wind (Figure 1c).

Through this example, we may understand that as a jet has removed enough angular momentum from a young star, the central object has span down making the magnetic rotator less efficient, reducing thus the magnetic rotator energy $E_{\rm MR}$. After passing through a critical stage, cylindrical collimation and knot formation seems to have disappeared completely and the flow is undergoing a continuous expansion with magnetic pinching not efficient any longer.

It should be emphasize that collimation is not the result of a strong magnetic field or of a strong magnetic rotator. Actually, the star could also spin up and still looses its jet by simply increasing enough its density anisotropy factor, δ. Obviously, the previous scenario may have been over-simplified as compared to reality because we have kept various parameters constant for simplicity. If the star reaches a stage comparable to the characteristics of the Sun today, the excess of equatorial density δ would indeed increase as measurements in coronal holes seem to suggest. So, the loss of the jet may be a combined effect of the various parameters, in particular λ and δ.

Preliminar results are now available that take into account a slight but positive increase of the pressure as we move from the axis to the external part of the jet ($\kappa > 0$). As a matter of fact, we expect κ to be positive for various energetical reasons (see previous section). Those results show that the jet remains always cylindrical, in this case. As ϵ becomes negative no sudden transition towards a radial configuration appears. Instead, we observe a smooth transition from the initial strongly collimated jet structure towards a wind that is still cylindrical but farther from the source with a larger asymptotic cross section.

The previous criterion for collimation does not seem to be really affected, at least quatitatively, in the sense that as the magnetic rotator becomes weaker, ϵ becoming negative, the flow undergoes a strong change in its degree of collimation. Practically, when $\epsilon < 0$, collimation seems to take place far from the source probably beyond the terminal shock with the interstellar medium. In other words and from a theoretical point of view, even the solar wind would be collimated, as already suggested by Heyvaerts & Norman (1989). However, full cylindrical collimation would not be observed because the wind would encounter the interstellar medium before collimation is achieved.

7. Critical Points and Self-similarity

So far we have summarized various results obtained with our θ-self-similar model for outflows from YSOs. Nevertheless, the major difficulty in solving consistently the MHD equations is the existence of critical or fixed points. Without going into the details of the various topologies, we would like to point out some of the reasons that make the problem so difficult.

Since Parker's original solar wind model (1958), it is well known that in order for these singularities to be avoided, specific boundary conditions have to be imposed on the source. Physically it means that the flow becomes stationary only if the source can adequately tune its boundary conditions. In classical equatorial winds (Weber & Davis 1967), such critical points appear when the poloidal velocity of the fluid is equal to the slow or the fast mode velocity. However, the situation is certainly not so simple in general axisymmetric outflows (Bogovalov, 1994).

In self-similar flows (see Tsinganos et al. 1995 for a complete demonstration), critical conditions correspond to the equation of slow and fast MHD waves but in the direction perpendicular to the self-similarity direction. Singularities appear now when the flow speed component in the perpendicular direction is equal to the velocity of the slow or the fast mode in this direction,

$$V_{\perp} = (V_{\mathrm{sm},\perp}; V_{\mathrm{fm},\perp}) \,. \tag{31}$$

In all axisymmetric systems, an extra singularity appears also at $M_a = 1$ when rotation is present. However, this relation is not affected by the actual direction of propagation of the waves. The condition $M_a = 1$ is equivalent to $V_{\perp} = V_{a,\perp}$ due to the incompressibility of pure Alfvén waves.

This modification of the critical points in self-similar flows was already pointed out by Bardeen & Berger (1978) but none of the other r-self-similar disc wind models (Blandford & Payne 1982, Rosso & Pelletier 1994, Contopoulos & Lovelace 1994) had to deal with this difficulty in their solutions. The first reason is that the slow magnetosonic point is usually assumed to be located below the starting point of the numerical calculation. The second reason is that none of the previous models shows a solution that really crosses the modified fast-magnetosonic point. In the case where $x < 1$ (*cf* Eq. 12) the critical solution is not collimated and reaches the fast mode critical point asymptotically. The only interesting solutions are the refocalising ones which do not fill all space. Although terminated, such solutions are thought to be physical if a shock front can form before complete refocalisation (Pelletier & Pudritz 1992). In the case $x > 1$, cylindrical collimated jets are obtained but the issue whether they cross the last critical point is still controversial, as we already mentioned in section 4.

In the frame of our θ-self-similar model, although there is no simple relation between pressure and density, critical points also correspond to MHD modes propagating in the r direction, perpendicularly to the self-similar direction,

$$V_r^4 - V_r^2(V_a^2 + V_s^2) + V_s^2 V_{a,r}^2 = 0\,. \tag{32}$$

The sound speed at a given point can be defined in terms of the local pressure variation with the density,

$$V_s^2 = \left.\frac{\mathrm{d}P}{\mathrm{d}\rho}\right|_{(R,A)} = -\frac{V_\star^2}{2}\left.\frac{\partial \Pi}{\partial M_a^2}\right|_{(R,A)} M^4 \frac{1+\kappa\frac{A}{A_\star}}{1+\delta\frac{A}{A_\star}}\,. \tag{33}$$

Of course, we cannot in general calculate this sound speed at any point. However, the structure of the dynamical equations (Sauty & Tsinganos, 1994) allows to calculate its value at the critical point (Tsinganos et al., 1995). Using the equation for the pressure gradient, we find the following value for the sound speed, at this particular point,

$$V_s^2 = V_\star^2 \frac{M^4}{R^4} \frac{1+\kappa\frac{A}{A_\star}}{1+\delta\frac{A}{A_\star}}\,. \tag{34}$$

Combining Eq. 34 with Eq. 32 leads to

$$(V_r^2 - V_{a,r}^2)(1+\kappa f) + (V_{a,\phi}^2 + V_{a,\theta}^2)\frac{\cos^2\theta}{\sin^2\theta} = 0\,, \tag{35}$$

which corresponds to the relation found by Sauty & Tsinganos (1994) in their topological analysis (Sauty 1993).

The conclusion is that the nature of the critical points in terms of MHD modes is very general and not restricted to polytropic winds. Moreover, the lack of spherical symmetry introduces modifications in the nature and the location of critical points that may have been underestimated so far in numerical models.

8. Towards the Construction of a Global Model?

We would like to conclude with a few remarks on the possibility of a global modelling of jets including the inner stellar source as well as the outer disc origin. One possibility would be, of course, to combine the present kind of "stellar" jet with an X-wind as suggested by Shu et al. (1994a,b and present proceedings). A tempting alternative would be to use simultaneously the various self-similar models developed so far. In particular, we may think of using a θ-self-similar model for the inner part of the jets from YSOs as we propose in the present paper while modelling the external part of the jet as a self-similar disc wind. Two configurations of the electrical current are to be considered in this case. Either, the electrical current is carried out from the inner jet and comes back into the disc *via* the disc wind, or current is everywhere flowing with the jet in which case it closes back outside the jet through the interstellar surrounding medium.

The first possibility means for the disc wind that $x < 1$. Globally the jet carries no net poloidal current. This configuration is favoured by MHD accretion-ejection models of the disc itself (Ferreira & Pelletier 1995, Ferreira 1994). However, the only super fastmagnetosonic solutions available are the terminated solutions with an *ad hoc* front shock. Although it is not excluded from observations, it is not obvious why such solutions that do not cross all critical points would be selected. Disc wind models could be responsible for the molecular flow while the optical jet would be extracted from the star and its close neighbourhood, fitting somehow the sketch of Pudritz & Norman (1986).

The second possibility implies $x > 1$. In this case the whole jet carries a net poloidal current. Disc wind models have been successful in producing cylindrically collimated jets that could be consistent with the inner cylindrical stellar jet we propose. However, an open question remains. While the stellar jet is automatically super fastmagnetosonic with θ-self-similar models, it is not clear whether r-self-similar can produce a super fastmagnetosonic jet (Rosso & Pelletier 1994) or not (Contopoulos & Lovelace 1994). As we already mentioned, the second possibility is more likely because the solution derived by Contopoulos & Lovelace (1994) was an exact one, unless they missed the small range of appropriate parameters to get through the fast magnetosonic point. After all, this may suggest that the inner jet (the optical part?) is strongly accelerated while the surrounding wind (the molecular flow?) is more like a breeze.

In conclusion, further investigations of the various self-similar models are still needed to complete our picture of the ejection phenomenon. We can still learn a lot from these "simplified" approaches before exhausting them. In addition, they may also be useful toward the goal of constructing more refined numerical models.

References

Bardeen J.M., Berger B.K., 1978, ApJ, 221, 105
Belcher J.W., MacGregor K.B., 1976, ApJ, 210, 498
Blandford R.D., Payne D.G., 1982, MNRAS, 199, 883
Bogovalov S.V., 1994, MNRAS, 270, 721
Bouvier J., Bertout C., Benz W., Mayor M., 1986, A&A, 165,110
Cantó J., 1980, A&A, 86, 327
Chan K.L., Henriksen R.H., 1980, ApJ, 241, 534
Contopoulos J., Lovelace R.V.E., 1994, ApJ, 429, 139
Ferreira J., 1994, thèse de doctorat de l'Université de Grenoble, Grenoble.

Ferreira J., Pelletier G., 1995, A&A, 295, 807
Hartmann L., MacGregor K.B., 1982, ApJ, 259, 180
Heyvaerts J., Norman C.A., 1989, ApJ, 347, 1055
Königl A., 1982, ApJ, 261, 115
Koupelis T., 1990, ApJ, 363, 79
Lovelace R.V.E., Berk H.L., Contopoulos J., 1991, ApJ, 379, 696
Lovelace, R.V.E., Contopoulos J., 1992, in: Ann. NY Acad. Sci. No. 675, 7th Florida Workshop on Nonlinear Astronomy: Astrophysical Disks, ed. S.F. Dermott et al., p. 286
Malbet, F., 1993, Environnement Circumstellaire des Etoiles Jeunes, thèse de doctorat. Université Paris VII, Paris
Mundt R., Ray T.P., Buehrke T., Raga A.C., 1990, A&A, 232, 37
Parker E.N., 1958, ApJ, 123, 664
Pelletier G., Pudritz R.E., 1992, ApJ, 394, 117
Pudritz R.E., Norman C.A., 1983, ApJ, 274, 677
Pudritz R.E., Norman C.A., 1986, ApJ, 301, 571
Pudritz R.E., 1991, The Physics of Disk Winds. In: Lada C.J., Kylafis N.D. (eds.) Proc. NATO ASI The Physics of Star Formation and Early Stellar Evolution, Kluwer, Dordrecht, p. 539
Raga A.C., Cantó J., 1989, 344, 404
Raga A.C., Cantó J., Biro S., 1993, MNRAS, 260, 163
Reipurth B., 1991, Herbig-Haro Objects. In: Lada C.J., Kylafis N.D. (eds.) Proc. NATO ASI The Physics of Star Formation and Early Stellar Evolution, Kluwer, Dordrecht, p. 497.
Rosso F., Pelletier G., 1994, A&A, 287, 325
Safier, 1993a, ApJ, 408, 115
Safier, 1993b, ApJ, 408, 148
Sakurai T., 1985, A&A, 152, 121
Sakurai T., 1987, PASJ, 39, 821
Sakurai T., 1990, Computer Physics Reports, 12, 247
Sauty C., 1993, thèse de doctorat de l'Université de Paris 7, Paris.
Sauty C., Tsinganos K., 1994, A&A, 287, 893
Shibata K., Uchida Y., 1987, PASJ, 39, 559
Shu F.H., Najita J., Ostriker E., Wilkin F., Ruden S., Lizano S., 1994a, ApJ, 429, 781
Shu F.H., Najita J., Ruden S., Lizano S., 1994b, ApJ, 429, 797
Trussoni E., Tsinganos K., 1993, A&A, 269, 589
Tsinganos K., 1982, ApJ, 253, 775
Tsinganos K., Sauty C., 1992a, A&A, 255, 405
Tsinganos K., Sauty C., 1992b, A&A, 257, 790
Tsinganos K., Sauty C., Trussoni E., Contopoulos J., Surlantzis G., 1995, (in preparation)
Tsinganos K., Trussoni E., 1991, A&A, 249, 156
Wardle M., Königl A., 1989, ApJ, 410, 218
Washimi H., Shibata S., 1993, MNRAS, 262, 936
Weber E.J., Davis L.Jr., 1967, ApJ, 148, 217

Stellar Jets: Spectral Diagnostics and Fluidodynamic Models

F. Bacciotti[1], C. Chiuderi[2], G. Hirth[3], A. Natta[4], and E. Oliva[4]

[1] Observatoire de la Côte d'Azur, B.P. 229, 06304 Nice Cedex 4, France
[2] Dipartimento di Astronomia e Scienza dello Spazio, Università di Firenze, Lg. E. Fermi 5, 50125 Firenze, Italy
[3] Max-Planck-Institut für Astronomie, Königstuhl 17, 69117 Heidelberg, Germany
[4] Osservatorio Astrofisico di Arcetri, Lg. E. Fermi 5, 50125 Firenze, Italy

Summary. We present a non-standard, straightforward procedure for a model-independent determination of crucial physical parameters of the linear section of stellar jets, such as the hydrogen ionization fraction x and the temperature T_e of the emitting gas. The method can be easily applied to every Herbig-Haro jet for which the brightest red lines have been measured, even if the lines are not calibrated and the ratios are not corrected for reddening. In the cases of HH 34 and HH 111 we find $x \sim 0.1$ and $T_e \sim 6000$. The momentum rates evaluated with the derived total number densities ($N \sim 10^4$ cm^{-3}) give strong support to the picture in which the luminous jet is only the fastest and most ionized component of a wider neutral flow that, in turn, has the capability of accelerating a molecular outflow. The ionization fraction cannot be explained in terms of equilibrium processes reflecting the local physical conditions of the medium, however the recombination length scales turn out to be comparable to the linear extension of the most luminous part of the jets: the observed degree of ionization is most likely the remnant of the heating and the excitation occurred in the initial accelerating region. This picture is supported by our diagnostics of RW Aur's optical jet, for which spatially resolved spectra have been taken: in this case the ionization fraction (around 10% again) is found to decrease slowly with distance from the star.

The combination of the time history of the ionization fraction, the proper radiative cooling, the recent observational evidence of a relative motion between the emitting gas and the pattern of bright knots, and the finding of multiple aligned bow-shocks in a single system, suggests that physical conditions consistent with the observed emission could be obtained through "mild" compressions of the central portion of the flow, caused by "damped" Kelvin-Helmoltz instabilities generated at the jet-ambient contact discontinuity. These compressions concentrate in smaller volumes the already available internal energy that is eventually radiated in optical and IR lines, unlike the weak shocks that actually increase the internal energy content by transforming the bulk kinetic energy into thermal random motions. A picture of this kind seems capable of resolving the long-standing problem of conciling the high supersonic velocity of the ionized material and the low excitation nature of the emission, and, at the same time, provides a very natural explanation for a number of observational constraints, not explicable in terms of the common shock excitation picture, among which the increase of the [SII]/Hα ratio along the jet and the disappearance of the optical emission after a characteristic scale length.

1. Introduction

A large number of Herbig-Haro (HH) jets shows a well-organized linear structure, extending over 0.1-0.5 *pc*, consisting of a narrow chain of travelling bright knots ("jet"), followed by an intermediate section where the emission disappears, and by one or more bow-shaped features aligned with the visible collimated section (see the good reviews by Mundt 1993, and Reipurth 1991, Reipurth & Heathcote 1993). Most of the available theoretical models rely on the firm belief that the observed lines form in the cooling region behind a shock. While this explanation seems quite reasonable for the bow-shaped terminal features, the interpretation of the linear jet section in terms of shocks faces serious problems, among which the individuation of the process capable of generating long-lived travelling shocks weak enough to produce the low excitation spectral features observed.

The main non-stationary models so far proposed involve either pulsations of the ejection velocity, that generate a series of miniature working surfaces travelling within the jet (Raga et al. 1990, Raga 1991, Kofman & Raga 1992, Raga & Kofman 1992, Stone & Norman 1993, Hartigan & Raymond 1993), or Kelvin-Helmholtz (K-H) shear instabilities at the boundary of the jet, that appear to be able to produce travelling internal shocks (Blondin et al. 1990, Coleman 1990, Appl & Camenzind 1992, Ray 1993, Yue & Winnewisser 1993, Bodo et al. 1993). In both cases, however, recent simulations show that the knotty structure is not well reproduced (Blondin et al. 1990, Stone & Norman 1993).

One of the greatest difficulties resides in the fact that available observations do not generally allow to estimate independently physical parameters of the flow, crucial for the specification of the dynamical and radiative properties, such as the excitation temperature T_e and the hydrogen total density n_H of the plasma.

In order to partly overcome these drawbacks, a new straightforward procedure has been developed, that allows to recover, in a model-independent way, the excitation temperature T_e and the hydrogen ionization fraction $x = n_{H^+}/n_H \approx n_e/n_H$ of the emitting plasma directly from intensity ratios of the brightest lines commonly observed in the spectra (Bacciotti,Chiuderi & Oliva 1995, hereafter BCO).

The method, briefly summed up here in Sect. 2, is model-independent in the sense that it relies only on cooling properties, and not on hypotheses on the heating agent.

In Sect. 3 we describe the application of the procedure to the cases of the HH 34 and HH 111 jets, and to the optical outflow of RW Aurigae, and we discuss some first physical implications of the results we obtain.

Finally in Sect. 4 we examine the possible consequences these results may have on hydrodynamic models for the morphology of optical jets. Our diagnostics in fact allow us to estimate the recombination and cooling time scales of the emitting gas: we shall show that if these are combined with important recent observational evidences, as the presence of several bow shaped features aligned along the jet axis (Reipurth 1993) and the detailed kinematic studies now available for several objects (Eislöffel 1993, Eislöffel & Mundt 1994), *it is possible to envisage a scenario for the formation of bright knots alternative to the common shock excitation picture.*

2. Procedure for the Determination of T_e and x

In most cases the brightest optical lines observed in bipolar jets are [OI]λ6300, H_α, [NII]λ6584 and [SII]$\lambda\lambda$6713,6731. Following standard tecniques we can easily determine the electron density through the [SII] doublet; our procedure allows, in addition, to determine T_e and $x = n_{H^+}/n_H \approx n_e/n_H$, from the available line ratios.

We consider low-excitation plasmas, being the low-excitation nature of the emission witnessed by the high value of the intensity ratio [SII]/Hα. The ionization of the various species should not be due to radiation, since no strong ionizing source is present in the vicinity of the object studied.

In this conditions it is reasonable to assume that the ionization equilibrium of nitrogen and oxygen is primarily determined by charge exchange with hydrogen, while sulphur is in any case completely ionized.

In addition we assume that the abundances of nitrogen, oxygen and sulphur are not far from their solar values (Walter et al. 1992, Peimbert et al. 1993), i.e. N/H=1.1 10^{-4}, O/H=8.5 10^{-4} and S/H=1.6 10^{-5}.

The emission coefficients of the lines under consideration are known function of n_e (whose value can be estimated from the [SII] doublet) and T_e. The ratios [SII]/Hα, [NII]/Hα and [SII]/[OI] depend therefore on T_e, which primarily determines the excitation rate of the forbidden lines, and x, the parameter which sets the relative abundances of H^+, N^+, and O^0. The derivation of T_e and x is based on the fact that the above line ratios have quite different behaviours with these parameters.

Handy approximate analytical expressions for the ionization fraction and excitation temperature are derived in BCO. It can be showed that the ratio N^+/H^+ is virtually independent on x, therefore the ratio [NII]/Hα can be used to estimate the temperature. The simple relation holds:

$$T_4 \simeq \frac{3.32}{ln\,[1.1\,10^2\,(H\alpha/[NII])\,T_4^{0.5}\,A_\odot(N)]} \tag{1}$$

where $T_4 = T_e/10^4$ K, and

$$A_\odot(N) = \frac{N/H}{1.1\,10^{-4}}$$

is the nitrogen abundance relative to its solar value.
Since the relative abundance S^+/S does not depend on x while H$\alpha \propto x$, the ionization fraction x can now be estimated from [SII]/Hα, using the [NII] temperature (Eq. 1). After some algebra one finds:

$$x \simeq 1.04\,T_4^{0.2}\frac{H\alpha}{[SII]}\left(\frac{[NII]}{H\alpha}\right)^{0.64}\cdot\frac{1}{1+n_e/n_c}A_\odot(S)A_\odot(N)^{-0.64} \tag{2}$$

where

$$A_\odot(S) = \frac{S/H}{1.6\,10^{-5}}$$

and $n_c \simeq 2000\ \mathrm{cm}^{-3}$ is the critical density of the [SII] doublet. The residual dependency on T_e is very slow and can be neglected using e.g. $T_4 = 0.7$. Eqs. (1), (2) allow therefore a rapid extimate of T_e and x. However it should be kept in mind that they loose their validity when $n_e \gg 2000\ \mathrm{cm}^{-3}$ (i.e. when the [SII] lines are collisionally deexcited) and when mechanisms other than charge exchange become important for the formation/destruction of N^+ and O^+ (e.g. when $x \geq 0.4$). More precise determinations of temperature and total density can be obtained numerically also including [OI] lines for which the analytical solution (i.e. the equivalent of Eq. (2)) is less handy due to the strong dependence of their collision strengths on T_e. The diagnostic diagrams that we derive in our work are all produced numerically, using a code developed by A. Raga (1992) and atomic parameters taken from Mendoza (1983), Giovanardi et al. (1987), Oliva et al. (1989), Hummer & Storey (1987) and from Butler & Dalgarno (1979) (see BCO for details). In this case one proceeds preparing contour plots on the same (x, T_e) frame of the three surfaces [SII]/[OI], [NII]/Hα and [SII]/Hα at the level indicated by the observations, taking care of observational uncertainty. The portion of the frame where the three obtained stripes cross each other gives then an indication on the x and T_e values, averaged along the line of sight. We notice that our procedure does not include a treatment for the photoionization of the species, therefore the method is not applicable in regions in the proximity of a strong shock front, such as the terminal bow-shock, where ionizing photons are likely produced.

3. Application to HH 34, HH 111 and to the Outflow of RW Aur

An application of this tecnique to the beam section of the well-known HH 34 and HH 111 jets (see BCO) has provided quite low values for both the ionization fraction and the temperature (Figures 1 and 2 respectively):

HH 34 jet:	$T_e = 5800 \pm 200$ K and $x = 0.067 \pm 0.010$
HH 111 jet:	$T_e = 5700 \pm 200$ K and $x = 0.11 \pm 0.02$

Since no source of ionizing photons is present, these results imply that the ionization state of the gas is not in equilibrium with local physical conditions. Also if moderate, in fact, such a value of the ionization fraction cannot be produced collisionally in a shock at a temperature of only 6000 K.

In order to find a solution to this apparent paradox, a plausible and innovative scenario was therefore proposed in BCO, as follows (see Figure 3). The ionization of the gas is produced in a strong shock that formes in the proximity of the star, in the nozzle that accelerates the flow, as for example in the simulations of Franck & Mellema (1994a, 1994b). This region is however heavily shielded from view by the dense circumstellar environment.

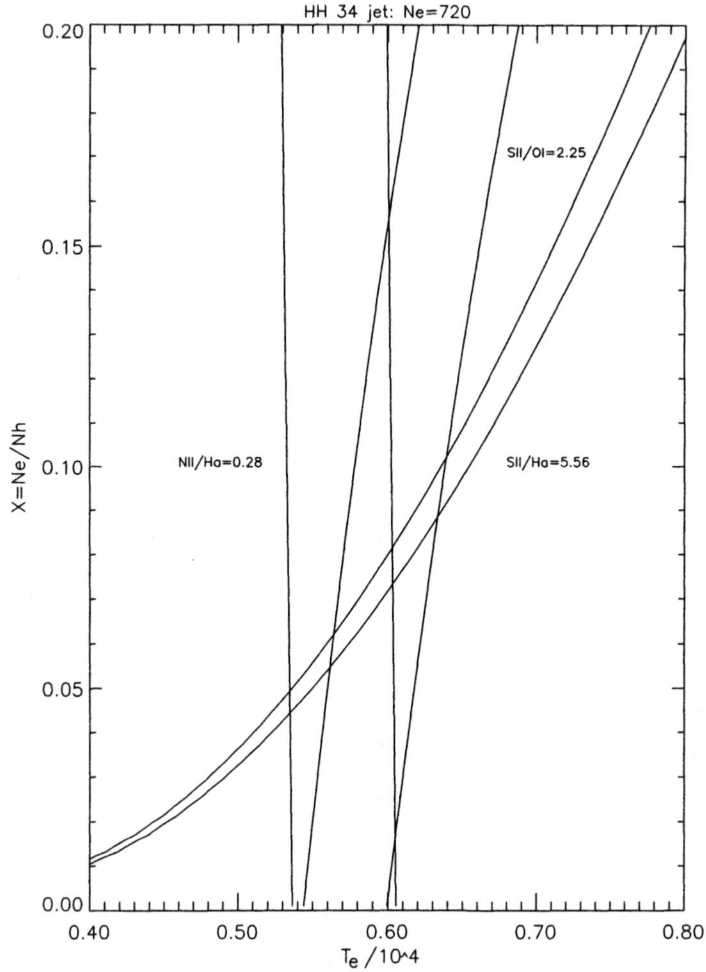

Fig. 1. Diagnostic diagram for the HH 34 jet, calculated for $N_e = 720 \text{ cm}^{-3}$

Immediately after, the gas starts to expand trying to reach pressure equilibrium with the more diffuse external cloud: here the ionization state decouples from local T_e variations, since the rapid expansion considerably inhibits recombination. As it will be discussed in the next section, the peculiar kinematics of the flow suggests that the subsequent emission observed in the beam section could be produced by "mild" compressions and not by an actual shock front. Also if beam shocks are present, however, they are in any case too weak to produce further substantial ionization. The ionization fraction therefore slowly decreases along the jet, the population of free electrons being regulated by recombination time scales, depending mainly on n_e.

In the cases of the HH 34 and HH 111 jets the halflife of the free electron population was found to be comparable with the the crossing time of the bright section of the jet: this implies that we should be able to detect a slow decrease

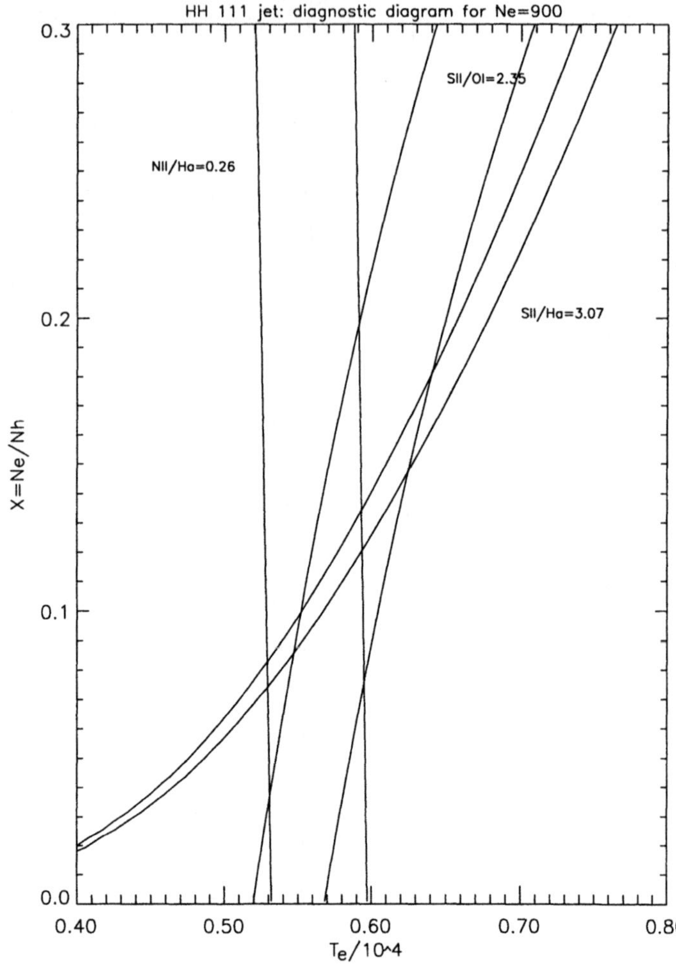

Fig. 2. Diagnostic diagram for the HH 111 jet, calculated for $N_e = 900\ \mathrm{cm}^{-3}$

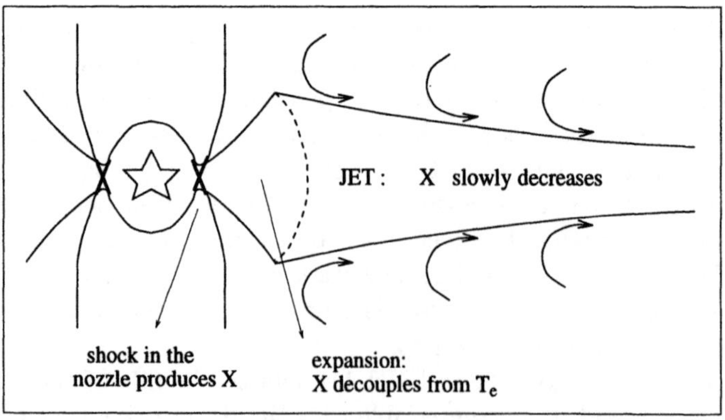

Fig. 3. Acceleration region of an optical jet

of x along the flow. This evidence could not been verified directly for HH 34 and HH 111, since spatially resolved spectra containing all the lines necessary to the diagnostics were not available.

The scenario proposed, however, has been confirmed by the results of the diagnostic analysis we performed for the case of the bipolar outflow of RW Aur (red lobe) for which detailed spatially resolved long-slit spectra were taken by one of us (G.A.H.): as it is showed by Figure 4, that collects the results obtained at many different positions along the axis, both the ionization fraction (again around 10%) and the electron density show a well defined slow decrease along the flow (Bacciotti, Hirth and Natta 1995).

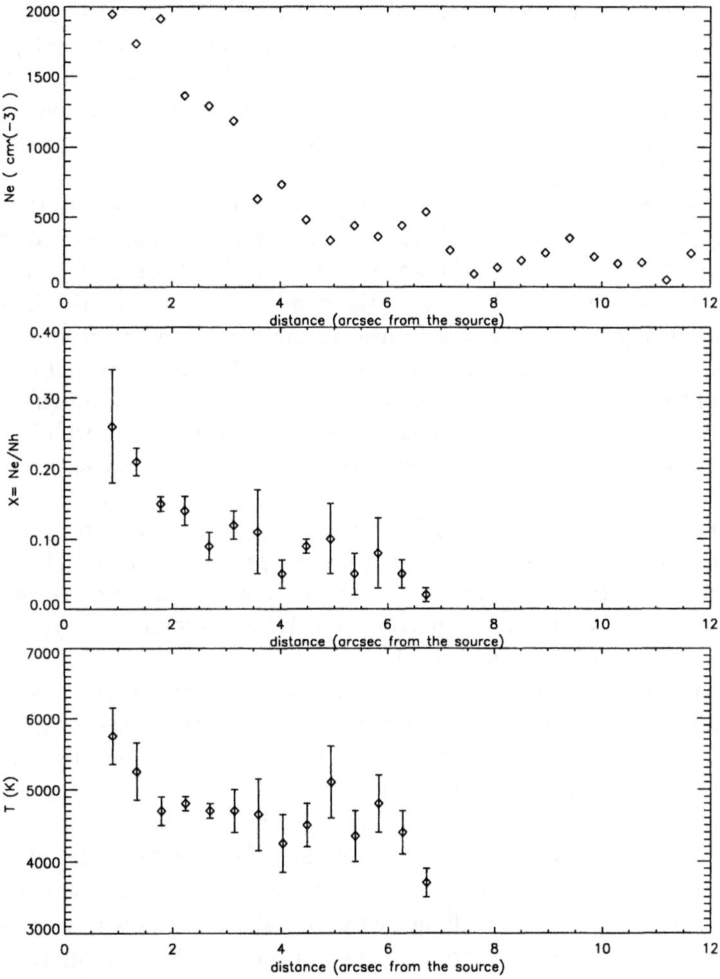

Fig. 4. Optical outflow of RW AUR (red lobe) : a) electron density; b) ionization fraction ; c) temperature

In order to better substantiate the picture proposed we mention that once recognized that at $T_e = 6000$ K Hα emission comes from recombination only, the slow decrease of x gives a very natural explanation for the growth of the [SII]/Hα ratio along the jet, and the different knot-to-knot luminosity contrast in [SII] and Hα profiles (see Bührke et al. 1988, Mundt et al. 1991).

This comes from the fact that the sulphur lines are strongly sensitive on T_e, while Hα is mostly dependent on $x = n_e/n_H$. As a result:

$$\frac{[SII]}{H\alpha} \propto T_e^{1/2} \frac{e^{-2.14/T_4}}{x}. \tag{3}$$

If in the axial portion of the flow there are no shocks able to further ionize the gas, the ionization fraction x can only decrease with time, and we then conclude that each successive knot should present an increasing value of the [SII]/Hα ratio, as observed. The same line of reasoning can be followed to justify the different knot-to-knot luminosity contrast observed in [SII] and in Hα images: the rise in temperature that produces the bright knot in [SII] opposes recombination; moreover [SII] $\propto xn_H^2$, while H$\alpha \propto x^2 n_H^2$.

We note that our approach is radically different with respect to that adopted in Hartigan, Morse & Raymond (1994): in this work the authors investigate the ionization state of the gas in the beam section of HH 34 and HH 111, by means of accurate radiative shock models, therefore assuming that the ionization is produced locally by shock fronts formed in the flow. The low excitation spectra observed in these objects necessarily imply very low shock velocities, which in turn produce an extremely low value of the ionization fraction, roughly one tenth of that determined by us for the same jets. Even if this approach is internally consistent, we believe that our diagnostic method has the advantage of relying only on cooling properties and not on the hydrodynamics of the flow. In fact the existence of a system of long-lived travelling shocks implied by the work of Hartigan, Morse & Raymond has not yet been proven definitively by any of the numerical simulations performed, except for those in which extremely regular and frequent pulsations of the ejection velocity are assumed, a hypothesis that we consider quite unrealistic (see BCO).

We conclude this paragraph presenting an estimate of the mass loss and momentum rates of the examined outflows. In all the cases studied we find $x \sim 0.1$, which implies that the total hydrogen number densities N in these flows are of the order of 10^4 cm^{-3}.

If an estimate of the radius of the jets is available, mass loss and momentum rates can be easily computed via the relations $\dot{M} = \pi r_j^2 \rho v_j$ and $\dot{P} = \dot{M} v_j = \pi r_j^2 \rho v_j^2$. It should be noticed however that these determinations could be not realistic if the value of the filling factor, usually not known, is substantially lower than one. For this reason it could be preferable to estimate the mass and momentum losses from an inspection of the absolute luminosity of a strong forbidden line, such as [OI] λ6300, but this is often difficult, due to the lack of knowledge of the ionization and excitation state of the species considered and to the strong dust absorption present in these regions.

Keeping in mind these consideration, we calculate $\dot{M}$ and $\dot{P}$ in the outflows considered through our determination of the ionization fraction, assuming a filling factor of order unity and a typical flow radius of $\sim 3\ 10^{15}$ cm (Raga et al. 1991, Reipurth et al. 1992, Hirth et al. 1994a,b). We obtain for the HH 34 jet $\dot{M} \sim 3\ 10^{-7}$ $M_{\odot}$ yr^{-1} and $\dot{P} \sim 5\ 10^{-5}$ $M_{\odot}$ yr^{-1} km s^{-1}, for the HH 111 jet $\dot{M} \sim 4\ 10^{-7}$ $M_{\odot}$ yr^{-1} and $\dot{P} \sim 1.5\ 10^{-5}$ $M_{\odot}$ yr^{-1} km s^{-1}, and finally for the red lobe of the outflow of RW Aurigae: $\dot{M} \sim 2\ 10^{-7}$ $M_{\odot}$ yr^{-1} and $\dot{P} \sim 1\ 10^{-5}$ $M_{\odot}$ yr^{-1} km s^{-1}.

If these estimates are realistic, room is left for the conclusion that the optical flows under study could be capable of driving a coaxial neutral and/or molecular flow, whose mass and momentum rates turn out to be typically $\dot{M} \sim 10^{-6}$ $M_{\odot}$ yr^{-1} and $\dot{P} \sim$ a few 10^{-5} $M_{\odot}$ yr^{-1} km s^{-1}. This in turn enforces the conjecture that the luminous jet we see at optical wavelengths should not be considered as an isolated flow of the physical size derived from the images, but that it represents only the fastest and most ionized component of a wider flow (see i.e. Masson & Chernin 1993, Raga & Cabrit 1993).

4. Are Beam Shocks Really Necessary to Produce the Observed Emission?

Let us now discuss the physical consequences our results may have on the models proposed. We first point out the importance of kinematical analyses.

4.1 Kinematics

Recent studies have indicated that it is possible to detect relative motions of the emitting material with respect to the pattern of disturbances traced by the position of the emission enhancements (Eislöffel, 1993, Eislöffel and Mundt, 1994) (see Figure 5a). It can be showed that this favours the interpretations involving K-H instabilities (Norman et al. 1985).

In addition, if each knot marks the location of a narrowing of the flow, the derived value of the relative velocity $v_{\rm rel}$ implies that each fluid particle has passed through at most one narrowing when it arrives at the end of the bright section of the jet (see Figure 6, where we sketch the time evolution of the jet).

Let us now turn to the physical implications of the observational evidence of aligned multiple bow-shock systems in a single Herbig-Haro flow (Reipurth and Heathcote 1991, 1992, 1993). They are probably the signature of recurrent eruptive events, each section of the flow advancing in the wake of the earlier shock (Figure 5b). Therefore the visible section of the jet, which is part of the last occurred outflow episode, advances in a medium already set in motion by the passage of preceding blast waves. This consideration leads us to suspect that the K-H perturbations, as being generated by the velocity shear at the boundary of the flow may be able to alter the shape of the central channel, but they may have not sufficient strength to produce significant shocks in the flow.

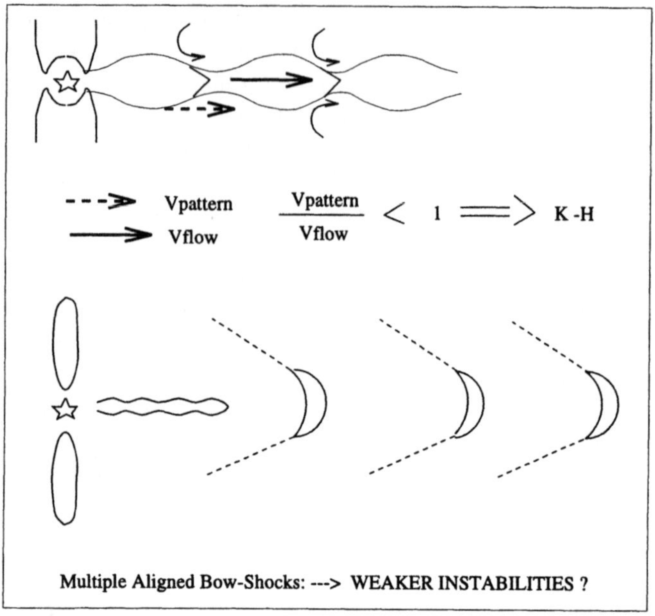

Fig. 5. Kinematics of the optical jet: a) relative motion between the emitting fluid and the pattern of disturbances; b) multiple aligned bow-shocks.

One may then conjecture that the emission observed along the axis of the flow could be related to the heating produced by the series of 'soft' compressions exerted on the jet's matter when it passes through the narrowings of the channel, and not to shocks of any kind.

In order to assess the reliability of this conjecture, that could eliminate the drawbacks the shock picture introduces in the modelzation of the nodular structure, we have studied the temporal evolution of the energy balance of a flow subject to slow recombination, under the effect of compression. In particular, we estimated if the amount of a compression a particle has to undergo in order to produce the observed emission is reasonable or not.

4.2 A Simple Model for Mild Compressions

For the sake of simplicity, we assume that the narrowing channel, in which the train of gas particles travels, has a conical shape. Due to the kinematical properties of the flow, we can concentrate on the portion of jet mass whose motion along the jet is affected by the action of a single compression (say, that labelled 'a' in Figure 6). We limit ourselves to the study of the fluid particles located on the axis of the flow, assigning to each of them the average thermodynamic properties of the jet's section the particle belongs to.

We consider an ideal inviscid and non-magnetized flow, and neglect gravitational effects, due to the low density of the gas. We study the following set of hydrodynamic equations:

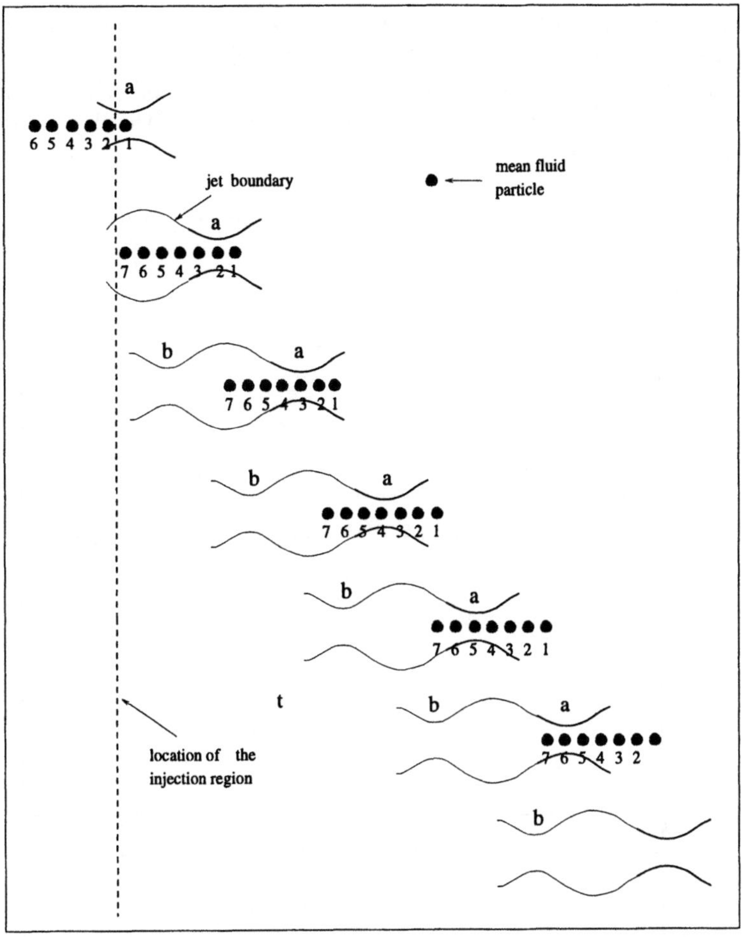

Fig. 6. Kinematics of the optical jet: snapshots at various time intervals

$$\frac{D\rho}{Dt} + \rho(\nabla \cdot \mathbf{v}) = 0 \qquad \rho\left(\frac{D\mathbf{v}}{Dt}\right) = -\nabla P + \mathbf{f}$$

$$\frac{1}{\gamma - 1}\left[\frac{DP}{Dt} - \frac{\gamma P}{\rho}\frac{D\rho}{Dt}\right] = -L(n_H, n_e, T_e)$$

$$\frac{Dn_e}{Dt} + n_e(\nabla \cdot \mathbf{v}) = -n_e^2 \alpha_B(H^0, T_e) \qquad P = (1.1 n_H + n_e) k_B T_e$$

where $D/Dt = \partial/\partial t + \mathbf{v}\cdot\nabla$ is the substantial time derivative, $\mathbf{f}$ is the compressive force per unit volume exerted by the perturbations at the jet's boundary, $\gamma = 5/3$ is the usual specific heats ratio, n_H and n_e are the hydrogen and electron number density respectively, and $\alpha_B(H^0, T_e)$ is the recombination coefficient valid in Case B, as in Hummer and Storey (1987). $L(n_H, n_e, T_e)$ are the energy losses appropriate to the temperatures involved. They include the recombination losses

L_{rec}, the free-free contribution of hydrogen L_{ff} and the radiative cooling L_{rad} of various atoms and ions. More details are given in BCO.

Note that in writing the above equations we have implicitly assumed that, contrary to the case of shocks, the bulk kinetic energy of the flow is not transformed in thermal energy.

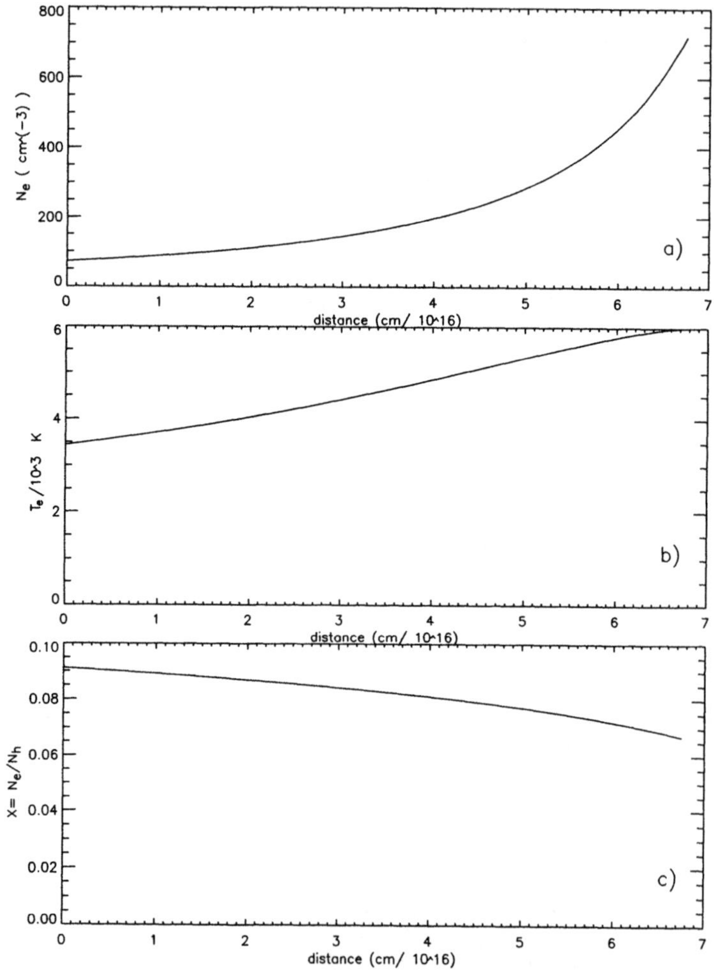

Fig. 7. Evolution of the electron density (**a**), the temperature (**b**) and the ionization fraction (**c**) of the test fluid particle during a single compression

Due to the assumed conical shape, the compression is parametrized by the ratio $\epsilon = R_i/R_{fin}$, where R_i and R_{fin} are the radii of the jet channel at the injection point and at the point of maximum compression respectively.

A numerical analysis involving the results of the previously described diagnostics shows that in both the cases of HH 34 and HH 111 the nature of the problem reduces the range of allowed compression rates between 3.5 and 5. In fact if $\epsilon < 3.5$ the compressive positive term cannot compensate the radiative losses, and the temperature cannot rise along the jet in the forward direction. If ϵ is greater than 5, on the other hand, the compression becomes too strong: in this case exceedingly high temperatures and densities are reached at the brightest knot.

Figures 7 and 8 illustrate the behaviour of various physical quantities under the action of the compression, taking a fiducial value $\epsilon = 3.7$. These results demonstrate that a sequence of mild compressions is able to reproduce the observed emission, without necessarily invoke the presence of shocks in the beam.

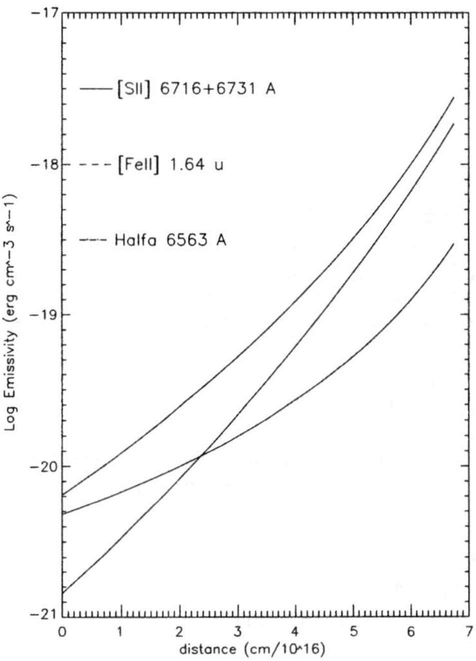

Fig. 8. Evolution of the emissivity of the test fluid particle during a single compression, in the lines [SII]$\lambda\lambda$6716, 6731, [FeII] 1.64 μ and H_α

Finally we point out that our interpretation offers a very natural explanation for the disappearing of the optical jet after a typical length scale of 10^{17} cm. In fact if there are no shocks in the beam that can produce new thermal energy at the expense of motion, a given particle that it is already passed (and heated by) a first narrowing, cannot produce the same amount of radiation in a second compression analogous to the first one, since it has lost most of its internal energy during the first compression episode. This can be demonstrated by a simple qualitative pressure-volume diagram (see BCO). According to this picture, the optical emission falls abruptly, at a location whose position depends only on the relative velocity of the emitting particles with respect to the system of compressions.

We conclude therefore that a phenomenological scenario that includes the peculiar kinematics of the flow, the action of damped Kelvin-Helmoltz perturbations, the time history of the ionization fraction and the proper cooling properties of the jet's matter provides a natural explanation of the observed jet morphologies and correctly predicts several other features.

The following step, consisting in a full numerical simulation of a large scale flow in conditions appropriate to the propagation of stellar jets, is now in progress.

Acknowledgement. The authors gratefully acknowledge the useful discussions they had with Jorge Canto', Alexandro Raga and Reinhard Mundt, as well as with Suzana Lizano, Steve Stahler, Guidetta Torricelli-Ciamponi, Giorgio Einaudi and Francesco Rubini.

References

Appl, S., and Camezind, M., 1992, ApJ 256, 354.
Bacciotti, F., Chiuderi, C. and Oliva, E., A&A, 296, 185
Bacciotti, F., Hirth, G.A. and Natta, A., 1995, A&A, in press
Blondin, J.M. Fryxell, B.A., and Königl, A., 1990, ApJ 360, 370.
Bodo, G., Trussoni, E., Massaglia, S., and Ferrari, A. 1993, in Proc. of the 6th International Workshop of the OAC on *Stellar Jets and Bipolar Outflows*, L.Errico and A. Vittone eds., Dordrecht, p. 307.
Butler, K., and Dalgarno, A., 1979, ApJ 234, 765.
Bührke, T., Mundt, R., and Ray, T., 1988, A&A 200, 99.
Coleman, C.S., 1990, MNRAS 244, 35.
Eislöffel, J., 1993, in Proc. of the 6th International Workshop of the OAC on *Stellar Jets and Bipolar Outflows*, L. Errico and A. Vittone eds., Dordrecht, p. 283.
Eislöffel, J., and Mundt, R., 1992, A&A 263, 292 (EM).
Eislöffel, J., and Mundt, R., 1994, A&A 284, 530.
Franck, A., and Mellema, G., 1994, A&A 289, 937
Franck, A., and Mellema, G., 1994, ApJ 430, 800
Giovanardi, C., Natta, A., and Palla, F., 1987, A & AS 70, 269.
Hartigan, P., Raymond, J., 1993, ApJ 409, 705.
Hartigan, P., Morse, J., Raymond, J., 1994, ApJ 436, 125.
Hirth, G.A., Mundt, R. and Solf, J., 1994a, A&A 285, 929.
Hirth, G.A., Mundt, R., Solf, J., and Ray, T., 1994b, ApJ 427, L99
Hummer, D.G., and Storey, P.J., 1987, MNRAS 224, 801.
Kofman, L., and Raga, A.C., 1992, ApJ 390, 359.
Masson, C.R., Chernin, L.M., 1993, ApJ 414, 230.
Mendoza, C. 1983, in IAU Symposium: "*Planetary Nebulae*", Vol. 103
Mundt, R., 1993, in Proc. of the 6th International Workshop of the OAC on *Stellar Jets and Bipolar Outflows*, L. Errico and A. Vittone eds., Dordrecht, p. 91.
Mundt, R., Ray, T.P., and Raga, A.C., 1991, A&A 252, 740.
Norman, M.L., Smarr, L., and Winkler, K.H., 1985, in *Numerical Astrophysics*, eds. J. Centrella, J. LeBlanc, and R. Bowes, (Boston, Jones & Bartlett), p. 88.
Oliva, E., Moorwood, A.F.M., and Danziger, I.J., 1989, A&A, 214, 307
Peimbert,M., Storey, P.J., and Torres-Peimbert, S., 1994 Rev. Mex. Astron. Astrofis. 29, 211.
Raga, A.C., 1991 in "*The Physics of Star Formation and Early Stellar Evolution*", eds. C.J. Lada and N.D. Kylafis, Kluwer Academic, Dordrecht, p. 247.
Raga, A.C., 1992, private communication.

Raga, A.C., and Cabrit, S., 1993, A&A 278, 267.
Raga, A.C., Cantó, J., Binette, L., and Calvet, N., 1990, ApJ 364, 601.
Raga, A.C., and Kofman, L., 1992, ApJ 386, 222.
Raga, A.C., Mundt, R., and Ray, T.P., 1991, A&A 252, 733.
Ray, T.P., 1993, in Proc. of the 6th International Workshop of the OAC on *Stellar Jets and Bipolar Outflows*, L. Errico and A. Vittone eds., Dordrecht, p. 241.
Reipurth, B., 1991 in "*The Physics of Star Formation and Early Stellar Evolution*", eds. C.J. Lada and N.D. Kylafis, Kluwer Academic, Dordrecht, p. 497.
Reipurth, B., and Heathcote, S.R., 1991, A&A 246, 511.
Reipurth, B., and Heathcote, S.R., 1992, A&A 257, 693.
Reipurth, B., and Heathcote, S., 1993, in "*Astrophysical jets*", D. Burgarella, M. Livio and C. O'Dea eds. (Cambridge University Press).
Reipurth, B., Raga, A.C., and Heathcote, S.R., 1992, ApJ 392, 145 (RRH).
Stone, J.M., and Norman, M.L., 1993, ApJ 413, 198.
Walter, D.K., Dufour, R.J., and Hester, J.J., 1992, ApJ 397, 196.
Yue, Z.Y., and Winnewisser, G., 1993, in Proc. of the 6th International Workshop of the OAC on *Stellar Jets and Bipolar Outflows*, L. Errico and A. Vittone eds., Dordrecht, p. 283.

Concluding Remarks

H. Elsässer

Max-Planck-Institut für Astronomie, Heidelberg-Königstuhl

This week's conference on "Disks and Outflows around Young Stars" has been a great pleasure for me and I would like to use these last minutes to express my gratitude: In the first place to thank you all for coming to Heidelberg from all parts of the world in order to attend this conference which has brought together most of the experts active in this field, and, secondly, for the great honor of dedicating this symposium to me. I feel great satisfaction seeing how lively this field is nowadays with respect to observations as well as theory.

Here at the Max-Planck-Institute for Astronomy the first steps into this area were taken about 20 years ago. With a newly constructed IR-camera using an image-tube we were able to take photographs in the near IR and did so to identify IR-sources found by means of rockets. One very stimulating result was the detection of the exciting star, not yet seen, of S 106 in the gap between the two lobes of this bipolar HII region. Carlos Eiroa in his doctoral thesis then derived the extinction of 20 magnitudes in V of this late O-type star obviously located in the center of a disk-like configuration. Extinction values higher than several magnitudes were something quite new in those years. S 106 was then our favorite object for a number of years. Guido Münch, Hans Hippelein, and Josef Solf studied the kinematics of the lobes and thus typical properties of a newly formed star with bipolar outflow became known. For me S 106 is still one of the most convincing cases in which one can see a disk more or less directly.

Since Reinhard Mundt joined the institute the jets have been another topic of central interest besides the ongoing activities of Thorsten Neckel and Jakob Staude. Now Steve Beckwith and his group is a further support.

Several years ago I had the feeling that the most exciting and fundamental problems had been solved. Therefore my main interest changed to extragalactic observations, first to interacting IR-galaxies. At present my coworkers and I are searching for galaxies in voids.

If I have learned one thing from this conference it is that my former impression was definitely wrong!

I wish you all success for your further work.

Before I close let me also thank the members of the Scientific Organizing Committee as well as of the Local Committee. You have done a tremendous job. Last but not least my thanks go to Steve Beckwith who was the initiator of this very successful conference.

Danke schön! – und: Auf Wiedersehen!

Table of Contents – CD-ROM

List of Participants

Péter Ábrahám	Budapest, Hungary
Jerry Acord	Bonn, Germany
Fred Adams	Ann Arbor, USA
Vassiliki Agapitou	London, England
Juan-Manuel Alcala	Garching, Germany
Immo Appenzeller	Heidelberg, Germany
Stefan Appl	Heidelberg, Germany
Pawel Artymowicz	Saltsjöbaden, Sweden
Francesca Bacciotti	Nice, France
Gianluigi Badd	Pino Torinese, Italy
Alexander Baranov	St. Petersburg, Russia
Eric Becklin	Los Angeles, CA, USA
Steven Beckwith	Heidelberg, Germany
Robbins Bell	Moffett Field, CA, USA
Stephen Bence	Cambridge, England
Claude Bertout	Grenoble, France
Nina Beskrovnaya	Pulkovo, St. Petersburg, Russia
Christoph Birk	Heidelberg, Germany
Stephan Blinder	Floirac, France
Peter Blondel	Amsterdam, The Netherlands
Karl-Heinz Böhm	Seattle, WA, USA
Stefan Bogun	Heidelberg, Germany
Vitor Bonifacio	London, England
Ian Bonnell	Cambridge, England
Sylvain Bontemps	Gif-sur-Yvette, France
Marie-Helene Boulard	Toulouse, France
Jerome Bouvier	Grenoble, France
Wolfgang Brandner	Garching, Germany
Max Camenzind	Heidelberg, Germany
John Carr	Columbus, OH, USA
Riccardo Cesaroni	Firenze, Italy
Sepideh Chakaveh	Göttingen, Germany
Simon Chapman	Cardiff, Great Britain
Carlos Chavarría-K.	Heidelberg, Germany
Claudio Codella	Firenze, Italy
Myles Corcoran	Dublin, Ireland
Livia Cordis	Hamburg, Germany
Elvira Covino	Naples, Italy
Douglas Creel	Los Angeles, CA, USA
Christopher Davis	Heidelberg, Germany
William Dent	Hilo, HI, USA
Catherine Dougados	Grenoble, France
Turlough Downes	Dublin, Ireland
Janet Drew	Oxford, England
Wolfgang Duschl	Heidelberg, Germany
Anne Dutrey	St. Martin d'Hères, France
Suzan Edwards	Northampton, MA, USA
Carlos Eiroa	Madrid, Spain

Jochen Eislöffel	Grenoble, France
Hans Elsässer	Heidelberg, Germany
Detlef Elstner	Potsdam, Germany
Christian Fendt	Lund, Sweden
Matilde Fernandez	Mexico, Mexico
Dirk Fiebig	Heidelberg, Germany
Thierry Foglizzo	Gif-sur-Yvette, France
Daniel Folha	London, England
Adam Frank	Minneapolis, MN, USA
Helmut Glänzer	Heidelberg, Germany
Andreas Glindemann	Heidelberg, Germany
Ana Gomez de Castro	Madrid, Spain
José Gómez	Granada, Spain
Elisabete Gouveia Dal Pino	Sao Paulo, SP, Brasilien
Malcolm Gray	Bristol, England
Vladimir Grinin	Pulkovo, St. Petersburg, Russia
Eike Guenther	London, England
Rolf Güsten	Bonn, Germany
Stephane Guilloteau	St. Martin d'Hères, France
Vasilii Gvaramadze	Tbilisi, Rep. of Georgia
Donald Hamilton	Heidelberg, Germany
Thomas Henning	Jena, Germany
Thomas Herbst	Heidelberg, Germany
Frederic Hessman	Göttingen, Germany
Annibal Hetem	Gif-sur-Yvette, France
Naomi Hirano	Kunitachi, Tokyo, Japan
Gerhard Hirth	Heidelberg, Germany
Melvin Hoare	Heidelberg, Germany
Klaus Hodapp	Honolulu, HI, USA
Michiel Hogerheijde	Leiden, The Netherlands
Ahmed Hujeirat	Heidelberg, Germany
Busaba Hutawarakorn	Jodrell Bank, Lower Withington, England
Nazar Ikhsanov	Pulkovo, St. Petersburg, Russia
Tim Jenness	Cambridge, England
Douglas Johnstone	Berkeley, CA, USA
Alexander Kakouris	Zografos, Greek
Uwe Kania	Heidelberg, Germany
Masayoshi Kiguchi	Higashi-Osaka, Japan
Arieh Königl	Chicago, IL, USA
Chris Koresko	Austin, TX, USA
Donald Korycansky	London, England
Joachim Krautter	Heidelberg, Germany
Wolfgang Kundt	Bonn, Germany
Sergei Lamzin	Heidelberg, Germany
Gregory Laughlin	Santa Cruz, CA, USA
Christoph Leinert	Heidelberg, Germany
Dietrich Lemke	Heidelberg, Germany
Rainer Lenzen	Heidelberg, Germany
Alain Lioure	Heidelberg, Germany
Kwok-Yung Lo	Garching, Germany
Victor Loskutov	St. Petersburg, Russia
Tigran Magakian	Byurakan, Ashtarak Distr., Armenien
Andrei Makalkin	Moscow, Russia
Karl Heinz Marien	Heidelberg, Germany

Christian Martin	Heidelberg, Germany
Mark McCaughrean	Heidelberg, Germany
Jennifer McDonald	Cambridge, England
Tom Megeath	Bonn, Germany
François Ménard	Grenoble, France
Alexander Menshchikov	Jena, Germany
Michael Meyer	Amherst, MA, USA
Luis Miranda	Madrid, Spain
Munetake Momose	Nobeyama, Nagano, Japan
Miguel Moreira	Lisboa, Portugal
Tigran Movsesyan	Byurakan, Ashtarak Distr., Armenien
Thomas Müller	Heidelberg, Germany
Guido Münch	Heidelberg, Germany
Reinhard Mundt	Heidelberg, Germany
Lee Mundy	College Park, MD, USA
Kaylene Murdoch	Oxford, England
Taishi Nakamoto	Tsukuba, Ibaraki, Japan
Takenori Nakano	Nobeyama, Nagano, Japan
Antonella Natta	Firenze, Italy
Thorsten Neckel	Heidelberg, Germany
Ralph Neuhäuser	Garching, Germany
Katsuo Ogura	Tokyo, Japan
Nagayoshi Ohashi	Nobeyama, Nagano, Japan
Michael Olberg	Onsala, Sweden
Elena Ortiz	Cantoblanco, Madrid, Spain
Martin Osterloh	Heidelberg, Germany
Gernot Paatz	Heidelberg, Germany
Deborah Padgett	Pasadena, USA
Rachael Padman	Cambridge, England
Francesco Palla	Firenze, Italy
Beate Patzer	Berlin, Germany
Kevin Pearson	Leicester, England
Antonio Pedrosa	Porto, Portugal
François Piché	Cambridge, England
Eckhart Pitz	Heidelberg, Germany
Mikhail Pogodin	Pulkovo, St. Petersburg, Russia
Gregor Pojmanski	Warsaw, Poland
Christina Popescu	Heidelberg, Germany
Lisa Prato	Stony Brook, NY, USA
Axel M. Quetz	Heidelberg, Germany
Tom Ray	Dublin, Ireland
Matthias von Rekowski	Bonn, Germany
Andrea Richichi	Heidelberg, Germany
Massimo Robberto	Pino Torinese, Italy
Douglas Roberts	Urbana, IL, USA
Günther Rüdiger	Potsdam, Germany
Anneila Sargent	Pasadena, CA, USA
Christoph Sauty	Meudon, France
Paul Scott	Cambridge, England
Karl-Friedrich Schuster	St. Martin dHeres, France
Frank Shu	Berkeley, CA, USA
Lionel Siess	Grenoble, France
Michael Simon	Stony Brook, NY, USA
Stephen Skinner	Sagamihara, Kanagawa, Japan
Michael Smith	Heidelberg, Germany

Josef Solf	Tautenburg, Germany
Cordula Sonnhalter	Würzburg, Germany
Jakob Staude	Heidelberg, Germany
Jürgen Steinacker	Jena, Germany
Michael Sterzik	Garching, Germany
James Stone	College Park, MD, USA
Steve Strom	Amherst, MA, USA
Koji Sugitani	Nagoya, Japan
Michel Tagger	Gif-sur-Yvette, France
Motohide Tamura	Mitaka, Tokyo, Japan
Jonathan Tedds	Edinburgh, Great Britain
Teresa Teixeira	London, England
Caroline Terquem	London, England
Leonardo Testi	Firenze, Italy
Eike Thamm	Jena, Germany
Clemens Thum	St. Martin d'Hères, France
Herman Tjin-A-Djie	Amsterdam, The Netherlands
Nicholas Tothill	Heidelberg, Germany
Vladimir Usov	Rejovot, Israel
Lourdes Verdes-Montenegro	Granada, Spain
Nikolai Voshchinnikov	Petrodvoietz, Russia
Malcolm Walmsley	Bonn, Germany
Jun-jie Wang	Beijing, China
Mark Wardle	Rochester, NY, USA
William Watson	Urbana, IL, USA
Norbert Weitzel	Heidelberg, Germany
Anthony Whitworth	Cardiff, Great Britain
Rainer Wichmann	Heidelberg, Germany
Helmut Wiesemeyer	Bonn, Germany
Georg Wilke	Heidelberg, Germany
Eugen Willerding	Bonn, Germany
Gerard Williger	Heidelberg, Germany
Thomas Wilson	Bonn, Germany
Ruslan Yudin	Pulkovo, St. Petersburg, Russia
Joao Yun	Lisboa, Portugal
Hans Zinnecker	Würzburg, Germany

1 Steve Strom
2 Douglas Roberts
3 Axel M. Quetz
4 Francesco Palla
5 Takenori Nakano
6 Suzan Edwards
7 Martin Osterloh
8 Frank Shu
9 Reinhard Mundt
10 Matilde Fernandez
11 Claudio Codella
12 unknown
13 Leonardo Testi
14 Elena Ortiz
15 Riccardo Cesaroni
16 Anneila Sargent
17 Frederic Hessman
18 Pawel Artymowicz
19 Eike Guenther
20 Antonella Natta
21 Janet Drew
22 Lisa Prato
23 Hans Elsässer
24 Norbert Weitzel
25 Michael Meyer
26 Jochen Eislöffel
27 Christoph Birk
28 unknown
29 Andreas Glindemann
30 unknown
31 Nicholas Tothill
32 Naomi Hirano
33 unknown
34 Rachael Padman
35 Michael Simon
36 Gregor Pojmanski
37 Koji Sugitani
38 Ian Bonnell
39 Immo Appenzeller
40 Katsuo Ogura
41 Andrea Richichi
42 Donald Korycansky
43 Vitor Bonifacio
44 Helmut Wiesemeyer
45 Michiel Hogerheijde
46 Daniel Folha
47 Günther Rüdiger
48 Antonio Pedrosa
49 Nagayoshi Ohashi
50 Tom Megeath
51 Jerry Acord
52 Kaylene Murdoch
53 Tim Jenness
54 Ralph Neuhäuser
55 unknown
56 Chris Koresko
57 Malcolm Walmsley
58 Stephen Bence
59 Klaus Hodapp
60 Deborah Padgett
61 Malcolm Gray
62 unknown
63 William Dent
64 Michael Olberg
65 José Gómez
66 Sylvain Bontemps
67 Karl Heinz Marien
68 Paul Scott
69 Jennifer McDonald
70 Ruslan Yudin
71 Douglas Creel
72 Fred Adams
73 Rainer Lenzen
74 Mark McCaughrean
75 Jürgen Steinacker
76 Elisabete Gouveia Dal Pino
77 Miguel Moreira
78 Detlef Elstner
79 Rainer Wichmann
80 Massimo Robberto
81 Jerome Bouvier
82 Alexander Men'shchikov
83 Matthias von Rekowski
84 Herman Tjin-A-Djie
85 Nina Beskrovnaya
86 Joao Yun
87 Annibal Hetem
88 Jun-jie Wang
89 Sergei Lamzin
90 Eike Thamm
91 Taishi Nakamoto
92 unknown
93 Arieh Königl
94 Christoph Leinert
95 Mikhail Pogodin
96 Kevin Pearson
97 Ana Gomez de Castro
98 Marie-Helene Boulard
99 unknown
100 Douglas Johnstone
101 Tigran Movsesyan
102 Thomas Henning
103 Turlough Downes
104 Thomas Herbst
105 Michel Tagger
106 Tigran Magakian
107 unknown
108 Myles Corcoran
109 François Piché
110 Eric Becklin
111 John Carr
112 Peter Blondel
113 Josef Solf
114 Beate Patzer
115 Luis Miranda
116 unknown
117 Munetake Momose
118 Anthony Whitworth
119 Alexander Kakouris
120 Guido Münch
121 Steven Beckwith
122 Carlos Eiroa
123 Hans Zinnecker
124 Robbins Bell
125 Jakob Staude
126 Thorsten Neckel

Photograph by Werner Neumann, MPIA

Contributing Authors

B/C = Book/CD-ROM

Lecture Notes in Physics

For information about Vols. 1–439
please contact your bookseller or Springer-Verlag

Vol. 440: H. Latal, W. Schweiger (Eds.), Matter Under Extreme Conditions. Proceedings, 1994. IX, 243 pages. 1994.

Vol. 441: J. M. Arias, M. I. Gallardo, M. Lozano (Eds.), Response of the Nuclear System to External Forces. Proceedings, 1994, VIII. 293 pages. 1995.

Vol. 442: P. A. Bois, E. Dériat, R. Gatignol, A. Rigolot (Eds.), Asymptotic Modelling in Fluid Mechanics. Proceedings, 1994. XII, 307 pages. 1995.

Vol. 443: D. Koester, K. Werner (Eds.), White Dwarfs. Proceedings, 1994. XII, 348 pages. 1995.

Vol. 444: A. O. Benz, A. Krüger (Eds.), Coronal Magnetic Energy Releases. Proceedings, 1994. X, 293 pages. 1995.

Vol. 445: J. Brey, J. Marro, J. M. Rubí, M. San Miguel (Eds.), 25 Years of Non-Equilibrium Statistical Mechanics. Proceedings, 1994. XVII, 387 pages. 1995.

Vol. 446: V. Rivasseau (Ed.), Constructive Physics. Results in Field Theory, Statistical Mechanics and Condensed Matter Physics. Proceedings, 1994. X, 337 pages. 1995.

Vol. 447: G. Aktaş, C. Saçlıoğlu, M. Serdaroğlu (Eds.), Strings and Symmetries. Proceedings, 1994. XIV, 389 pages. 1995.

Vol. 448: P. L. Garrido, J. Marro (Eds.), Third Granada Lectures in Computational Physics. Proceedings, 1994. XIV, 346 pages. 1995.

Vol. 449: J. Buckmaster, T. Takeno (Eds.), Modeling in Combustion Science. Proceedings, 1994. X, 369 pages. 1995.

Vol. 450: M. F. Shlesinger, G. M. Zaslavsky, U. Frisch (Eds.), Lévy Flights and Related Topics in Physics. Proceedigs, 1994. XIV, 347 pages. 1995.

Vol. 451: P. Krée, W. Wedig (Eds.), Probabilistic Methods in Applied Physics. IX, 393 pages. 1995.

Vol. 452: A. M. Bernstein, B. R. Holstein (Eds.), Chiral Dynamics: Theory and Experiment. Proceedings, 1994. VIII, 351 pages. 1995.

Vol. 453: S. M. Deshpande, S. S. Desai, R. Narasimha (Eds.), Fourteenth International Conference on Numerical Methods in Fluid Dynamics. Proceedings, 1994. XIII, 589 pages. 1995.

Vol. 454: J. Greiner, H. W. Duerbeck, R. E. Gershberg (Eds.), Flares and Flashes, Germany 1994. XXII, 477 pages. 1995.

Vol. 455: F. Occhionero (Ed.), Birth of the Universe and Fundamental Physics. Proceedings, 1994. XV, 387 pages. 1995.

Vol. 456: H. B. Geyer (Ed.), Field Theory, Topology and Condensed Matter Physics. Proceedings, 1994. XII, 206 pages. 1995.

Vol. 457: P. Garbaczewski, M. Wolf, A. Weron (Eds.), Chaos – The Interplay Between Stochastic and Deterministic Behaviour. Proceedings, 1995. XII, 573 pages. 1995.

Vol. 458: I. W. Roxburgh, J.-L. Masnou (Eds.), Physical Processes in Astrophysics. Proceedings, 1993. XII, 249 pages. 1995.

Vol. 459: G. Winnewisser, G. C. Pelz (Eds.), The Physics and Chemistry of Interstellar Molecular Clouds. Proceedings, 1993. XV, 393 pages. 1995.

Vol. 460: S. Cotsakis, G. W. Gibbons (Eds.), Global Structure and Evolution in General Relativity. Proceedings, 1994. IX, 173 pages. 1996.

Vol. 461: R. López-Peña, R. Capovilla, R. García-Pelayo, H. Waelbroeck, F. Zertuche (Eds.), Complex Systems and Binary Networks. Lectures, México 1995. X, 223 pages. 1995.

Vol. 462: M. Meneguzzi, A. Pouquet, P.-L. Sulem (Eds.), Small-Scale Structures in Three-Dimensional Hydrodynamic and Magnetohydrodynamic Turbulence. Proceedings, 1995. IX, 421 pages. 1995.

Vol. 463: H. Hippelein, K. Meisenheimer, H.-J. Röser (Eds.), Galaxies in the Young Universe. Proceedings, 1994. XV, 314 pages. 1995.

Vol. 464: L. Ratke, H. U. Walter, B. Feuerbach (Eds.), Materials and Fluids Under Low Gravity. Proceedings, 1994. XVIII, 424 pages, 1996.

Vol. 465: S. Beckwith, J. Staude, A. Quetz, A. Natta (Eds.), Disks and Outflows Around Young Stars. Proceedings, 1994. XII, 361 pages, 1996.

Vol. 466: H. Ebert, G. Schütz (Eds.), Spin – Orbit-Influenced Spectroscopies of Magnetic Solids. Proceedings, 1995. VII, 287 pages, 1996.

Vol. 467: A. Steinchen (Ed.), Dynamics of Multiphase Flows Across Interfaces. 1994/1995. XII, 267 pages. 1996.

Vol. 468: C. Chiuderi, G. Einaudi (Eds.), Plasma Astrophysics. 1994. VII, 326 pages. 1996.

Vol. 469: H. Grosse, L. Pittner (Eds.), Low-Dimensional Models in Statistical Physics and Quantum Field Theory. Proceedings, 1995. XVII, 339 pages. 1996.

Vol. 470: E. Martínez-González, J. L. Sanz (Eds.), The Universe at High-z, Large-Scale Structure and the Cosmic Microwave Background. Proceedings, 1995. VIII, 254 pages. 1996.

Vol. 471: W. Kundt (Ed.), Jets from Stars and Galactic Nuclei. Proceedings, 1995. X, 290 pages. 1996.

New Series m: Monographs

Vol. m 1: H. Hora, Plasmas at High Temperature and Density. VIII, 442 pages. 1991.

Vol. m 2: P. Busch, P. J. Lahti, P. Mittelstaedt, The Quantum Theory of Measurement. XIII, 165 pages. 1991. Second Edition: XIII, 181 pages. 1996.

Vol. m 3: A. Heck, J. M. Perdang (Eds.), Applying Fractals in Astronomy. IX, 210 pages. 1991.

Vol. m 4: R. K. Zeytounian, Mécanique des fluides fondamentale. XV, 615 pages, 1991.

Vol. m 5: R. K. Zeytounian, Meteorological Fluid Dynamics. XI, 346 pages. 1991.

Vol. m 6: N. M. J. Woodhouse, Special Relativity. VIII, 86 pages. 1992.

Vol. m 7: G. Morandi, The Role of Topology in Classical and Quantum Physics. XIII, 239 pages. 1992.

Vol. m 8: D. Funaro, Polynomial Approximation of Differential Equations. X, 305 pages. 1992.

Vol. m 9: M. Namiki, Stochastic Quantization. X, 217 pages. 1992.

Vol. m 10: J. Hoppe, Lectures on Integrable Systems. VII, 111 pages. 1992.

Vol. m 11: A. D. Yaghjian, Relativistic Dynamics of a Charged Sphere. XII, 115 pages. 1992.

Vol. m 12: G. Esposito, Quantum Gravity, Quantum Cosmology and Lorentzian Geometries. Second Corrected and Enlarged Edition. XVIII, 349 pages. 1994.

Vol. m 13: M. Klein, A. Knauf, Classical Planar Scattering by Coulombic Potentials. V, 142 pages. 1992.

Vol. m 14: A. Lerda, Anyons. XI, 138 pages. 1992.

Vol. m 15: N. Peters, B. Rogg (Eds.), Reduced Kinetic Mechanisms for Applications in Combustion Systems. X, 360 pages. 1993.

Vol. m 16: P. Christe, M. Henkel, Introduction to Conformal Invariance and Its Applications to Critical Phenomena. XV, 260 pages. 1993.

Vol. m 17: M. Schoen, Computer Simulation of Condensed Phases in Complex Geometries. X, 136 pages. 1993.

Vol. m 18: H. Carmichael, An Open Systems Approach to Quantum Optics. X, 179 pages. 1993.

Vol. m 19: S. D. Bogan, M. K. Hinders, Interface Effects in Elastic Wave Scattering. XII, 182 pages. 1994.

Vol. m 20: E. Abdalla, M. C. B. Abdalla, D. Dalmazi, A. Zadra, 2D-Gravity in Non-Critical Strings. IX, 319 pages. 1994.

Vol. m 21: G. P. Berman, E. N. Bulgakov, D. D. Holm, Crossover-Time in Quantum Boson and Spin Systems. XI, 268 pages. 1994.

Vol. m 22: M.-O. Hongler, Chaotic and Stochastic Behaviour in Automatic Production Lines. V, 85 pages. 1994.

Vol. m 23: V. S. Viswanath, G. Müller, The Recursion Method. X, 259 pages. 1994.

Vol. m 24: A. Ern, V. Giovangigli, Multicomponent Transport Algorithms. XIV, 427 pages. 1994.

Vol. m 25: A. V. Bogdanov, G. V. Dubrovskiy, M. P. Krutikov, D. V. Kulginov, V. M. Strelchenya, Interaction of Gases with Surfaces. XIV, 132 pages. 1995.

Vol. m 26: M. Dineykhan, G. V. Efimov, G. Ganbold, S. N. Nedelko, Oscillator Representation in Quantum Physics. IX, 279 pages. 1995.

Vol. m 27: J. T. Ottesen, Infinite Dimensional Groups and Algebras in Quantum Physics. IX, 218 pages. 1995.

Vol. m 28: O. Piguet, S. P. Sorella, Algebraic Renormalization. IX, 134 pages. 1995.

Vol. m 29: C. Bendjaballah, Introduction to Photon Communication. VII, 193 pages. 1995.

Vol. m 30: A. J. Greer, W. J. Kossler, Low Magnetic Fields in Anisotropic Superconductors. VII, 161 pages. 1995.

Vol. m 31: P. Busch, M. Grabowski, P. J. Lahti, Operational Quantum Physics. XI, 230 pages. 1995.

Vol. m 32: L. de Broglie, Diverses questions de mécanique et de thermodynamique classiques et relativistes. XII, 198 pages. 1995.

Vol. m 33: R. Alkofer, H. Reinhardt, Chiral Quark Dynamics. VIII, 115 pages. 1995.

Vol. m 34: R. Jost, Das Märchen vom Elfenbeinernen Turm. VIII, 286 pages. 1995.

Vol. m 35: E. Elizalde, Ten Physical Applications of Spectral Zeta Functions. XIV, 228 pages. 1995.

Vol. m 36: G. Dunne, Self-Dual Chern-Simons Theories. X, 217 pages. 1995.

Vol. m 37: S. Childress, A.D. Gilbert, Stretch, Twist, Fold: The Fast Dynamo. XI, 410 pages. 1995.

Vol. m 38: J. González, M. A. Martín-Delgado, G. Sierra, A. H. Vozmediano, Quantum Electron Liquids and High-T_c Superconductivity. X, 299 pages. 1995.

Vol. m 39: L. Pittner, Algebraic Foundations of Non-Commutative Differential Geometry and Quantum Groups. XII, 469 pages. 1996.

Vol. m 40: H.-J. Borchers, Translation Group and Particle Representations in Quantum Field Theory. VII, 131 pages. 1996.

Vol. m 41: B. K. Chakrabarti, A. Dutta, P. Sen, Quantum Ising Phases and Transitions in Transverse Ising Models. X, 204 pages. 1996.

GPSR Compliance

The European Union's (EU) General Product Safety Regulation (GPSR) is a set of rules that requires consumer products to be safe and our obligations to ensure this.

If you have any concerns about our products, you can contact us on ProductSafety@springernature.com

In case Publisher is established outside the EU, the EU authorized representative is:

Springer Nature Customer Service Center GmbH
Europaplatz 3
69115 Heidelberg, Germany

Batch number: 10620143

Printed by Printforce, the Netherlands